Alpine and Polar Treelines in a Changing Environment

Alpine and Polar Treelines in a Changing Environment

Special Issue Editor

Gerhard Wieser

MDPI • Basel • Beijing • Wuhan • Barcelona • Belgrade • Manchester • Tokyo • Cluj • Tianjin

Special Issue Editor
Gerhard Wieser
Division of Alpine Timberline
Ecophysiology Federal Research
and Training Centre for Forests,
Natural Hazards and Landscape
(BFW)
Austria

Editorial Office
MDPI
St. Alban-Anlage 66
4052 Basel, Switzerland

This is a reprint of articles from the Special Issue published online in the open access journal *Forests* (ISSN 1999-4907) (available at: https://www.mdpi.com/journal/forests/special_issues/Alpine_Polar_Treelines).

For citation purposes, cite each article independently as indicated on the article page online and as indicated below:

LastName, A.A.; LastName, B.B.; LastName, C.C. Article Title. *Journal Name* **Year**, *Article Number*, Page Range.

ISBN 978-3-03928-630-0 (Pbk)
ISBN 978-3-03928-631-7 (PDF)

Cover image courtesy of Gerhard Wieser.

Contents

About the Special Issue Editor

Gerhard Wieser is an associate professor whose research focus is the ecophysiology of the alpine treeline. He studied botany and microbiology in Innsbruck where he received his Ph.D. in 1983. In 1987, he joined the Federal Research and Training Centre for Forests, Natural Hazards, and Landscape in Innsbruck, where he is headed of the Division Alpine Timberline Ecophysiology.

Editorial

Alpine and Polar Treelines in a Changing Environment

Gerhard Wieser

Division of Alpine Timberline Ecophysiology, Federal Research and Training Centre for Forests, Natural Hazards and Landscape (BFW), Rennweg 1, Innsbruck A-6020, Austria; Gerhard.Wieser@uibk.ac.at; Tel.: +43-512-573933-5120

Received: 23 February 2020; Accepted: 25 February 2020; Published: 26 February 2020

Concerns have been raised with respect to the state of high-altitude and high-latitude treelines, as they are anticipated to undergo considerable modifications due to global change, especially due to climate warming [1–4]. Given that high-elevation treelines are temperature-limited vegetation boundaries [5–8], they are considered to be sensitive to climate warming. As a consequence, in a future, warmer environment, an upward migration of treelines is expected because low air and root-zone temperatures constrain their regeneration and growth. Despite the ubiquity of climate warming, treeline advancement is not a worldwide phenomenon: While some treelines have been advancing rapidly, others have responded sluggishly or have remained stable [9]. This variation in responses is attributed to the potential interaction of a continuum of site-related factors, which may lead to the occurrence of locally conditioned temperature patterns. Furthermore, competition amongst species and below-ground resources has been suggested as an additional aspect to explain the variability of the movement of treelines. This Special Issue is dedicated to the discussion of treeline responses to changing environmental conditions in different areas of the globe. A short outline of the individual contributions will be presented in the following paragraphs.

Effects of climate change on conifers within the treeline ecotone of the Central Austrian Alps is reviewed by Wieser et al. [10]. They outline tree growth conditioned by elevation and possible effects of climate change on carbon gain and water relations, which are derived from space-for-time studies and manipulative experiments. In addition, based on long-term observational records, possible future tendencies of tree growth in a warmer environment are discussed. Oberhuber et al. [11] analyzed the growth trends of conifers along environmental transects in the Central European Alps and explain the missing adequate growth response to climate warming, competition for resources in increasingly denser stands at subalpine sites, and by frost desiccation injuries of evergreen tree species at the Krummholz-limit. Wieser et al. [12] investigated the effects of artificial topsoil drought on the water use of *Picea abies* and *Larix decidua* Mill. saplings at the treeline in the Austrian Alps. Their study revealed that a three-year water shortage in this layer of the ground did not considerably reduce water loss in both species investigated. Mayr et al. [13] investigated winter embolism and recovery in *Pinus mugo* L. and state that future changes in snow cover regimes may significantly affect embolism and refilling processes. With respect to *Pinus cembra* L. saplings, Oberhuber et al. [14] demonstrated that a weak apical control might serve as protection against severe climatic conditions above the treeline during the winter.

High resolution maps of climatological data are essential for modeling the potential impacts of climate change on forests. For the Swiss Alps, Zischg et al. [15] presented such maps of temperature, relative humidity, radiation, and "föhn" winds. Apart from altitude, these maps also take into account micro-relief, slope, and aspect. Chiang et al. [16] proposed that aside from temperature, potential changes in light quality and quantity also play an important role in the phenology and growth of woody plants in boreal and temperate climates.

González-Rodriguez et al. [17] assessed the photosynthetic performance of *Pinus canariensis* Chr. Sm. Ex DC in Buch in Tenerife, Canary Islands to cope with seasonal changes in environmental conditions, especially summer drought. They found that, at the treeline, *Pinus canariensis* displays a drought avoidance strategy due to a great plasticity in gas exchange, antioxidants, and pigments. Pakharkova et al. [18] demonstrated that *Pinus sibirica* Du Tour and *Abies sibirica* L. showed different patterns of photosynthetic pigments along an elevational gradient in the West Sayan ridge, Siberia. As the decline in photosynthetic pigments with increasing elevation was more pronounced in *Pinus sibirica* needles than in *Abies sibirica* needles, the authors conclude that under conditions of future climate warming, *Pinus sibirica* trees will have an advantage in colonizing zones above the present treeline. Shiryaev et al. [19] investigated the long-term influence of climate change on the spatiotemporal dynamics on mycobiota in the Polar Ural. Their results showed that the composition of the fungal community followed alterations in vegetation, which changed from a forest-tundra to a boreal forest in the past 60 years.

Yu et al. [20] investigated the microenvironmental effects on the physiological performance of *Betula ermanii* at and beyond the treeline on Changbai Mountain in Northeast China. Their results showed that mature trees at these levels did not differ substantially in their ecophysiological performance due to microclimatic amelioration at microsites above the treeline. Leaf and soil $\delta^{15}N$ patterns were studied along elevational gradients in tree- and shrublines in three different climatic zones in Wolong Nature Reserve in Southwest China by Wang et al. [21], who reported that $\delta^{15}N$ leaf and $\delta^{15}N$ soil values were higher in subtropical forests compared to dry and wet-temperate forests.

McGlone et al. [22] presented a study on the postglacial treeline history of Sub-Antarctic Campbell Island, south of New Zealand. Their results pointed out that the treeline position in the southern hemisphere is noticeably affected not only by temperature, but also by cloudiness and seasonality. As a consequence, they concluded that a continuous increase in warming may not necessarily cause an advancement of the treeline in oceanic regions. A dendroclimatic assessment of Ponderosa pine radial growth along elevational transects in Western Montana, USA was presented by Montpellier et al. [23]. Their research findings suggest that Ponderosa pine trees at lower elevations may be better adapted to withstand warm and dry periods, while trees at high elevations are better suited to cool and wet conditions.

Seedling regeneration is an important feature discussing treeline dynamics in a future, changing environment. Based on a worldwide survey, Johnson and Yeakley [24] showed that seedling regeneration varied with respect to climate zone and microsite. Their results suggest that due to climate change, seedling regeneration will mainly benefit in cold and wet locations. The review of Holtmeier and Broll [25] provides a literature overview of treeline research from its onset to the present. They detected a reiterative pattern: a moderate number of ideas that, at present, are considered novel, that originated several decades ago, and tend to confirm prior knowledge. Additionally, they also outline further research questions.

Finally, I would like to express my gratitude to all the authors for their timely, high-quality contributions providing insights into high-altitude and high-latitude treelines within the context of global change. Furthermore, I thank all the anonymous reviewers for maintaining the quality standard of the Special Issue. I also appreciate the fruitful co-operation with the MDPI *forests* team, especially A. Zhang, during all stages of the project.

Conflicts of Interest: The author declares no conflicts of interest.

References

1. Wieser, G.; Tausz, M. Trees at their upper limit: Treelife limitation at the Alpine Timberline. In *Plant Ecophysiology*; Springer: Berlin/Heidelberg, Germany, 2007; Volume 5.
2. Körner, C. *Alpine Treelines: Funtional Ecology of the Global High. Elevation Tree Limits*; Springer: Berlin/Heidelberg, Germany, 2012.

3. Holtmeier, F.-K.; Broll, G. Treeline advance—Driving processes and adverse factors. *Landsc. Online* **2007**, *1*, 1–33. [CrossRef]

4. Smith, W.K.; Germino, M.J.; Johnson, D.M.; Reinhardt, K. The altitude of alpine treeline: A bellwether of climate change. *Bot. Rev.* **2009**, *75*, 163–190. [CrossRef]

5. Körner, C. A re-assessment of high elevation tree line positions and their explanation. *Oecologia* **1998**, *115*, 445–459. [PubMed]

6. Esper, J.; Cook, E.R.; Schweingruber, F.H. Low-frequency signals in long tree-ring chronologies for reconstructing past temperature variability. *Science* **2002**, *295*, 2250–2253. [CrossRef] [PubMed]

7. Körner, C.; Paulsen, J. A world-wide study of high altitude treeline temperatures. *J. Biogeogr.* **2004**, *31*, 713–732. [CrossRef]

8. Rossi, S.; Deslauriers, A.; Gričar, J.; Seo, J.-W.; Rathgeber, C.B.K.; Anfodillo, T.; Morin, H.; Levanic, T.; Oven, P.; Jalkanen, R. Critical temperatures for xylogenesis in conifers of cold climates. *Glob. Ecol. Bioeogr.* **2008**, *17*, 696–707. [CrossRef]

9. Harsch, M.A.; Hulme, P.E.; Huntley, B. Are treelines advancing? A global meta-analysis of treeline response to climate warming. *Ecol. Lett.* **2009**, *12*, 1040–1049. [CrossRef]

10. Wieser, G.; Oberhuber, W.; Gruber, A. Effects of Climate Change at Treeline: Lessons from Space-for-Time Studies, Manipulative Experiments, and Long-Term Observational Records in the Central Austrian Alps. *Forests* **2019**, *10*, 508. [CrossRef]

11. Oberhuber, W.; Bendler, U.; Gamper, V.; Geier, J.; Hölzl, A.; Kofler, W.; Krismer, H.; Waldboth, B.; Wieser, G. Growth Trends of Coniferous Species along Elevational Transects in the Central European Alps Indicate Decreasing Sensitivity to Climate Warming. *Forests* **2020**, *11*, 132. [CrossRef]

12. Wieser, G.; Oberhuber, W.; Gruber, A.; Oberleitner, F.; Hasibeder, R.; Bahn, M. Artificial Top Soil Drought Hardly Affects Water Use of *Picea abies* and *Larix decidua* Saplings at the Treeline in the Austrian Alps. *Forests* **2019**, *10*, 777. [CrossRef]

13. Mayr, S.; Schmid, P.; Rosner, S. Winter Embolism and Recovery in the Conifer Shrub (*Pinus mugo* L.). *Forests* **2019**, *10*, 941. [CrossRef]

14. Oberhuber, W.; Geisler, T.; Bernich, F.; Wieser, G. Weak Apical Control of Swiss Stone Pine (*Pinus cembra* L.) May Serve as a Protection against Environmental Stress above Treeline in the Central European Alps. *Forests* **2019**, *10*, 744. [CrossRef]

15. Zischg, A.; Gubelmann, P.; Frehner, M.; Huber, B. High Resolution Maps of Climatological Parameters for Analyzing the Impacts of Climatic Changes on Swiss Forests. *Forests* **2019**, *10*, 617. [CrossRef]

16. Chiang, C.; Olsen, J.; Basler, D.; Bånkestad, D.; Hoch, G. Latitude and Weather Influences on Sun Light Quality and the Relationship to Tree Growth. *Forests* **2019**, *10*, 610. [CrossRef]

17. González-Rodriguez, Á.M.; Brito, P.; Lorenzo, J.; Jiménez, M. Photosynthetic Performance in *Pinus canariensis* at Semiarid Treeline: Phenotype Variability to Cope with Stressful Environment. *Forests* **2019**, *10*, 845. [CrossRef]

18. Pakharkova, N.; Borisova, I.; Sharafutdinov, R.; Gavrikov, V. Photosynthetic Pigments in Siberian Pine and Fir under Climate Warming and Shift of the Timberline. *Forests* **2020**, *11*, 63. [CrossRef]

19. Shiryaev, A.; Moiseev, P.; Peintner, U.; Devi, N.; Kukarskih, V.; Elsakov, V. Arctic Greening Caused by Warming Contributes to Compositional Changes of Mycobiota at the Polar Urals. *Forests* **2019**, *10*, 1112. [CrossRef]

20. Yu, D.; Wang, Q.; Wang, X.; Dai, L.; Li, M. Microsite Effects on Physiological Performance of *Betula ermanii* at and beyond an Alpine Treeline Site on Changbai Mountain in Northeast China. *Forests* **2019**, *10*, 400. [CrossRef]

21. Wang, X.; Jiang, Y.; Ren, H.; Yu, F.; Li, M. Leaf and Soil δ^{15}N Patterns Along Elevational Gradients at Both Treelines and Shrublines in Three Different Climate Zones. *Forests* **2019**, *10*, 557. [CrossRef]

22. McGlone, M.; Wilmshurst, J.; Richardson, S.; Turney, C.; Wood, J. Temperature, Wind, Cloud, and the Postglacial Tree Line History of Sub-Antarctic Campbell Island. *Forests* **2019**, *10*, 998. [CrossRef]

23. Montpellier, E.; Soulé, P.; Knapp, P.; Maxwell, J. Dendroclimatic Assessment of Ponderosa Pine Radial Growth along Elevational Transects in Western Montana, USA. *Forests* **2019**, *10*, 1094. [CrossRef]

24. Johnson, A.; Yeakley, J. Microsites and Climate Zones: Seedling Regeneration in the Alpine Treeline Ecotone Worldwide. *Forests* **2019**, *10*, 864. [CrossRef]
25. Holtmeier, F.; Broll, G. Treeline Research—From the Roots of the Past to Present Time. A Review. *Forests* **2020**, *11*, 38. [CrossRef]

 *forests*

 MDPI

Review

Effects of Climate Change at Treeline: Lessons from Space-for-Time Studies, Manipulative Experiments, and Long-Term Observational Records in the Central Austrian Alps

Gerhard Wieser [1,*], Walter Oberhuber [2] and Andreas Gruber [2]

[1] Department of Alpine Timberline Ecophysiology, Federal Research and Training Centre for Forests, Natural Hazards and Landscape (BFW), Rennweg 1, A-6020 Innsbruck, Austria

[2] Department of Botany, Leopold-Franzens-Universität Innsbruck, Sternwartestraße15, A-6020 Innsbruck, Austria; walter.oberhuber@uibk.ac.at (W.O.); andreas.gruber@itworks.co.at (A.G.)

* Correspondence: gerhard.wieser@uibk.ac.at; Tel.: +43-512-573933-5120

Received: 20 May 2019; Accepted: 10 June 2019; Published: 14 June 2019

Abstract: This review summarizes the present knowledge about effects of climate change on conifers within the treeline ecotone of the Central Austrian Alps. After examining the treeline environment and the tree growth with respect to elevation, possible effects of climate change on carbon gain and water relations derived from space-for-time studies and manipulative experiments are outlined. Finally, long-term observational records are discussed, working towards conclusions on tree growth in a future, warmer environment. Increases in CO_2 levels along with climate warming interact in complex ways on trees at the treeline. Because treeline trees are not carbon limited, climate warming (rather than the rising atmospheric CO_2 level) causes alterations in the ecological functioning of the treeline ecotone in the Central Austrian Alps. Although the water uptake from soils is improved by further climate warming due to an increased permeability of root membranes and aquaporin-mediated changes in root conductivity, tree survival at the treeline also depends on competitiveness for belowground resources. The currently observed seedling re-establishment at the treeline in the Central European Alps is an invasion into potential habitats due to decreasing grazing pressure rather than an upward-migration due to climate warming, suggesting that the treeline in the Central Austrian Alps behaves in a conservative way. Nevertheless, to understand the altitude of the treeline, one must also consider seedling establishment. As there is a lack of knowledge on this particular topic within the treeline ecotone in the Central Austrian Alps, we conclude further research has to focus on the importance of this life stage for evaluating treeline shifts and limits in a changing environment.

Keywords: treeline; climate change; ecosystem manipulation; space-for-time substitution; long-term trends; Central Austrian Alps

1. Introduction

Alpine treelines are obvious vegetation boundaries. In the Central Austrian Alps, treelines generally form an ecotone between the closed forest below and the treeless alpine zone above [1–8]. Due to abiotic climatic severity within this transition zone, trees become flagged and stunted, which leads to scrub-like trees higher up at the krummholz limit. Therefore, researchers commonly define treelines as the upper elevational limit of trees greater than 2 m in height [9,10]. Such a height ensures that tree crowns are well coupled to atmospheric conditions measured at a standard height (2 m) in weather stations. Further up to the krummholz limit, stunted bush-like trees experience microclimatic conditions that dominate the next higher altitudinal vegetation belt [11] and are characterized by low-growing vegetation (e.g., dwarf-shrubs, grassland, and meadows).

Generally, on a global and on a continental scale, the formation and the maintenance of treelines seem to be correlated with air or soil temperatures, although the altitudinal position of a treeline may vary with respect to site conditions. A world-wide survey indicates that in temperature limited ecosystems, a growing mean season air temperature of 5.5 to 7.0 °C constrains tree growth [12,13]. Moreover, in a global survey, a growing season mean soil temperature of 6.7 ± 0.8 °C in 10 cm soil depth matches the upper elevational limit of tree growth [14,15]. As in nature, such mean temperatures generally do not exist, and they should be considered as an indicator of heat deficiency rather than an underlying factor [2,3,16–18].

Presently, people have raised concern about treelines, because they are anticipated to experience considerable modifications due to global change, especially climate warming [11,19–24]. Global mean surface temperature has increased by about 0.6 ± 0.2 °C during the last century, and global change models predict a further increase by 1.4–5.8 °C for upcoming decades [25]. As observed changes appear to be most pronounced at high altitudes [26,27], and considering that alpine (high elevation) treelines are undoubtedly associated with heat deficiency, treeline ecotones are ideally suited for climate change monitoring [28].

Yet, knowledge of tree response to warming in treeline environments is critical for understanding potential alterations that will likely occur in a changing environment. Inference about future climate change typically relies on one of three approaches: space-for-time substitution, manipulative experiments, and long-term observational records.

- The space-for-time approach [15,29,30] uses variations of environmental conditions along altitudinal gradients, where warmer temperatures at lower elevations represent a likely future climate, while lower temperatures at higher elevations represent the present. Such variations in environmental conditions offer a great possibility for comparative research on ecophysiological adaptations to environmental alterations [24,29,31] with minimal confounding biogeographic influence and maximal interpretability [32]. Elevational transects are also considered as powerful tools to investigate climate-driven changes in tree growth [33,34].
- In-situ manipulative warming and rain shelter experiments are common methodologies for assessing the effects of rapid climate change [35]. They can be quite effective in simulating climate warming [36–39] and top soil drought [40,41] in high elevation forests. Compared to the space-for-time approach, such techniques provide an explicit control in simulating climate warming or artificial soil drought.
- Time series data [35] of tree growth and stable isotopes coupled with time series data of climate may facilitate a mechanistic understanding of climate-related influences on physiological processes, such as leaf gas exchange and stem wood formation, in response to recent climate warming and increasing CO_2 concentration [42].

This review summarizes the current knowledge about potential climate change effects on treeline-associated conifers. The focus is on the Central Austrian Alps, as geographical variations in mesoclimates and interferences by continental, oceanic, and Mediterranean influences hamper drawing a single scenario for the entire European Alps [24,43]. After examining the treeline environment and tree growth with respect to elevation, potential effects of climate change on carbon gain and water relations derived from space-for-time studies and manipulative experiments are examined. Finally, long-term observational records are discussed to evaluate effects of climate change on tree growth at their upper limit.

2. The Treeline Environment

The treeline environment in the Central Tyrolean Alps is characterized by harsh climatic conditions where short growing seasons alternate with periods of dormancy during the winter [44]. Figure 1 provides an example for seasonal changes in thermal conditions and precipitation within the treeline ecotone on Mount Patscherkofel in the Central Tyrolean Alps for the period 1961–1990. Due to recent climate warming, the mean annual air temperature during the last 28 years (1991–2018) increased on average by 1.1 °C (i.e., to 3.1 °C) as compared to the previous 30 years (2.0 °C), while a considerable trend in air temperature was absent from 1925 to 1981 (Figure 1).

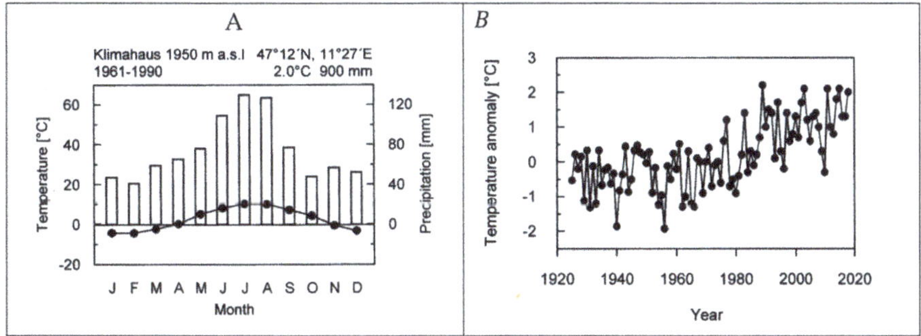

Figure 1. (**A**) Climate diagram (mean monthly temperature in lines and mean monthly precipitation in bars) and (**B**) temporal variation in mean annual air temperature during the period 1925–2018 relative to the 1961–1990 average at the treeline on Mount Patscherkofel. Compiled after [45] and [46].

The temperature increase since 1982 apparently was most pronounced during spring (by 1.5 °C) and summer (by 1.5 °C) compared with autumn (by 0.3 °C) and winter (by 0.8 °C), implying early snowmelt in spring. Thus, the growing season increased by 4 ± 1 weeks during the last three decades [44]. Higher temperatures also intensified evaporative demand [45]. In contrast to temperature, total annual precipitation did not change significantly during the last 94 years (45).

Within the treeline ecotone soil depth, length of the growing season and the air temperature declined with increasing altitude, while soil temperature increased along the same elevational gradient [47,48]. The observed lower soil temperature at the forest limit can primarily be attributed to a closed canopy. The latter prevents radiative warming and soil heat flux of the rooting zone [8,15,47,49,50]. However, this is not the case in open stands at the tree and the krummholz limit. Moreover, in addition to altitude, slope angle and relief strongly determine the microclimate at any site within the treeline ecotone. Differences in radiation and wind velocity due to topographical features strongly influence soil development, soil temperature, snow cover duration, soil water availability, and, consequently, seedling establishment [49,51–56].

3. Tree Growth at Treeline

There is correlative and extensive evidence that temperature strongly affects tree growth and tissue formation at the treeline [29,36,50,57,58]. Height growth of *Pinus cembra* trees in the Tyrolean Alps has been shown to be highly correlated with growing season mean air temperature [58]. Moreover, radial stem and shoot increment of *P. cembra*, *Picea Abies*, and *Larix decidua* cease at temperatures lower than 5–7.5 °C [37,59,60]. Soil temperature also has been considered as a substantial factor restricting root growth [61] and influencing above ground metabolism [62]. Low soil nutrient availability (particularly nitrogen [63,64]) has also been attributed to terminating tree growth at the treeline [5,63], where decomposition and mineralization are limited by low soil temperatures [65–67]. Additionally, competition for below ground resources also may noticeably influence tree growth [68,69].

3.1. Height Growth

At the treeline in the Central Tyrolean Alps, researchers in [70] and [71] observed that cumulative height growth of naturally growing *P. cembra* trees was considerably lower in the kampfzone (i.e., the upper treeline belt) at 2194 m above sea level (a.s.l.) as compared to the upper end of the closed forest in 1995 m a.s.l (Figure 2). In subalpine (1730–2080 m a.s.l.) afforestations in the Sellrain and in the

Schmirn Valley, Tyrol, Austria, height growth of young *P. cembra* [72,73], *P. abies*, and *L. decidua* [74] trees exceeding a height > 0.5 m also declined with elevation.

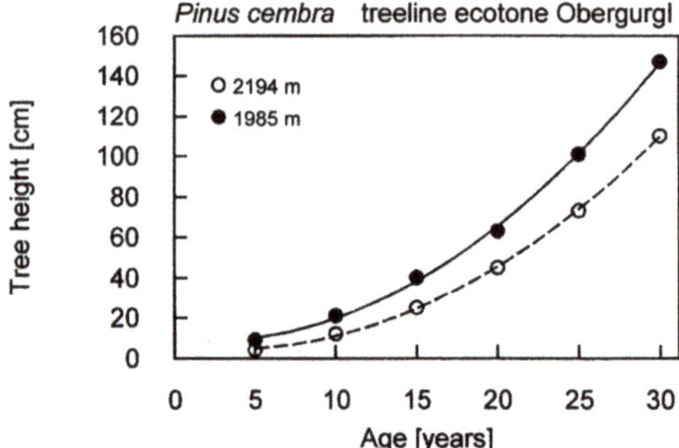

Figure 2. Cumulative height increment of *Pinus cembra* trees in relation to tree age at the forest limit [1985 m above sea level (a.s.l.); solid circles, solid line] and the krummholz limit (2194 m a.s.l.; open circles, dashed line) in Obergurgl, Ötztal, Tyrol, Austria. After [71].

Data in Table 1 show that the observed reduction in cumulative height growth with increasing elevation corresponds to a decline in the length of the growing season and an increase in wind velocity [72]. The effect of wind on height growth is more marked with increasing elevation because stands open up and trees are isolated from each other [5,18,74] and probably also because of a corresponding increase in wind damage with increasing elevation [3,18,72,75].

Table 1. Mean annual height increment of young *Pinus cembra* trees, mean growing season length, and mean wind velocity during the growing season at various altitudes in a subalpine afforestation in the Sellrain valley, Tyrol, Austria. Relative values are given as percentage of data from 1730 m a.s.l. After [72].

	Absolute Values			Relative Values (%)		
	1730 m	1800 m	1900 m	1730 m	1800 m	1900 m
Annual height increment (cm)	24.1	21.8	11.8	100	90	49
Growing season (days)	112	107	99	100	96	88
Wind velocity (m s^{-1})	1.2	1.3	3.1	100	108	258

In addition to elevation, exposure and microtopography strongly determine the microclimate at any site within the treeline ecotone and hence also influence height growth. For example, in the Sellrain valley, the cumulative height growth of 25-year-old *P. cembra* trees growing on a north-exposed slope 2000 m a.s.l was only 1.5 m as compared to 5.1 m in even aged *P. cembra* trees growing at 1900 m a.s.l on a south facing slope [72]. Similar differences in height growth with respect to microtopography have also been reported for *P. cembra* and *P. abies* trees in the Ötztal valley [71] and in the Schmirn valley [74], respectively. Height growth of seedlings, by contrast, is barely affected by topography and/or elevation [73,74]. This is because seedlings experience a microclimate comparable to that of short-stature plants that profit from life form-driven passive solar energy use, which facilitates canopy and soil heating [11].

3.2. Diameter Growth

Stem diameter growth also declines with increasing elevation, as shown for young *P. cembra* [73], *P. abies*, and *L. decidua* [74] trees in the Schmirn Valley, Tyrol, Austria. In general, the onset of cambial activity depends on the timing of snowmelt and the rise in air and soil temperature [47,76–78]. Reference [79] noted that cambial activity of *P. cembra* at the treeline on Mount Patscherkofel began one week later at the krummholz limit at 2180 m a.s.l. than at the forest limit and at the treeline at 1950 and 2110 m a.s.l., respectively. Maximum radial increment as well as the termination of annual ring formation were independent of altitude and occurred at the same time throughout the treeline ecotone [47,79]. At the beginning of the growing season, higher numbers of cambial cells were found at sites with an open canopy (i.e., the treeline and the krummholz limit) as compared to the closed forest limit. Yet, root zone temperatures were also significantly higher in open stands as compared to the forest limit, indicating that soil temperature may influence tree growth at the treeline.

However, studies on the effect of artificial soil warming on tree growth in boreal forests [80,81] and in the Swiss [82] and the Austrian Alps [83] yielded ambiguous results, indicating species-specific responses with results ranging from no stimulation in growth to a strong growth stimulation. Three years of soil warming did not cause any response in stem diameter increment of 25-year-old *P. cembra* trees in an afforestation at 2150 m a.s.l. treeline in the Sellrain valley [38,84]. In the kampfzone on Mount Patscherkofel south of Innsbruck (2.180 m a.s.l.), root zone cooling and root zone warming hardly affected diameter growth of *P. cembra* [36]. The results of this study indicated that *P. cembra* reacted to soil cooling with a decline and to soil warming with an increase in radial stem increment when compared to control trees with soil temperature left un-manipulated. Observed differences in diameter growth, however, were not statistically significant [36], probably because—in addition to soil temperature—varying soil nutrient contents with respect to microtopography are also known to influence radial growth at the treeline [85].

In contrast to root-zone temperature manipulation, nitrogen fertilization and understory removal significantly increased radial growth of 25-year-old *P. cembra* trees in a subalpine afforestation at 2150 m a.s.l. [84]. Removal of belowground competition also improved seedling growth within the treeline ecotone [71] and tree growth in subalpine forests [68,69]. Moreover, restricted seedling and tree establishment above the current treeline has also been attributed to competition with neighboring low stature understory vegetation. [23,86–89]. However, evaluating long-term modifications in species interactions is still under debate [90,91] and awaits clarification for the Central European Alps. Neither understory removal nor nitrogen addition nor soil temperature manipulation considerably influenced seasonal dynamics in radial growth of *P. cembra* at the treeline [36,84].

Table 2. Growing season mean whole tree non-structural carbohydrate (NSC) pools (% dry matter) and the contribution of needle, branch, stem, and root NSC accumulation patterns (%) to whole tree NSC pools of *Pinus cembra* trees along an altitudinal transect between 1750 and 2175 m a.s.l. on Mount Patscherkofel. After [92].

Elevation (m)	Whole Tree NSC Pool (% Dry Weight)	Needle (%)	Branch (%)	Stem (%)	Root (%)
2175	4.7 ± 1.2	61	16	11	12
2100	4.0 ± 0.8	51	12	21	16
1950	3.9	49	7	26	18
1750	3.8 ± 1.1	47	7	25	21

Nevertheless, a decline in radial growth with increasing elevation was dominantly caused by a corresponding decline in temperature [93]. As already pointed out by [5], at cooler temperatures, there is a tendency for photosynthates to be transformed into non-structural carbohydrates (NSC) and their components (soluble sugars and starch rather than cellulose), which in turn may limit diameter growth. Studies on mobile carbohydrates in *P. cembra* trees along a south facing 425 m elevation transect from the closed forest at 1750 m a.s.l. up to the krummholz limit at 2175 m a.s.l. [92] indicated that growing season mean whole tree NSC pools increased with elevation (Table 2). The observed increase in growing season mean whole tree NSC may be attributed to an increase in leaf mass with elevation [71,94], as needles contained the largest NSC fraction of the whole tree (Table 2).

4. Water and Carbon Relations within the Treeline Ecotone

4.1. Water Relations

Ample precipitation (Figure 1) and moderate evaporative demand usually cause soil water availability to be sufficient to meet the tree water demand within the treeline ecotone. Furthermore, minimum tree water potentials during the growing season have been found to stay considerably above a critical value, causing xylem cavitation risks [95,96]. In general, transpiration of conifers at the treeline shows a very pronounced response to increasing evaporative demand in terms of solar radiation and vapor pressure deficit under appropriate water supply [97,98]. Multiple regression analyses showed that, at the forest limit and at the treeline, solar radiation had a similar effect on tree water loss as vapor pressure deficit [97]. At the krummholz limit, by contrast, vapor pressure deficit had a bigger effect on tree water loss than irradiance. In addition, the absence of a closed canopy at the krummholz limit triggered an intensification in fundamental atmospheric processes [7,98].

Slope angle is a determinant of topo- and microclimatic related differences in solar radiation [98] and wind velocity [7,98]. At the krummholz limit, higher wind velocities than those at the forest limit and at the treeline (Table 1) cause a wind-induced clustering of the needles [99,100]. Consequently, the response of the stomata is impaired due to partial stomatal narrowing [99,100] (Figure 3).

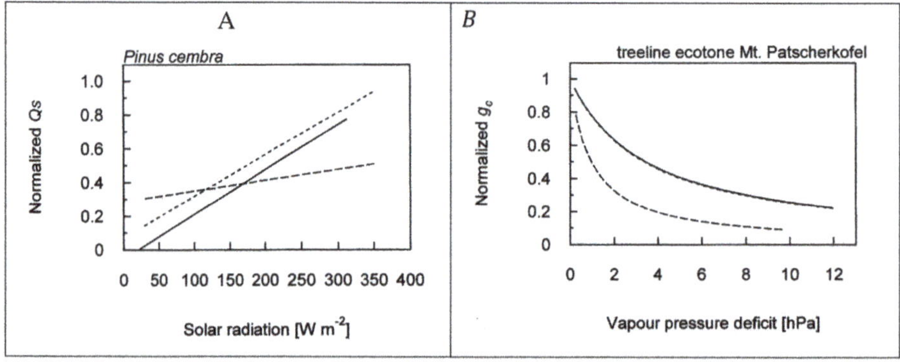

Figure 3. The relationship between (**A**) solar radiation and daily mean normalized sap flow density (Qs) and (**B**) between vapor pressure deficit and normalized daily mean crown conductance (g_c) of *Pinus cembra* at the krummholz limit (dashed lines), the treeline (dotted lines), and the forest limit (solid lines). After [98].

Soil temperature also seemed to control the water transport along the soil–plant–atmosphere continuum. Passive soil warming significantly improved sap flow density of *P. cembra* by 11% to 19% above the level of control trees during the second and the third year of treatment, respectively [38]. This effect appeared to result from a warming-induced reduction in water viscosity, an increased permeability of root membranes [101–103], and aquaporin-mediated changes in root conductivity [104,105], as tree fine-root production and mycorrhization did not respond to warming [106]. Hardly affected were leaf-level transpiration and conductance for water vapor, thus water-use efficiency stayed unchanged, as confirmed by needle $^{\delta13}C$ analysis [38].

Nevertheless, presuming a future environment with higher temperatures coupled with a decline in relative humidity and thus a considerable increase in evaporative demand [45] may reduce the water supply of trees at wind-exposed ridges and windward sun-exposed slopes with thin soil layers, particularly at their upper distribution limit. The latter is already affected today by temporary top-soil drought [52,98]. However, experimental studies on the effects of top-soil drought indicated that three years of rain exclusion did not considerably reduce sap flow density of *P. abies* and *L. decidua* saplings at the treeline in the Stubai Valley. The lack of a significant sap flow–soil water content correlation in both tree species indicated sufficient water supply, suggesting that whole tree water loss of young trees at the treeline primarily depends on evaporative demand [40]. Furthermore, roots of the treeline-associated conifers are able to penetrate into rocky undergrounds, allowing for the utilization of water sources in deep and wet soil layers, as shown for *L. decidua* trees using hydrogen stable isotope analysis [107].

4.2. Carbon Relations

Trees at the treeline do not show signs of greater limitations in net CO_2 uptake rate compared to trees at lower elevations [8], and their maximum net CO_2 uptake rate at ambient CO_2 is similar or even higher compared to their relatives at lower elevation sites [7]. As shown for current-year *P. cembra* needles in Figure 4, the temperature optimum of net CO_2 uptake shifted towards lower temperatures with increasing elevation, while maximum net CO_2 uptake rate at ambient CO_2 (A_{max}) tended to increase with elevation. The higher efficiency of carbon uptake per unit leaf area at the krummholz limit as compared to the treeline and the forest limit (Figure 4) may be attributed to elevation related differences in needle morphology [32,49] with respect to 100-needle dry weight and specific leaf area (Table 3).

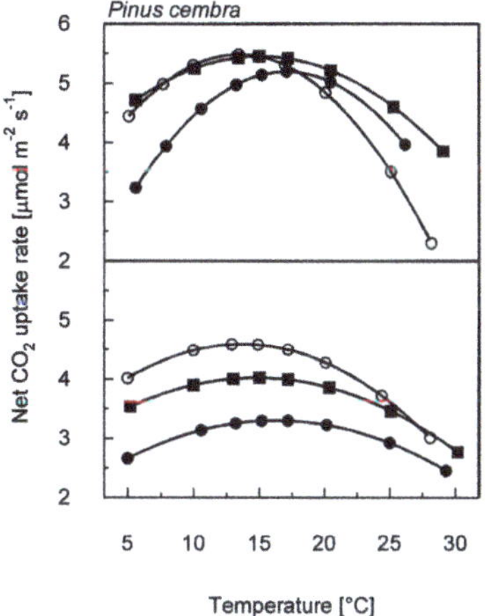

Figure 4. The relationship between temperature and net photosynthesis of fully developed, current-year *Pinus cembra* needles at the krummholz limit [open circles, tree line (closed squares), and the forest limit (solid circles) assessed in summer (top) and fall (bottom)]. After [48] and [108].

Table 3. The 100-needle dry weight and the specific leaf area (SLA) of current-year *Pinus cembra* needles at the krummholz limit (2180 m a.s.l.), the treeline (2100 m a.s.l.), and the forest limit (1950 m a.s.l.) on Mount Patscherkofel. After [108].

Elevation (m)	100-Needle Dry Weight (g)	SLA (cm^2 g^{-1})
2180	1.04 ± 0.97	36.5 ± 1.0
2100	1.31 ± 0.72	35.4 ± 0.8
1950	1.29 ± 0.67	43.8 ± 0.7

Conversely, *P. cembra* needles did not show elevational differences in foliar nitrogen concentration [32]. Thus, the combination of high nitrogen content with low specific leaf area [considered as a morphological feature of sun type needles (~thicker needles)] may enlarge the quantity of photosynthetic enzymes with increasing elevation. For plants in marginal habitats where assimilation is restricted to a short growing season, a higher amount of photosynthetic enzymes is suggested to provide a highly cost-effective system [109].

In general, air temperature hardly limits net CO_2 uptake, as more than 80% of full photosynthetic capacity is reached between 5 and 20 °C (Figure 4). During the growing season, net CO_2 uptake is primarily restricted by low irradiance and the accompanied low air and needle temperatures. Thus, the temperature optimum of net photosynthesis shifts towards higher values when photon flux density is high and towards lower values when photon flux density is low [110–112].

Dark respiration also adapts to the average temperature conditions that prevail in a local habitat. It is well known that, at any given temperature, trees growing at low temperature respire at higher rates as compared to trees growing in a warmer environment. When accounting for actual in situ temperatures, respiratory carbon losses of trees in the krummholz belt are similar or even lower than at the forest limit ([108]; Table 4). This genotypic and acclimative adaptation to lower temperatures [113,114] is also mirrored in a higher temperature sensitivity with increasing elevation (Figure 4, Table 4), permitting a higher metabolic activity at cooler conditions.

Table 4. Average air temperature during summer (T_{air}), corresponding night-time dark respiration (R_d), and the temperature sensitivity of respiration (Q_{10}) of current-year *Pinus cembra* needles at the krummholz limit (2180 m a.s.l.), the treeline (2100 m a.s.l.), and the forest limit (1950 m a.s.l.) assessed at the prevailing temperature conditions in summer 2007 on Mount Patscherkofel. After [108].

Elevation (m)	Tair (°C)	R_d (μmol m^{-2} s^{-1})	Q_{10}
2175	9.1	0.27	2.5
2100	10.0	0.28	2.3
1950	11.1	0.31	2.0

4.3. Long Term Trends in Carbon and Water Relations

At the treeline in the Central Austrian Alps, the maximum net CO_2 uptake rate at ambient CO_2 (A_{max}) of sun-exposed twigs from the upper canopy of mature, field grown *P. cembra*, *P. abies*, and *L. decidua* trees increased significantly between 1934 and 2012 (Figure 5). Temporal changes in A_{max} of *P. cembra* were also mirrored in tree growth and tree-ring $\delta^{13}C$, as basal area increment was significantly positively correlated with increasing A_{max}, whereas tree-ring $\delta^{13}C$ was significantly negatively correlated with A_{max}. Furthermore, in *P. cembra*, A_{max} tended to increase with tree-ring derived intrinsic water-use efficiency (Figure 5) and hence also with tree-ring derived intercellular CO_2 partial pressure, similar to a classical A/Ci curve [45]. The observed increase in A_{max} of *P. cembra* can be ascribed to both a rise in atmospheric CO_2 concentration and to the observed increase in T_{air}. In addition, temperature optimum of A_{max} for *P. cembra* increased from 12.5 °C in 1956 [111] to 17.1 °C in 2007 [48]. Thus, the observed increase in temperature optimum of A_{max} of 4.6 °C during the period 1956–2007 matched the observed increase in mean growing season air temperature of 0.9 °C per decade. In parallel to A_{max}, transpiration of *P. cembra* increased from 0.42 mmol m^{-2} s^{-1} in 1934 [115] to 1.12 mmol m^{-2} s^{-1} in 2012 [38]. As a result, the instantaneous water use efficiency of photosynthesis did not change considerably and averaged 4.3 ± 0.7 μmol mol^{-1} between 1934 and 2012, suggesting an increase in stomatal conductance for water vapor during the past 89 years [45]. Temporal patterns in $\Delta^{13}C$ and $\Delta^{18}O$ also suggest a parallel increase in CO_2-fixation and stomatal conductance of conifers at the treeline during the past 40 years [42]. The stability in tree-ring derived instantaneous water-use efficiency of 70-year-old *P. cembra*, *P. abies*, and *L. decidua* trees was accompanied by an increase in basal area increment, suggesting that young trees benefit from both climate warming and rising ambient CO_2 [42]. To the contrary, in 120-year-old *P. cembra* trees, nearby basal area increment remained largely stable during the past 35 years, indicating that trees had matured after 95 years of growth [45]. Thus, it seems that in environments under non-limiting water availability, as is the case at the treeline in the Central Austrian Alps [44], increasing temperature counteracts the diminishing effects of rising atmospheric CO_2 concentration on leaf conductance in adult trees [116,117].

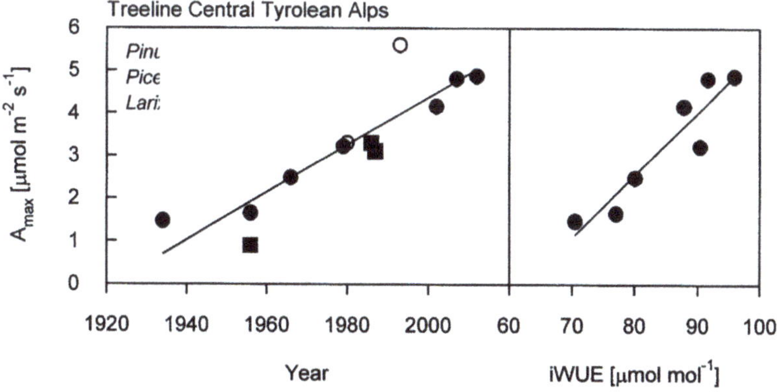

Figure 5. Temporal variation in maximum net CO_2 uptake rate at ambient CO_2 (A_{max}) of *Pinus cembra* (solid circles), *Picea abies* (solid squares), and *Larix decidua* (open circles) between 1925 and 2013 (left), and the relationship between A_{max} and tree-ring derived intrinsic water-use efficiency (iWUE) obtained for *Pinus cembra* at the treeline in the Central Tyrolean Alps. Compiled after [38,45,48,110,115,118–124].

5. Treeline Fluctuations

Radiocarbon dating of soil charcoal fragments indicates that the treeline has changed its position during the Holocene several times due to anthropogenic impacts [125]. In the Central Alps, the treeline reached its maximum elevation of 2700 m a.s.l. between 8000 and 5000 years BP [126,127]. Humans have influenced the timberline ecotone of the Tyrolean Alps since 7000 years BP [128–130], as nearly all accessible slopes were deforested to gain pastures [131] and timber for mining, firewood, and construction wood, particularly during the Middle Ages [21]. Presently, the actual treeline in the Central Tyrolean Alps is 150–300 m below its potential climatic maximum reached during the Holocene [132,133]. On siliceous parent material, the presence of Podzols under dwarf shrub communities above the present forest limit points towards the existence of a forest canopy in the past [134]. Nevertheless, topography and local climate have influenced human impacts; in particular, slopes with southern exposure became completely deforested [125]. Therefore, it is uncertain if the present actual treeline in the Central Tyrolean Alps mirrors equilibrium between climate and tree-specific ecophysiological features [25]. Furthermore, potential changes in ecosystem dynamics and functioning have also been attributed to recent changes in land-use management [135].

As growth [50] and reproduction [5,86,136] of trees at the treeline are controlled by temperature, an upslope migration of tree species has been predicted under climate warming ([21] and further references therein). Yet, the elevational movement of timberlines is dependent initially on new seedling establishment in favorable microsites that appear to be generated by ecological facilitation [86]. Even so, treelines influenced by pastoral use for centuries, as in the Central Austrian Alps, for example, respond differently to climate warming than undisturbed treelines [21]. Thus, the observed increase in seedling establishment in potential habitats beyond the forest limit of the Central Tyrolean Alps [132,133] is probably a consequence of decreasing grazing pressure [24] and changes in land-use management.

A huge percentage of seedling re-establishment at the treeline in the Swiss Alps [137] also resulted from invasion into potential habitats rather than an upward migration, as shown in pine seedlings by [70] and [138], leading [11] to suppose that the treeline in the Central European Alps behaves in a "conservative" way. Nevertheless, tree advancement to higher elevations is primarily dependent on seedling establishment in favorable microsites [86]. In addition, seedlings have to cope with a dense dwarf shrub and/or a closed grass cover and thus with impoverished mycorrhizal symbiosis; consequently, they also encounter competition for below-ground resources. Treeline migrations, however, may also influence carbon metabolism, soil microbial activity, and even ecosystem fluxes within the treeline ecotone ([18] and further references therein). Thus, understanding seedling physiology is a prerequisite for estimating future elevational shifts in treeline associated conifers within the Central Tyrolean Alps.

6. Conclusions and Climate Change Perspectives

Changes in temperature and soil water availability as well as the increasing atmospheric CO_2 concentration interact in complex ways in trees and forests at the treeline. As trees within the treeline ecotone are not carbon limited [8,13], climate warming—rather than the rising atmospheric CO_2 level—causes alterations in the ecological functioning of the treeline ecotone in the Central Austrian Alps [10,18]. Additional carbon acquired by treeline trees upon warming is available for both below (water nutrients) and above (light) demands with respect to resource sequestration [139] and competitiveness. The latter is known to curtail tree survival at the treeline [140]. Although the water uptake from soils will be improved by further climate warming due to an increased permeability of root membranes and aquaporin-mediated changes in root conductivity, the water balance may be perturbed at wind and sun-exposed sites with shallow soils due to increasing evaporative demand, as is indicated already today in the krummholz zone [98].

One has to take into account that treelines are vulnerable to anthropogenic influences such as land use changes and management practices [135]. The current observed seedling re-establishment at the treeline in the Central European Alps is an invasion into potential habitats due to decreasing grazing pressure rather than an upward-migration [133,137]. This kind of habitat reoccupation suggests the treeline in the Central European Alps behaves in a conservative way [11].

Moreover, in a future, warmer environment, tree population dynamics within the treeline ecotone will be controlled by severe weather events such as early and late frost events or soil drought during the summer rather than by a gradual increase in mean surface temperature [141]. Finally, seedling establishment should be considered for understanding the altitude of the treeline [86]. However, as there is a lack of knowledge in this particular topic within the treeline ecotone in the Central Austrian Alps, further research is needed regarding the importance of this life stage for evaluating treeline shifts and limits in a future, changing environment.

Author Contributions: G.W., W.O. and A.G. contributed equally to the manuscript.

Funding: This research received no external funding.

Conflicts of Interest: The authors declare no conflict of interest.

References

1. Däniker, A. Biologische Studien über Baum- und Waldgrenzen, insbesondere über die klimatischen Ursachen und deren Zusammenhänge. *Vierteljahresschrift der Naturforschenden Gesellschaft Zürich* **1923**, *68*, 1–102.
2. Holtmeier, F.-K. *Geoökologische Beobachtungen und Studien an der subarktischen und Alpinen Waldgrenze in Vergleichender Sicht*; Steiner: Wiesbaden, Germany, 1974.
3. Holtmeier, F.-K. *Mountain Timberlines. Ecology, Patchiness, and Dynamics*; Advances in Global Change Research; Kluwer Academic: Dordrecht, The Netherlands; Boston, MA, USA; London, UK, 2009; Volume 36.
4. Wardle, P. Alpine timberlines. In *Arctic and Alpine Environments*; Ives, J.D., Barry, R., Eds.; Methuen Publishing: London, UK, 1974; pp. 371–402.
5. Tranquillini, W. Physiological ecology of the alpine timberline. In *Tree Existence at High Altitudes With special Reference to the European Alps*; Ecological Studies; Springer: Berlin, Germany, 1979; Volume 31.
6. Slatyer, R.O.; Noble, I.R. Dynamics of treelines. In *Landscape Boundaries: Consequences for Biotic Diversity and Ecological Flows*; Ecological Studies; Hansen, A., DiCastri, F., Eds.; Springer: Berlin, Germany; New York, NY, USA, 1992; Volume 92, pp. 346–359.
7. Wieser, G.; Tausz, M. *Trees at Their Upper Limit: Treelife Limitation at the Alpine Timberline*; Plant Ecophysiology; Springer: Berlin, Germany, 2007; Volume 5.
8. Körner, C. *Alpine Treelines. Funtional Ecology of the Global High Elevation Tree Limits*; Springer: Basel, Switzerland, 2012.

9.   Kullman, L. 20th century climate warming and tree limit rise in the southern Scandes in Sweden. *Ambio* **2001**, *30*, 2–80. [CrossRef] [PubMed]

10. Wieser, G. Lessons from the timberline ecotone in the Central Tyrolean Alps: A review. *Plant Ecol. Divers.* **2012**, *5*, 127–139. [CrossRef]

11. Grace, J.; Berninger, F.; Nagy, L. Impact of climate change on the treeline. *Ann. Bot.* **2002**, *90*, 537–544. [CrossRef] [PubMed]

12. Sveinbjörnsson, B. North American and European treelines: External factors and internal processes controlling position. *Ambio* **2000**, *29*, 388–395. [CrossRef]

13. Körner, C.; Hoch, G. A test of treeline theory on a montane permafrost island. *Arct. Antarct. Alpine Res.* **2006**, *38*, 113–119. [CrossRef]

14. Körner, C.; Paulsen, J. A world-wide study of high altitude treeline temperatures. *J. Biogeogr.* **2004**, *31*, 713–732. [CrossRef]

15. Körner, C. Climatic treelines: Conventions, global patterns, causes. *Erdkunde* **2007**, *61*, 316–324. [CrossRef]

16. Tuhkanen, S. Climatic parameters and indices in plant geography. *Acta Phytogenetica Suecica* **1980**, *67*, 1–110.

17. Ohsawa, M. An interpretation of latitudinal patterns of forest limits in south and East Asian mountains. *J. Ecol.* **1990**, *78*, 262–339. [CrossRef]

18. Wieser, G.; Holtmeier, F.-K.; Smith, W.K.; William, K. Treelines in a changing global environment. In *Trees in a Changing Environment*; Plant Ecophysiology; Tausz, M., Grulke, N., Eds.; Springer: Dordrecht, The Netherland, 2014; Volume 9, pp. 221–263.

19. Holtmeier, F.-K. *Die Höhengrenze der Gebirgswälder. Arbeiten aus dem Institut für Landschaftsökologie, Band 8*; Westfälische Wilhelms-Universität: Münster, Germany, 2000.

20. Holtmeier, F.K.; Broll, G. Sensitivity and response of northern hemisphere altitudinal and polar treelines to environmental change at landscape and local scales. *Glob. Ecol. Biogeogr.* **2005**, *14*, 395–410. [CrossRef]

21. Holtmeier, F.-K.; Broll, G. Treeline advance—Driving processes and adverse factors. *Landsc. Onl.* **2007**, *1*, 1–33.

22. Walther, G.-R.S.; Pott, R. Climate change and high mountain vegetation shifts. In *Mountain Ecosystems. Studies in Treeline Ecology*; Broll, G., Keplin, B., Eds.; Springer: Berlin, Germany, 2005; pp. 77–95.

23. Smith, W.K.; Germino, M.J.; Johnson, D.M.; Reinhardt, K. The altitude of alpine treeline: A bellwether of climate change. *Bot. Rev.* **2009**, *75*, 163–190. [CrossRef]

24. Wieser, G.; Matyssek, R.; Luzian, R.; Zwerger, P.; Pindur, P.; Oberhuber, W.; Gruber, A. Effects of atmospheric and climate change at the timberline of the Central European Alps. *Ann. For. Sci.* **2009**, *66*, 402. [CrossRef] [PubMed]

25. Intergovernmental Panel of Climate Change (IPCC). *Climate Change. The Physical Science Basis*; Cambridge University Press: Cambridge, UK, 2013.

26. Böhm, R.; Auer, I.; Brunetti, M.; Maugeri, M.; Nanni, T.; Schöner, W. Regional temperature variability in the European Alps: 1760–1998 from homogenized instrumental time. *Int. J. Climatol.* **2001**, *1801*, 1779–1801. [CrossRef]

27. Beniston, M. Mountain climates and climate change: An overview of processes focusing on the European Alps. *Pure Appl. Geophys.* **2005**, *162*, 1587–1606. [CrossRef]

28. Becker, A.; Körner, C.; Brun, J.-J.; Guisan, A.; Tappeiner, U. Ecological and land use studies along elevational gradients. *Mount. Res. Dev.* **2007**, *27*, 58–65. [CrossRef]

29. Körner, C. Carbon limitation in trees. *J. Ecol.* **2003**, *94*, 4–17. [CrossRef]

30. Blois, J.L.; Williams, J.W.; Fitzpatrick, M.C.; Jackson, S.T.; Ferrier, S. Space cab substitute for time in predicting climate-change effects on biodiversity. *Proc. Natl. Acad. Sci. USA* **2015**, *110*, 9374–9379. [CrossRef]

31. Larcher, W. Ergebnisse des IBP-Projektes "Zwergstrauchheide Patscherkofel". *Sitzungsbericht Österreichische Akademie der Wissenschaften, Math. Naturwiss. Kl. I.* **1977**, *186*, 301–371.

32. Körner, C. The use of altitude in ecological research. *Trends Ecol. Evol.* **2007**, *22*, 569–574. [CrossRef] [PubMed]

33. King, G.; Gugleri, F.; Fonti, P.; Frank, D.C. Tree growth response along an elevational gradient: Climate or genetics? *Oecologia* **2013**, *173*, 1587–1600. [CrossRef] [PubMed]

34. Jochner, M.; Bugmann, H.; Nötzli, M.; Bigler, C. Tree growth response to changing temperatures across space and time: A fine-scale analysis at the treeline in the Swiss Alps. *Trees* **2018**, *32*, 645–660. [CrossRef]

35. Elmendorf, S.C.; Henry, G.H.R.; Hollister, R.D.; Fossa, M.A.; Gould, W.A.; Hermanutz, L.; Hofgraad, A.; Jonsdottri, I.S.; Levesque, E.; et al. Experiment, monitoring, and gradient methods used to infer climate

change effects on plant communities yield constant patterns. *Proc. Natl. Acad. Sci. USA* **2015**, *112*, 448–452. [CrossRef] [PubMed]

36. Gruber, A.; Wieser, G.; Oberhuber, W. Opinion paper: Effects of simulated soil temperature on stem diameter increment of *Pinus cembra* at the alpine timberline: A new approach based on root zone roofing. *Eur. J. For. Res.* **2010**, *129*, 141–144. [CrossRef]

37. Hagedorn, F.; Martin, M.; Rixen, C.; Rusch, S.; Bebi, P.; Zürcher, A.; Siegwolf, R.T.S.; Wipf, S.; Escape, C.; Roy, J. Short-term responses of ecosystem carbon fluxes to experimental soil warming at the Swiss alpine treeline. *Biogeochemistry* **2010**, *97*, 7–19. [CrossRef]

38. Wieser, G.; Grams, T.E.E.; Matyssek, R.; Oberhuber, W.; Gruber, A. Soil warming increased whole-tree water use of *Pinus cembra* at the treeline in the Central Tyrolean Alps. *Tree Physiol.* **2015**, *35*, 279–288. [CrossRef]

39. Dawes, M.A.; Schleppi, P.; Hättenschwiler, P.; Hagedorn, F. Soil warming opens the nitrogen cycle at the alpine treeline. *Global Change Biol.* **2017**, *23*, 421–434. [CrossRef]

40. Wieser, G.; Oberhuber, W.; Gruber, A.; Oberleitner, F.; Hasibeder, R.; Bahn, M. Artificial top soil drought hardly affects water use of young *Picea abies* and *Larix decidua* trees at treeline in the Austrian Alps. *Forests* **2019**. under review.

41. Bahn, M.; Erb, K.; Harris, E. *ClimLUC—Climate Extremes and Land-Use Change: Effects on Ecosystem Processes and Services*; Austrian Academy of Sciences Press: Vienna, Austria, 2018. [CrossRef]

42. Wieser, G.; Oberhuber, W.; Gruber, A.; Leo, M.; Matyssek, R.; Grams, T.E.E. Stable water use efficiency under climate change of three sympatric conifer species at the Alpine treeline. *Front. Plant Sci.* **2016**, *7*, 799. [CrossRef]

43. Theurillat, J.P.; Guisan, A. Potential impact of climate change on vegetation in the European Alps: A review. *Clim. Change* **2001**, *50*, 77–109. [CrossRef]

44. Wieser, G. Climate at the alpine timberline. In *Trees at Their Upper Limit. Treelife Limitation at the Alpine Timberline*; Plant Ecophysiology; Wieser, G., Tausz, M., Eds.; Springer: Berlin, Germany, 2007; Volume 5, pp. 19–36.

45. Wieser, G.; Oberhuber, W.; Waldboth, B.; Gruber, A.; Matyssek, R.; Siegwolf, R.T.W.; Gramd, T.E.E. Long-term trends in leaf level gas exchange mirror tree-ring derived intrinsic water-use efficiency of *Pinus cembra* at treeline during the last century. *Agric. For. Meteorol.* **2018**, *248*, 251–258. [CrossRef]

46. HISTALP. Available online: http://www.zamg.ac.at/histalp/inxex.php (accessed on 11 April 2019).

47. Gruber, A.; Wieser, G.; Oberhuber, W. Intra-annual dynamics in stem CO_2 efflux in relation to cambial activity and xylem development in *Pinus cembra*. *Tree Physiol.* **2009**, *29*, 641–649. [CrossRef] [PubMed]

48. Wieser, G.; Oberhuber, W.; Walder, L.; Spieler, D.; Gruber, A. Photosynthetic temperature adaptation of *Pinus cembra* within the timberline ecotone of the Central Austrian Alps. *Ann. For. Sci.* **2010**, *67*, 201. [CrossRef] [PubMed]

49. Aulitzky, H. Die Bodentemperaturverhältnisse in der Kampfzone oberhalb der Waldgrenze und im subalpinen Zirben-Lärchenwald. *Mitteilungen der Forstlichen Bundesversuchsanstalt Mariabrunn* **1961**, *59*, 153–208.

50. Oberhuber, W. Limited by growth processes. In *Trees at Their Upper Limit. Tree Life Limitation at the Alpine Timberline*; Plant Ecophysiology; Wieser, G., Tausz, M., Eds.; Springer: Berlin, Germany, 2007; Volume 5, pp. 131–143.

51. Aulitzky, H. Über die Windverhältnisse einer zentralalpinen Hangstation in der subalpinen Stufe. *Mitteilungen der Forstlichen Bundesversuchsanstalt Mariabrunn* **1961**, *59*, 209–230.

52. Aulitzky, H. Grundlagen und Anwendungen des vorläufigen Wind-Schnee-Ökogramms. *Mitteilungen der Forstlichen Bundesversuchsanstalt Mariabrunn* **1963**, *60*, 763–834.

53. Kronfuss, H. Schneelage und Ausaperung an der Waldgrenze. *Mitteilungen der Forstlichen Bundesversuchsanstalt Wien* **1967**, *75*, 207–241.

54. Kronfuss, H. Räumliche Korrelation zwischen der Windstärke in Bodennähe und der Schneedeckenandauer. *Centralblatt für das Gesamte Forstwesen* **1970**, *87*, 99–116.

55. Kronfuss, H.; Stern, R. *Strahlung und Vegetation*; Angewandte Pflanzensoziologie; Publishet Forstliche Bundesversuchsanstalt: Wien, Austria, 1978; Volume 24.

56. Larcher, W. Winter stress in high mountains. In *Establishment and Tending of Subalpine Forest: Research and Management*; Turner, H., Tranquillini, W., Eds.; Berichte Eidgenössische Anstalt für das forstliche Versuchswesen Birmensdorf: Zürich, Switzerland, 1985; Volume 270, pp. 11–19.

57. Loris, K. Dickenwachstum von Zirbe, Fichte und Lärche an der alpinen Waldgrenze/Patscherkofel. Ergebnisse der Dendrometermessungen 1976/79. *Mitteilungen der Forstlichen Bundesversuchsanstalt Wien* **1981**, *142*, 417–441.

58. Kronfuss, H. Der Einfluß der Lufttemperatur auf das Höhenwachstum der Zirbe. *Centralblatt für das Gesamte Forstwesen* **1994**, *111*, 165–181.

59. Kanninen, M. Shoot elongation in Scots Pine: Diurnal variations and response to temperature. *J. Exp. Bot.* **1985**, *36*, 1760–1770. [CrossRef]

60. Hässler, R.; Streule, A.; Turner, H. Shoot and root growth of young *Larix decidua* in contrasting microenvironments neat the alpine timberline. *Phyton* **1999**, *39*, 47–52.

61. Tranquillini, W.; Unterholzner, R. Das Wachstum zweijähriger Lärchen einheitlicher Herkunft in verschiedenen Höhenlagen. *Centralblatt für das Gesamte Forstwesen* **1968**, *85*, 43–59.

62. Havranek, W.M. Gas exchange and dry matter allocation in larch at the alpine timberline on Mount Patscherkofel. In *Establishment and Tending of Subalpine Forest: Research and Management*; Turner, H., Tranquillini, W., Eds.; Berichte Eidgenössische Anstalt für das forstliche Versuchswesen Birmensdorf: Zürich, Switzerland, 1985; Volume 270, pp. 135–142.

63. Sveinbjörnsson, B.; Nordell, O.; Kauhanen, H. Nutrient relations of mountain birch growth at and below the elevational tree-line in Swedish Lapland. *Funct. Ecol.* **1992**, *6*, 213–220. [CrossRef]

64. LeBauer, D.S.; Treseder, K.K. Nitrogen limitation of net primary productivity in terrestrial ecosystems is globally distributed. *Ecology* **2008**, *89*, 371–379. [CrossRef]

65. Nadelhoffer, J.K.; Giblin, A.E.; Shaver, G.R.; Laiúndre, J.A. Effects of temperature and substrate quality on element mineralization in six Arctic soils. *Ecology* **1991**, *72*, 242–253. [CrossRef]

66. Rustad, L.E.; Campbell, J.L.; Marion, G.M.; Norby, R.J.; Mitchell, M.J.; Hartley, A.E.; Cornelissen, J.H.C.; Gurevitch, J. GCTE-NEWS. A meta analysis of the response of soil respiration, net nitrogen mineralization, and aboveground plant growth to experimental ecosystem warming. *Oecologia* **2001**, *126*, 543–562. [CrossRef]

67. Melillo, J.M.; Steudler, P.A.; Aber, J.D.; Newkirk, K.; Lux, H.; Bowles, F.P.; Catricala, C.; Magill, A.; Ahrens, T.; Morrisseau, S. Soil warming and carbon-cycle feedback to the climate system. *Science* **2002**, *298*, 2173–2176. [CrossRef]

68. Platt, K.H.; Allen, R.B.; Coomes, D.A.; Wiseer, S.K. Mountain beech seedling responses to removal of below-ground competition and fertiliser addition. *N. Z. J. Ecol.* **2004**, *28*, 289–293.

69. Song, H.; Cheng, S.; Zhang, Y. The growth of two species of subalpine conifer sapliplings in response to soil warming and inter-competition in Mt. Gongga on the south-eastern fringe of the Qinghai-Tibetan plateau, China. *World J. Eng. Technol.* **2016**, *4*, 398–412. [CrossRef]

70. Paulsen, J.; Weber, U.M.; Körner, C. Tree growth near treeline: Abrupt or gradual reduction with altitude? *Arct. Antarct. Alpine Res.* **2000**, *32*, 14–20. [CrossRef]

71. Oswald, H. Verteilung und Zuwachs der Zirbe (*Pinus cembra*) de subalpinen Stufe an einem zentralalpinen Standort. *Mitteilungen der Forstlichen Bundesversuchsanstalt Mariabrunn* **1963**, *60*, 437–499.

72. Kronfuss, H.; Havranek, W.M. Effects of elevation and wind on the growth of *Pinus cembra* L. in a subalpine afforestation. *Phyton* **1999**, *39*, 99–106.

73. Li, M.H.; Yang, J. Effects of microsite on growth of *Pinus cembra* in the subalpine zone of the Austrian Alps. *Ann. For. Sci.* **2004**, *61*, 319–325. [CrossRef]

74. Li, M.H.; Yang, J.; Kräuchi, N. Growth response of *Picea abies* and *Larix decidua* to elevation in subalpine areas of Tyrol, Austria. *Can. J. For. Res.* **2003**, *33*, 653–662. [CrossRef]

75. Holtmeier, F.-K.; Broll, G. Altitudinal and polar treelines in the northern hemisphere - causes and response to climate change. *Polarforschung* **2010**, *79*, 139–153.

76. Vaganov, E.A.; Hughes, M.K.; Kirdyanov, A.V.; Schweingruber, F.H.; Silkin, P.P. Influence of snowfall and melting time on tree growth in subarctic Eurasia. *Nature* **1999**, *400*, 149–151. [CrossRef]

77. Lupi, C.; Morin, H.; Deslauries, A.; Rossi, S. Xylogenesis in black spruce: Does soil temperature matter? *Tree Physiol.* **2012**, *32*, 74–82. [CrossRef]

78. Rossi, S.; Deslauries, A.; Anfodillo, T.; Morin, H.; Saracino, A.; Motta, R.; Borghetti, M. Conifers in cold environments synchronyze maximum growth rate of tree-ring formation with day length. *New Phytol.* **2006**, *170*, 301–310. [CrossRef]

79. Gruber, A.; Zimmenmann, J.; Wieser, G.; Oberhuber, W. 2. Effects of climate variables on intra-annual stem radial increment in *Pinus cembra* (L.) along the timberline ecotone. *Ann. For. Sci.* **2009**, *66*, 503. [CrossRef] [PubMed]

80. Strömgren, M.; Linder, S. Effects of nutrition and soil warming on stemwood production in a boreal Norway spruce stadns. *Glob. Ghange Biol.* **2002**, *8*, 1194–1204.

81. Dao, M.C.E.; Rossi, R.; Walsh, D.; Morin, H.; Houle, D. A 6-yer-long manipulation with soil warming and canopy nitrogen additions does not affect xylem phenology and cell production of mature black spruce. *Front. Plant Sci.* **2015**, *6*, 877. [CrossRef] [PubMed]

82. Dawes, M.A.; Philipson, C.D.; Fonti, P.; Bebi, P.; Hättenschwiler, S.; Hagedorn, F.; Rixen, C. Soil warming and CO_2 enrichment induce biomass shifts in alpüine treeline vegetation. *Glob. Change Biol.* **2015**, *21*, 2005–2021. [CrossRef]

83. Loranger, H.; Zotz, G.; Bader, M.Y. Early establishment of trees at the alpine treeline: Idiosyncratic species responses to temperature-moisture interactions. *AOB Plants* **2016**, *8*, plw053. [CrossRef] [PubMed]

84. Gruber, A.; Oberhuber, W.; Wieser, G. Nitrogen addition and understory removal but not soil warming increases radial growth of Pinus cembra at treeline in the Central Austrian Alps. *Front. Plant Sci.* **2018**, *9*, 711. [CrossRef] [PubMed]

85. Anschlag, K.; Broll, G.; Holtmeier, F.-K. Mountain birchseedlings in the treeline ecotone, subarctic Finland: Variationin above-and below grond growth depending onmicrotopography. *Arct. Antarct. Alpine Res.* **2008**, *40*, 609–616. [CrossRef]

86. Brodersen, C.R.; Germino, M.J.; Johnson, D.M.; Reinhardt, K.; Smith, W.K.; Resler, L.M.; Bader, M.Y.; Sala, A.; Kueppers, L.M.; Broll, G. Seedling survival at timberline is critical to conifer mountain forest elevation and extent. *Front. For. Global Change* **2019**, *2*, 9. [CrossRef]

87. Batllori, E.; Camarero, J.J.; Gutierrez, E. Current regeneration patterns at the treeline in the Pyrenees indicate similar recruitment processes irrespective to the past disturbance regime. *J. Biogeogr.* **2010**, *37*, 1938–1950.

88. Elliott, G.P. Influences of 20th century warming at the upper treeline contingent on local-scale interactions: Evidence from a latitudinal gradient in the Rocky Mountains, USA. *Glob. Ecol. Biogeogr.* **2011**, *20*, 46–57. [CrossRef]

89. Grau, O.; Ninot, J.M.; Blanco-Moreno, J.M.; van Logtestijn, R.S.P.; Cornelissen, J.H.C.; Callaghan, T.V. Shrub-tree interactions and environmental changes drive treeline dynamics in the Subarctic. *Oikos* **2012**, *121*, 1680–1690. [CrossRef]

90. Liang, E.; Wang, Y.; Pizo, S.; Lu, X.; Camarero, J.J.; Zhu, H.; Zhu, L.; Ellison, A.M.; Ciaias, P.; Penuelas, J. Species interactions slow warming-induced upward shifts of treelines on the Tibetan Plateau. *Proc. Natl. Acad. Sci. USA* **2016**, *113*, 4380–4385. [CrossRef] [PubMed]

91. Camarero, J.J.; Linares, J.C.; Garcia-Cervigon, A.L.; Batllori, E.; Martinez, I.; Gutierrez, E. Back to the future: The responses of alpine treelines to climate warming are constrained by the current ecotone structures. *Ecosystems* **2017**, *20*, 683–700. [CrossRef]

92. Gruber, A.; Pirkebner, D.; Oberhuber, W.; Wieser, G. Spatial and seasonal variations in mobile carbohydrates i Pinus cembre in the treeline ecotone of the Central Austrian Alps. *Eur. J. For. Res.* **2011**, *130*, 173–179. [CrossRef]

93. Oberhuber, W. Influence of climate on radial growth of *Pinus cembra* within the alpine timberline ecotone. *Tree Physiol.* **2004**, *24*, 219–301. [CrossRef] [PubMed]

94. Bernoulli, M.; Körner, C. Dry matter allocation in treeline trees. *Phyton* **1999**, *39*, 7–12.

95. Mayr, S. *Limits in water relations. In Trees at Their Upper Limit. Treelife Limitation at the Alpine Timberline*; Wieser, G., Tausz, M., Eds.; Springer: Dordrecht, The Netherlands, 2007; pp. 145–162.

96. Anfodillo, T.; Rento, S.; Carraro, V.; Furlanetto, L.; Urbinati, C.; Carrer, M. Tree water relations and climatic variations at the alpine timberline: Seasonal changes of sap flux and xylem water potential in *Larix decidua* Miller, *Picea abies* (L.) Karst. and *Pinus cembra* L. *Ann. For. Sci.* **1998**, *55*, 159–172. [CrossRef]

97. Wieser, G.; Gruber, A.; Oberhuber, W. Sap flow characteristics and whole-tree water use of *Pinus cembra* across the treeline ecotone of the central Tyrolean Alps. *Eur. J. For. Res.* **2014**, *133*, 287–295. [CrossRef]

98. Aulitzky, H. The microclimatic conditions in a subalpine forest as basis for the management. *Geo J.* **1984**, *8*, 277–281. [CrossRef]

99. Caldwell, M. The effect of wind on stomatal aperture, photoynthesis, and transpiration of *Rhododendron ferrugineum* L. and *Pinus cembra* L. *Centralblatt für das gesamte Fortswesen* **1970**, *87*, 193–201.

100. Caldwell, M. Pant gas exchange at high wind speeds. *Plant Physiol.* **1970**, *46*, 535–537. [CrossRef] [PubMed]

101. Goldstein, G.H.; Brubaker, L.B.; Hinckley, T.M. Water relations of white spruce (*Picea glauca* (Moench(Voss)) at tree line in north central Alaska. *Can. J. For. Res.* **1985**, *15*, 1080–1087. [CrossRef]

102. Wan, X.; Zwiazek, J.J. Mercuric chloride effects on root water transport in aspen seedlings. *Plant Physiol.* **1999**, *121*, 939–946. [CrossRef] [PubMed]

103. Wan, X.; Zwiazek, J.J.; Lieffers, V.J.; Landhäusser, S.M. Hydraulic conductance in aspen (*Populus tremuloides*) seedlings exposed to low root temperatures. *Tree Physiol.* **2001**, *21*, 691–696. [CrossRef] [PubMed]

104. McElrone, A.J.; Bichler, J.; Pockman, W.T.; Addington, R.N.; Linder, C.R.; Jackson, R.B. Aquaporin-mediated changes in hydraulic conductivity of deep tree roots accessed via caves. *Plant Cell Environ.* **2007**, *30*, 1411–1421. [CrossRef] [PubMed]

105. Ruggiero, C.; Angelino, G.; Maggio, A. Developmental regulation of water uptake in wheat. *J. Plant Physiol.* **2007**, *164*, 1170–1178. [CrossRef] [PubMed]

106. Rainer, G.; Kuhnert, R.; Unterholzer, M.; Dresch, P.; Gruber, A.; Peintner, U. Host-specialist dominated ectomycorrhizal communities of *Pinus cembra* are not affected by temperature manipulation. *J. Fungi* **2015**, *1*, 55–75. [CrossRef] [PubMed]

107. Valentini, R.; Anfodillo, T.; Ehlringer, J. Water sources utilization and carbon isotope composition ($\delta^{13}C$) of co-occurring species along an altitudinal gradient in the Italian Alps. *Can. J. For. Res.* **1994**, *24*, 1575–1578. [CrossRef]

108. Walder, L.M. Der Gaswechsel der Zirbe (*Pinus cembra* L.) im Waldgrenzökoton am Patschertkofel. Diploma Thesis, Botany, University Innsbruck, Tyrol, Austria, June 2007.

109. Birmann, K.; Körner, C. Nitrogen status of conifer needles at the alpine treeline. *Plant Ecol. Divers.* **2009**, *2*, 233–241. [CrossRef]

110. Pisek, A.; Winkler, E. Assimilationsvermögen und Respiration der Fichte (*Picea excelsa* LINK) in verschiedenen Höhenlagen und der Zirbe (*Pinus cembra* L.) an der Waldgrenze. *Planta* **1958**, *51*, 518–543. [CrossRef]

111. Pisek, A.; Larcher, W.; Moser, W.; Pack, I. Kardinale Temperaturbereiche der Photosynthese und Grenztemperaturen des Lebens der Blätter verschiedener Spermatophyten. III. Temperaturabhängigkeit und optimaler Temperaturbereich der Nettophotosynthese. *Flora* **1969**, *Ab. B 158*, 608–630.

112. Wieser, G. Carbon dioxide gas exchange of cembran pine (*Pinus cembra*) at the alpine timberline during winter. *Tree Physiol.* **1997**, *17*, 473–477. [CrossRef] [PubMed]

113. Havranek, W.M.; Tranquillini, W. Physiological processes during winter dormancy and their ecological significace. In *Ecophysilogy of Coniferous Forest*; Smith, W.K., Hinckley, T.M., Eds.; Academic Press: San Diego, CA, USA, 1995; pp. 95–124.

114. Larcher, W. *Ökophysiologie der Pflanzen: Leben, Leistung und Stressbewältigung der Pflanzen in ihrer Umwelt*; UTB für Wissenschaft: Ulmer Stuttgart, Germany, 2001.

115. Cartellieri, E. Jahresgang von osmotischem Wert, Transpiration und Assimilation einiger Ericaceen der alpinen Zwergstrauchheide und von *Pinus cembra*. *Jahrbuch für Wissenschaftliche Botanik* **1935**, *82*, 460–506.

116. Marchin, R.E.; Broadhead, A.A.; Bostic, L.E.; Dunn, R.R.; Hoffmann, W.A. Stomatal acclimation to vapor pressure deficit doubles transpiration of small tree seedlings with warming. *Plant Cell Environ.* **2016**, *39*, 2221–2234. [CrossRef] [PubMed]

117. Saurer, M.; Spahni, R.; Frank, D.C.; Joos, F.; Leuenberger, M.; Loader, N.J.; Mc Carrol, D.; Gagen, M.; Poulter, B.; Sierwolf, R.T.S.; et al. Spatial variability and temporal trends in water-use efficiency of European forest. *Global Change Biol.* **2014**, *20*, 332–336. [CrossRef] [PubMed]

118. Koch, W.; Klein, W.; Walz, H. Neuartige Gaswechsel-Messanlage für Pflanzen in Laboratorium und Freiland. *Siemens Zeitschrift* **1968**, *42*, 392–404.

119. Havranek, W.M. Stammatmung, Dickenwachstum und Photosynthese einer Zirbe (*Pinus cembra*) an der Waldgrenze. *Mitteilungen der Forstlichen Bundesversuchsanstalt Wien* **1981**, *142*, 443–467.

120. Wieser, G.; Häsler, R.; Götz, B.; Koch, W.; Havranek, W.M. Role of climate, crown position, tree age and altitude in calculated ozone flux into needles of *Picea abies* and *Pinus cembra*: A synthesis. *Environ. Pollut.* **2000**, *110*, 415–422. [CrossRef]

121. Wieser, G. Environmental control of carbon dioxide gas exchange in needles of a mature *Pinus cembra* tree at the alpine timberline during the growing season. *Phyton* **2004**, *44*, 145–153.

122. Havranek, W.M.; Wieser, G.; Bodner, M. Ozone fumigation of Norway spruce at timberline. *Ann. Sci. For.* **1989**, *46s*, 581–585. [CrossRef]

123. Benecke, U.; Schulze, E.-D.; Matyssek, R.; Havranek, W.M. Environmental control of CO_2-assimilation and leaf conductance in Larix decidua Mill. I. a comparison of contrasting natural environments. *Oecologia* **1981**, *50*, 54–61. [CrossRef] [PubMed]

124. Volgger, E. Zur Ozonempfindlichkeit der europäischen Lärche (Larix decidua Mill.) an der Waldgrenze. Diploma Thesis, Botany, University of Innsbruck, Innsbruck, Austria, 1995.

125. Stöhr, D. Soils - heterogeneous at a microscale. In *Trees at Their Upper Limit. Tree Life Limitation at the Alpine Timberline*; Plant Ecophysiology; Wieser, G., Tausz, M., Eds.; Springer: Berlin, Germany, 2007; Volume 5, pp. m37–m56.

126. Carcaillet, C. Changes in landscape structure in the northwestern Alps over the last 7000 years: Lessons from soil charcoal. *J. Veg. Sci.* **2000**, *11*, 705–714. [CrossRef]

127. Tinner, W.; Theurillat, J.P. Uppermost limit, extent, and fluctuations of the timberline and treeline ecocline in the Swiss Central Alps during the past 11500 yr. *Arct. Antarct. Alpine Res.* **2003**, *35*, 158–169. [CrossRef]

128. Bortenschlager, S. Beiträge zur Vegetationsgeschichte Tirols. I. Inneres Ötztal und unteres Inntal. *Berichte des Naturwissenschaftlich-MedizinischenVereins in Innsbruck* **1984**, *71*, 19–56.

129. Patzelt, G. Modellstudie Ötztal—Landschaftsgeschichte in Hochgebirgsraum. *Mitteilungen der Österreichischen Geographischen Gesellschaft* **1996**, *138*, 53–70.

130. Sturmböck, M. *Die spät- und Postglaziale Vegetationsgeschichte des Nordwestlichen Südtirols. Dissertationes Botanicae*; Cramer: Berlin, Germany, 1999; Volume 299.

131. Leidlmair, A. *Landeskunde Österreich. Landesnatur, Kulturlandschaft, Bevölkerung, Wirtschaft, Die Bundesländer*; Paul List (Verlag): München, Germany, 1983.

132. Luzian, R.; Pindur, P. *Prähistorische Lawinen. Nachweis und Analyse holozäner Lawinenereignisse in den Zillertaler Alpen, Österreich*; Mitteilungen der Kommission für Quartärforschung der Österreichische Akademie der Wissenschaft: Wien, Austria, 2007; Volume 16.

133. Zwerger, P.; Pindur, P. Waldverbreitung und Waldentwicklung im Oberen Zemmgrund. In *Prähistorische Lawinen. Nachweis und Analyse holozäner Lawinenereignisse in den Zillertaler Alpen*; Luzian, R., Pindur, P., Eds.; BFW Berichte: Wien, Austria, 2007; Volume 141, pp. 69–97.

134. Neuwinger-Raschendorfer, I. Bodenfeuchtemessungen. *Mitteilungen der Forstlichen Bundesversuchsanstalt Mariabrunn* **1963**, *59*, 257–264.

135. Cernusca, A. Aims and tasks of ECOMONT. In *Land-Use Changes in European Mountain Ecosystems. ECOMONT—Concepts and Results*; Cernusca, A., Tappeiner, U., Bayfield, N., Eds.; Blackwell: Berlin, Germany, 1999; pp. 13–35.

136. Smith, W.K.; Germino, M.J.; Hanckock, T.E.; Johnson, D.M. Another perspective on altitudinal limits of alpine timberlines. *Tree Physiol.* **2003**, *23*, 1101–1112. [CrossRef]

137. Gehring-Fasel, J.; Gusian, A.; Zimmermann, N.E. Tree line shifts in the Swiss Alps: Climate change or land abandonment. *J. Veg. Sci.* **2007**, *18*, 571–582. [CrossRef]

138. Hättenschwiler, S.; Körner, C. Responses to recent climate warming of *Pinus sylvestris* and *Pinus cembra* within their montane transition zone in the Swiss Alps. *J. Veg. Sci.* **1995**, *6*, 357–368. [CrossRef]

139. Mooney, H.A.; Winner, W.E. Partitioning response of plants to stress. In *Response of Plants to Multiple Stresses*; Mooney, H.A., Winner, W.E., Pell, E.J., Eds.; Academic Press: San Diego, CA, USA, 1991; pp. 129–141.

140. Callaway, R.M. Competition and facilitation on elevation gradients in subalpine forests in the northern Rock Mountains, USA. *Oikos* **1998**, *82*, 561–573. [CrossRef]

141. Seidl, R.; Thom, D.; Kautz, M.; Martin-Benito, D.; Peltoniemi, M.; Vacchiano, G.; Wild, J.; Ascoli, D.; Petr, M.; Honkaniemi, J.; et al. Forest disturbances under climate change. *Nat. Clim. Change* **2017**, *7*, 395–2017. [CrossRef] [PubMed]

 *forests*

Article

Growth Trends of Coniferous Species along Elevational Transects in the Central European Alps Indicate Decreasing Sensitivity to Climate Warming

Walter Oberhuber [1], Ursula Bendler [1], Vanessa Gamper [1], Jacob Geier [1], Anna Hölzl [1], Werner Kofler [1], Hanna Krismer [2], Barbara Waldboth [1] and Gerhard Wieser [3,*]

[1] Department of Botany, Leopold-Franzens-Universität Innsbruck, Sternwartestraße 15, A-6020 Innsbruck, Austria; walter.oberhuber@uibk.ac.at (W.O.); ursula.bendler@student.uibk.ac.at (U.B.); vanessa.gamper@student.uibk.ac.at (V.G.); jacob.geier@student.uibk.ac.at (J.G.); anna.hoelzl111@gmail.com (A.H.); werner.kofler@uibk.ac.at (W.K.); barbara.waldboth@student.uibk.ac.at (B.W.)

[2] Department of Forest-and Soil Sciences, Institute of Silviculture, University of Natural Resources and Life Sciences, Peter-Jordan-Straße 82, A-1190 Vienna, Austria; hanna.krismer@tirol.gv.at

[3] Division of Alpine Timberline Ecophysiology, Federal Research and Training Centre for Forests, Natural Hazards and Landscape (BFW), Rennweg 1, A-6020 Innsbruck, Austria

* Correspondence: gerhard.wieser@uibk.ac.at; Tel.: +43-512-5739-335120

Received: 21 November 2019; Accepted: 20 January 2020; Published: 22 January 2020

Abstract: Tree growth at high elevation in the Central European Alps (CEA) is strongly limited by low temperature during the growing season. We developed a tree ring series of co-occurring conifers (Swiss stone pine, Norway spruce, European larch) along elevational transects stretching from the subalpine zone to the krummholz limit (1630–2290 m asl; n = 503 trees) and evaluated whether trends in basal area increment (BAI) are in line with two phases of climate warming, which occurred from 1915–1953 and from 1975–2015. Unexpectedly, results revealed that at subalpine sites (i) intensified climate warming in recent decades did not lead to a corresponding increase in BAI and (ii) increase in summer temperature since 1915 primarily favored growth of larch and spruce, although Swiss stone pine dominates at high elevations in the Eastern CEA, and therefore was expected to mainly benefit from climate warming. At treeline, BAI increases in all species were above the level expected based on determined age trend, whereas at the krummholz limit only deciduous larch showed a minor growth increase. We explain missing adequate growth response to recent climate warming by strengthened competition for resources (nutrients, light, water) in increasingly denser stands at subalpine sites, and by frost desiccation injuries of evergreen tree species at the krummholz limit. To conclude, accurate forecasts of tree growth response to climate warming at high elevation must consider changes in stand density as well as species-specific sensitivity to climate variables beyond the growing season.

Keywords: elevational transect; basal area increment; climate warming; conifers; European Alps; growth trend

1. Introduction

It is well established that tree growth at high elevation and in boreal regions is mainly temperature-limited [1–4] and temperatures below 5 °C during the growing season impair tissue formation in woody plants [5,6]. Consistently, numerous dendroclimatological studies have revealed that radial stem growth in the Central European Alps (CEA) is limited by low summer temperature e.g., [7,8]. Climate observations reveal a rise of air temperature in recent decades and global change models predict even further temperature increases in the future [9]. This phenomenon is especially pronounced in the European Alps, showing almost twice increase of air temperature compared to the global rate [10,11]. Although climate warming was found to increase radial growth at high elevation and in boreal regions [12], radial growth did not consistently track recent temperature increase [13].

A low-frequency offset between radial tree growth that does not correspond to the increase in summer temperatures is known as a divergence problem [14]. Although in the European Alps no unusual late 20th century divergence problem was found by [12], several authors [15,16] have found that recent temperature increase had a negative effect on tree growth at high elevation. Similarly, [13] reported a loss of sensitivity of temperature-limited temperate and boreal tree-ring series starting in the latter half of the 20th century. Understanding the growth response of high elevation forests to climate warming is of great importance because forests are a major part of the global carbon cycle and significant alterations are expected in upcoming decades, see [17] and further references therein.

Several authors have encouraged the application of large-scale dendroclimatological studies to assess tree growth response to atmospheric changes [18]. However, at the local scale tree growth could be biased by nonclimatic factors like resource availability (soil nutrients, water, light) and topography [19–22]. Therefore, to evaluate tree growth response in CEA to climate warming since the early 20th century, we used a recently acquired collection of tree-ring width data from 503 trees distributed at several elevational transects across the subalpine zone, treeline, and krummholz limit. Radial growth measurements were converted to basal area increments (BAI), aggregated according to elevation (subalpine, treeline, and krummholz limit), and resulting BAI time series were compared with homogenized climate data from representative instrumental stations of the CEA.

Instrumental temperature records show two phases of climate warming from early to mid-20th century and from mid-1970s until 2015 [10], which are also evident on larger geographic scales [9,13,23–25]. We therefore focused on elevational-dependent changes in radial growth trends of the three dominant coniferous tree species in the Eastern CEA, i.e., Norway spruce (*Picea abies* [L. Karst.] L.), European larch (*Larix decidua* Mill.), and Swiss stone pine (*Pinus cembra* L.), and evaluated whether the observed growth trends in BAI are consistent with the increase in summer air temperature over these two phases of climate warming. We expected that BAI of all species, but especially *P. cembra*, which is the dominant conifer at high elevation within the study area, would show increasing BAI consistent with distinct climate warming during the study period (1915–2015) amounting to >2 °C.

2. Material and Methods

2.1. Study Sites

The study sites are located along the Eastern CEA (Table 1) in the area of the Tuxer and Stubaier Alps (Tyrol, Austria) and Ötztaler Alps (border area of Austria and Italy) and encompass elevational transects including the subalpine zone (1630–2290 m asl), treeline (1950–2130 m asl) and the krummholz-limit (2130–2270 m asl). Treeline and the krummholz-limit are defined as uppermost patches of trees of at least 2 m height, and crippled individuals occurring above treeline and less than 2 m in height, respectively. Although most continental climate conditions enable subalpine forests to reach highest elevations at sites SCH1 and SCH2 (Table 1), anthropogenic influence (pasturing) impairs development of a natural treeline and krummholz-limit. Selected tree species include European larch (*Larix decidua*), Norway spruce (*Picea abies*) and Swiss stone pine (*Pinus cembra*), which are the dominant and widespread tree species at high elevation in the Eastern CEA [26].

Table 1. Site characteristics of selected stands along elevational transects in the Central European Alps. Geographic coordinates are given for the lowest site at each sampled transect (Lat = latitude; Lon = longitude; ΔE = elevational difference between highest and lowest stand).

Site	Lat	Lon	Elevation (m asl)			ΔE	Aspect	Slope (°)
			Subalpine	Treeline	Krummholz	(m)		
Patscherkofel (PTK)	47.206	11.452	1960–2010	2080–2130	2140–2180	220	WSW–W	10–30
Viggarspitze (VIG)	47.213	11.480	2030–2050	2120–2060	2220–2270	240	SSW	10–40
Morgenköpfl (MOR)	47.179	11.469	1880–2020	2060–2120	2130–2220	340	SW–NW	10–30
Kaserstattalm (KAS)	47.115	11.290	1640–1800	1950–1990	-	350	SE–SW	20–30
Elferlift (ELF)	47.096	11.305	1630–1810	-	-	180	NW–NE	20–30
Lazauntal (SCH1)	46.756	10.779	2040–2290	-	-	250	E	30–45
Finailtal (SCH2)	46.743	10.827	2120–2250	-	-	130	W	30–50

Within the Eastern CEA, the geology of the Tuxer and Ötztaler Alps is dominated by siliceous parent material. In the Stubaier Alps, the bedrock at selected sites (Elferlifte, Kaserstattalm) is of calcareous origin [27]. Local disturbance at some sites (primarily above the edge of the closed subalpine forest, i.e., at treeline and the krummholz limit) is restricted to grazing. Annual/summer temperature means and annual/summer precipitation totals during the study period 1915–2015 amounted to 0.5/7.7 °C and 1133/433 mm, respectively [10].

2.2. Tree-Ring Analysis

A uniform sampling design was applied at all selected sites. Along elevational transects, two radii per tree were obtained from opposite sides of the stem using an increment borer. Trees free of major stem or crown anomalies due to wind or snow breakage were randomly selected and cored parallel to the contour to avoid biases induced by formation of reaction wood. At subalpine sites, cores were extracted from dominant mature trees at breast height (1.3 m). Due to a decrease in tree height towards higher elevation, cores were taken between 0.5 m and 0.2 m height at treeline and the krummholz limit, respectively (n = 503 trees; Table 2).

Table 2. Age structure and sample depth of selected stands along elevational transects (mean age (year) ± standard deviation (SD)/n sampled trees). Abbreviations of sites as in Table 1; *Pice* = *Pinus cembra* L., *Lade* = *Larix decidua*, *Pcab* = *Picea abies* L. Karst.

	Subalpine			Treeline			Krummholz-Limit		
	Pice	*Lade*	*Pcab*	*Pice*	*Lade*	*Pcab*	*Pice*	*Lade*	*Pcab*
PTK	124 ± 43/8	133 ± 45/5	146 ± 38/10	37 ± 15/15	32 ± 19/14	26 ± 18/13	15 ± 5/10	12 ± 3/10	23 ± 17/10
VIG	138 ± 39/8	128 ± 28/6	85 ± 11/5	22 ± 7/10	29 ± 17/11	-	14 ± 4/9	10 ± 3/12	-
MOR	118 ± 6/10	103 ± 12/10	-	25 ± 7/11	31 ± 20/13	-	17 ± 3/9	17 ± 5/12	-
KAS	-	133 ± 23/10	142 ± 23/24	14 ± 3/53	18 ± 4/51	16 ± 4/46	-	-	-
ELF	-	175 ± 47/25	150 ± 43/35	-	-	-	-	-	-
SCH1	270 ± 54/24								
SCH2	-	128 ± 35/14	-	-	-	-	-	-	-
mean ± SD	199 ± 42	133 ± 32	131 ± 29	25 ± 8	27 ± 15	21 ± 11	15 ± 4	13 ± 4	23 ± 17
n total	50	70	74	89	89	59	28	34	10

Increment cores were air-dried, mounted on wooden holders, and the surface was prepared with a razor blade. For contrast enhancement of tree-ring boundaries, white chalk powder was rubbed into the tracheid lumen. Ring widths were measured to the nearest 1 μm using a light microscope (Olympus SZ61, Tokyo, Japan) fitted with a LINTAB measuring system (Frank Rinn, Heidelberg, Germany). Correct dating of measured time series of ring width were statistically checked with the TSAP-Win

Scientific and COFECHA [28] software and mean tree-ring width chronologies were developed for each site and species.

Correctly dated tree-ring width chronologies were converted into basal area increment (BAI) values excluding the bark and assuming a circular shape of the stem with the following formula:

$$BAI = \pi \, (R^2_n - R^2_{n-1}),$$ (1)

where R is the radius of the tree inside the bark and n is the year of tree ring formation. Bark thickness and stem diameter at coring height were measured at the time of sampling. BAI data were not standardized to preserve the long-term growth trend over the study period. BAI instead of ring-width time series were chosen because BAI is more closely related to biomass increment than ring width and BAI reduces tree size and age effects on growth trends [18,29–31].

To be able to distinguish climate warming-induced growth increases from inherent age/size related growth trends, species-specific biological growth curves were developed by aligning annual increments of individual trees to the biological (cambial) age instead of the calendar year and calculating age/size-related growth trends by fitting linear functions to the data [31]. Number of missing rings, to the pith in the increment cores, were estimated via the curvature and width of the inner rings. Species-specific age/size related trends in BAI were then plotted in a BAI time series developed at treeline and krummholz limit (mean tree age <30 years Table 2). At treeline and the krummholz limit, radial growth of saplings may be strongly restricted for numerous years due to pronounced competition with dwarf shrubs and grasses for nutrients and light. As soon as saplings are well established, radial growth distinctly increases [32]. The onset of the expected age/size-related growth trend was scheduled to that specific year. We are aware that this dating occurred somewhat arbitrarily, but a change in timing by ±2 years has no crucial effect on data interpretation.

2.3. Climate Data

Mean summer (June through August) and yearly temperature (°C) and precipitation (mm) data for the Greater Alpine Region (Histalp_region AT-West and HighLevel) were obtained from the HISTALP database (http://www.zamg.ac.at/histalp/) [10]. Climate data are mean monthly temperature and precipitation sums collected from several sites located at high (mean elevation 2091 m asl) and low elevation (mean elevation 773 mThis is o.k. asl), respectively, which are representative for the study region. During the last one hundred years, two phases of climate warming, i.e., from 1915–1953 ($W_{ph}I$) and from 1975–2015 ($W_{ph}II$) are obvious in climate records (Figure 1). Mean temperature during 5-year-long periods at the start and end of these phases were calculated. Additionally, temperature difference between start and end of the study period (1915–2015; $W_{ph}III$) was determined based on selected 5-year means (Figure 1).

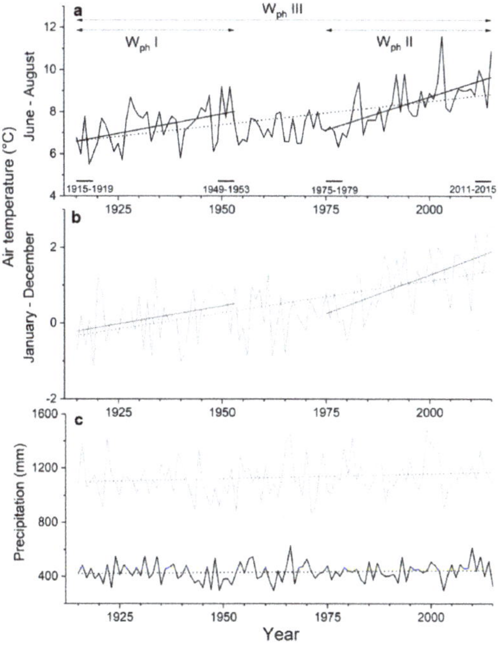

Figure 1. Mean air temperature (**a,b**) and precipitation (**c**) during June–August (black lines) and January–December (grey lines) in the Alpine region for the period 1915–2015. In (a), increase in air temperature was divided into two phases: phase I ($W_{ph}I$) from 1915–1953 and phase II ($W_{ph}II$) from 1975–2015. Period III ($W_{ph}III$) encompasses the whole study period 1915–2015. Linear air temperature and precipitation trends are shown in solid lines for $W_{ph}I$ and $W_{ph}II$ and dotted lines for $W_{ph}III$. Five-year periods used to calculate mean values of air temperature and basal area increment during selected periods ($W_{ph}I$-III) are indicated.

2.4. Statistical Analysis

Student's independent sample *t*-test was used to determine significant differences among selected time periods with respect to air temperature and BAI. Homogeneity of variances was asserted using Levene's test. Age trends in BAI time series were fit by linear regression analysis. All tests were performed using STATISTICA 12 (StatSoft Inc., Tulsa, OK, USA).

3. Results

During the study period 1915–2015 yearly (January-December) and summer (June–August) air temperatures showed an increasing trend, which intensified since the late 1970s (Figure 1). During the selected warming phases, WphI and WphII, yearly/summer air temperature increased by +1.04/+1.7 °C and +1.62/+2.52 °C, respectively, indicating accelerated climate warming in recent decades (Figure 2).

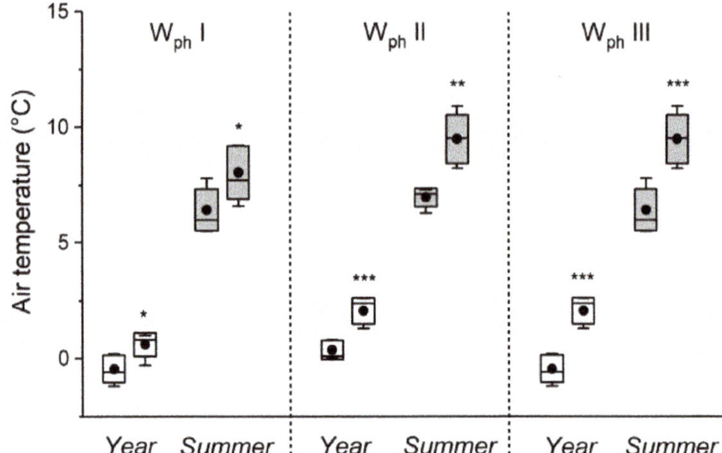

Figure 2. Comparison of yearly and summer (June–August) air temperatures between the beginning and end of phase I (W_{ph}I), phase II (W_{ph}II), and during the whole study period, i.e., beginning of phase I and end of phase II (W_{ph}III; cf. Figure 1). A box and whiskers plot shows the median (horizontal line) and mean values (closed circle) ± standard deviation (box); whiskers extend to the 10th and 90th percentiles. Statistical significant differences between mean temperatures at the beginning and end of the warming phases are indicated by asterisks (* $p \leq 0.05$, ** $p \leq 0.01$, *** $p \leq 0.001$).

Comparing the first (1915–1919) and last (2011–2015) five-year mean values, yearly and summer air temperature increased by +2.5 °C and +3.1 °C, respectively (Figure 3). In contrast, yearly and summer precipitation showed a slight but statistically not significant increase ($p > 0.05$) over the whole study period (Figure 1c).

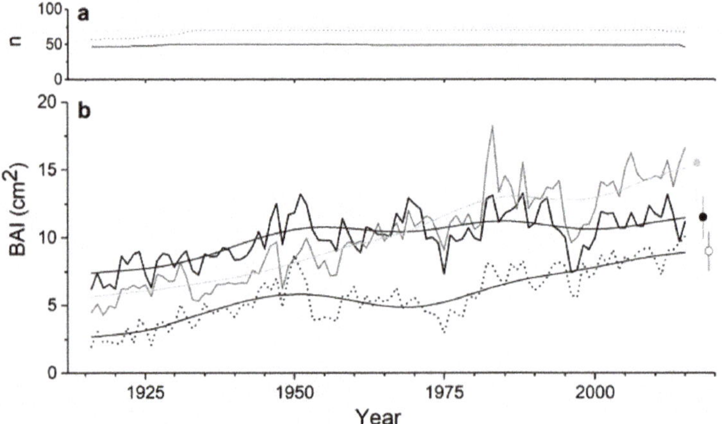

Figure 3. Sample depth (**a**) and basal area increment (BAI; **b**) of selected tree species from subalpine sites (*Pinus cembra* = continuous black line; *Picea abies* = continuous grey line; *Larix decidua* = dotted black line). Data were smoothed based on a fast Fourier transform low-pass filter (thin lines; number of points set to 10). Mean values ± standard error are indicated.

In the subalpine zone, mean age of all tree species at coring height was >130 years, whereas at treeline and at the krummholz limit mean age of tree species averaged to *c.* 25 and c. 17 years, respectively (Table 2). Mean stem diameter of *P. cembra*, *L. decidua*, and *P. abies* decreased from 40 ± 7 cm, 38 ± 8 cm, and 39 ± 6 cm in the subalpine zone, to 11 ± 3 cm, 11 ± 3 cm, and 10 ± 2 cm at the treeline, and 5 ± 1 cm, 5 ± 1 cm, and 4 ± 1 cm at the krummholz limit, respectively.

The number of trees included in the mean BAI stayed nearly constant during the study period 1915–2015 (Figure 3a). Increasing growth is obvious in all species during W_{ph}I. Afterwards, growth of *P. cembra* increased at a lower rate, while BAI of *P. abies* and *L. decidua* also strongly increased during the second warming phase (W_{ph}II; Figures 3b and 4).

Although BAI increase of all species was statistically significant in W_{ph}I-III (Figure 4), BAI increase significantly declined during W_{ph}II compared to W_{ph}I in *P. cembra*, and remained constant in *L. decidua* and *P. abies* (Table 3). During the whole study period (1915–2015, W_{ph}III) BAI increase was highest in *P. abies* and lowest in *P. cembra* (Table 3).

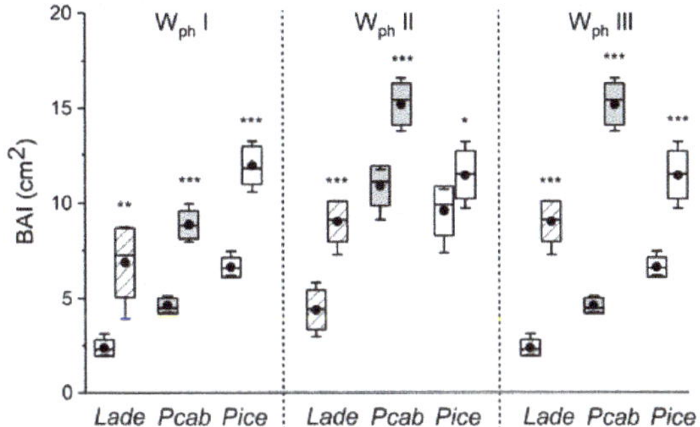

Figure 4. Comparison of mean basal area increment (BAI) of selected tree species (*Lade* = *Larix decidua*; *Pcab* = *Picea abies*; *Pice* = *Pinus cembra*) from subalpine sites at beginning and end of increasing air temperature during phase I (W_{ph}I), phase II (W_{ph}II) and during the whole study period (W_{ph}III; cf. Figure 1). Box and whiskers plot showing median (horizontal line) and mean (closed circle) ± standard deviation (box); whiskers extend to the 10th and 90th percentiles. Statistical significant differences between BAI at the beginning and end of warming phases are indicated by asterisks (* $p \leq 0.05$, ** $p \leq 0.01$, *** $p \leq 0.001$).

Table 3. Increase in basal area increment (cm^2) during selected phases of climate warming (W_{ph}I-III). Mean values ± standard deviations are shown. Statistically significant differences of mean values between species and W_{ph}I and W_{ph}II are indicated by different letters ($p \leq 0.05$).

Species	W_{ph}I	W_{ph}II	W_{ph}III
Pinus cembra	5.35 ± 0.84 [a]	1.89 ± 1.78 [b]	4.84 ± 1.33 [a]
Larix decidua	4.49 ± 1.91 [a]	4.65 ± 0.54 [a]	6.66 ± 1.05 [b]
Picea abies	4.24 ± 0.82 [a]	4.28 ± 1.23 [a]	10.57 ± 1.22 [c]

When BAI is plotted against cambial age, age/size trends become obvious (Figure 5). In the first few decades of juvenile growth BAI shows a linear increase. Linear functions fitted to BAI of juvenile growth indicate that developed age/size trends extend up to c. 40 years in *P. cembra* and *L. decidua* and c. 35 years in *P. abies*. Slopes of the regression lines indicate that juvenile growth was highest and lowest in *P. cembra* and *L. decidua*, respectively. BAI increase during the last decade is clearly above the determined age trend in each species at treeline (Figure 6), but at the krummholz limit BAI increase was higher than the expected age trend in some years in *L. decidua* only (Figure 6).

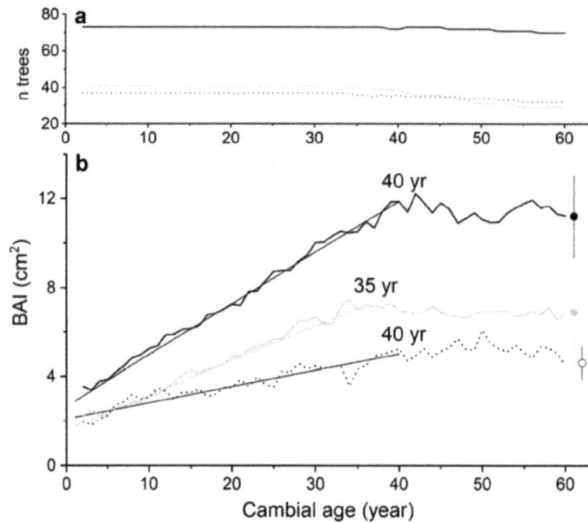

Figure 5. Sample depth (**a**) and basal area increment (BAI; **b**) of selected tree species from subalpine sites at cambial age 1–60 years (*Pinus cembra* = continuous black line; *Picea abies* = continuous grey line; *Larix decidua* = dotted black line). Variance among samples are indicated by mean standard error. Age trends were fit by linear regression analysis and are indicated by thin lines. *Pinus cembra*: y = 0.224x + 2.897, R^2 = 0.995; *Picea abies*: y = 0.151x + 1.762, R^2 = 0.975; *Larix decidua*: y = 0.071x + 2.165, R^2 = 0.877.

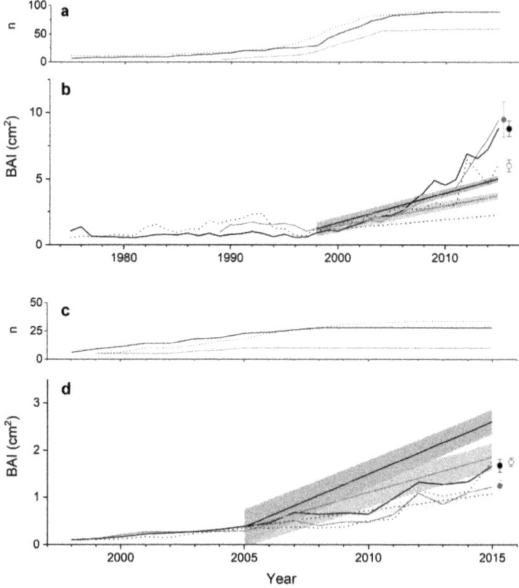

Figure 6. Sample depth and basal area increment (BAI) of tree species at treeline (**a**,**b**) and krummholz limit (**c**,**d**). *Pinus cembra* = continuous black line; *Picea abies* = continuous grey line; *Larix decidua* = dotted black line. Variance among samples is indicated by mean standard error. Species-specific age trends in BAI were plotted based on regression equations given in Figure 5. 95% confidence intervals are indicated. Onset of age trend in BAI was assumed to occur at the time when continuous growth increase was detected (for details see Materials and Methods).

4. Discussion

To strengthen the common climate signal and cancel out the local nonclimate signals, such as disturbance and topography, we pooled growth data (n = 503 trees) from several high-elevation forest sites. Long-term growth trends in the subalpine zone (1630–2290 m asl) revealed a dominating increase in BAI of mature forests (mean age >130 years) from 1915–2015, which is consistent with intense climate warming within the CEA [10]. However, results revealed stable BAI increase in the subalpine zone during $W_{ph}I$ and $W_{ph}II$ in *L. decidua* and *P. abies*, and decreasing BAI increase in *P. cembra*, although climate warming was much more distinct during the recent period (+1.6 °C in $W_{ph}I$ vs. +2.5 °C in $W_{ph}II$ during June through August, i.e., the main growing period; Figure 2). Age/size trends are not a conclusive explanation for this finding, because c. 90% of age/size related BAI increase is reached when trees are c. 30 years old (Figure 4), indicating that age/size effects are only marginally involved in BAI increase during $W_{ph}I$. As suggested by [33], the strong increase in wood increment in the early 20th century in the Eastern CEA is partly attributable to changes in forest management, i.e., reduction or abandonment of forest grazing, pollarding, or litter raking. Although we cannot refute changing land-use practices during $W_{ph}I$ on BAI, it is reasonable to assume that forest management practices were less pronounced on steep slopes at high elevation, which prevail within the study area (Table 1).

A loss of thermal response of radial tree growth in recent decades has been reported for *P. cembra* from one site within the study area on Mt. Patscherkofel [15] and from other temperature-limited sites [13,16]. Several hypotheses have been put forward to explain the missing adequate growth response of temperature-limited trees to recent climate warming [13,14], but the exact cause is still unknown. Several authors [15,34,35] attributed reduced radial growth in the European Alps to late-summer drought under increasing temperature. There is evidence that warm-season vapor pressure deficit and maximum daily temperature are closely related to forest drought stress [36]. Within the study area, however, increasing moisture stress in concordance with higher temperature is unlikely, because ample precipitation and moderate evaporative demand are regarded to cause soil water availability to be sufficiently high throughout the growing season [37]. Furthermore, although summer warming in the study region has been most prominent since late 1970s, it was not accompanied by a drying trend with decreased precipitation (Figure 1c). Missing matches between summer temperature increase and radial growth might be due to sensitivity of radial growth to several other climate parameters, such as previous autumn temperature and winter precipitation [8,35,38,39], which might partly compensate for more favorable growing conditions during summer. Alternatively, stand dynamics might cause increasing competition among trees for light and nutrients, which leads to decreasing sensitivity of growth to climate warming [40]. In the subalpine zone, lowest sensitivity to recent climate warming was found in late successional *P. cembra* showing highest mean age. Although tree age effects on growth response to climate was reported by [41,42], [35] reported that the climate signal in *P. cembra* is maximized in older trees and [43] found little evidence for decreasing climate sensitivity in old *P. cembra*, indicating that tree age per se is not the cause of reduced temperature sensitivity in this species.

Sampling issues were also reported to seriously affect growth trend evaluation [44–46]. A common sampling bias in dendrochronology is the 'big tree selection bias', i.e., time series are constructed from similar sized large trees of different ages [45,47]. This sampling design can lead to misleading estimates of growth trends within a stand [48]. We avoided this bias by evaluating growth trends only throughout periods with constant sample depth, i.e., BAI time series were developed from trees of comparable age rather than size. Another potential sampling bias is the 'slow-grower survivorship bias', i.e., an underrepresentation of fast-growing trees in the earlier portion of the developed time series, which yields to an apparent increase in growth over time. Due to the fact that only living trees were sampled in this study, we cannot unequivocally exclude the occurrence of this bias in our data set. However, considering this bias would lead to higher BAI in earlier parts of the record, causing a decrease in growth response to climate warming during $W_{ph}I$ and the whole study period (1915–2015; $W_{ph}III$).

Of the three selected species, most distinct growth response to climate warming was expected for *P. cembra*, which dominates at highest elevations in the Eastern CEA [26]. However, in the subalpine zone the largest growth increase over the whole study period was detected in *P. abies* (Table 3), suggesting that elevation may be associated with other environmental factors in addition to temperature [49] or that in old-growth *P. cembra* stands the benefit to growth brought on by a longer growing season may have been low in comparison with some other factors, such as the increasing respiration losses caused by warmer summer temperatures. However, the latter argument is contradicted by [50], who reported that plants have the capacity to adjust to a warming environment via physiological acclimation of respiration to warming. Strongest growth response of *P. abies* during the last century might be related to different disturbance history of stands, which may have affected the response to warming through the changing canopy structure [40]. Alternatively, species-related growth responses to climate were reported by several authors [39,51,52].

Although trees at high elevation are generally saturated with carbohydrates [53–57] found support for the carbohydrate limitation hypothesis in deciduous *L. decidua*. They determined that *L. decidua* augmented its growth as a response to the addition of CO_2, indicating that this species is potentially carbon limited at treeline. Hence, the observed increase in BAI of *L. decidua* in this study indicates that this deciduous species benefited not only from climate warming, but probably also from effects of CO_2 fertilization on growth [37].

At treeline, increase in BAI was higher than the calculated age/size trend in all species, which is attributable to intensive climate warming in recent decades, and highest sensitivity of tree growth to temperature occurred when approaching the upper limit of tree existence [4]. At the krummholz limit, however, only deciduous *L. decidua* showed an increase in BAI which exceeded the expected age/size trend. Missing growth response of evergreen conifers at the krummholz limit is most likely due to winter desiccation, i.e., damage to needles and branches caused by ice crystal abrasion of blowing snow and late winter water losses, which cannot be replaced because of frozen soil [58,59].

5. Conclusions

We found significant elevational and species-related trends in BAI of widespread coniferous species of the Eastern CEA, but no consistent growth response that is in parallel with the rapid rise of summer temperature in recent decades. Our results therefore revealed that although enhanced tree growth occurred at high elevation in response to climate warming, growth increase in the subalpine zone has declined in *P. cembra* and remained constant in *P. abies* and *L. decidua* over the last decades, despite the more intensified warming taking place since late 1970s. This finding indicates that tree growth response to climate warming is possibly nonlinear, and that the underlying ecological and physiological mechanisms, i.e., stand dynamics and climatic influences beyond the growing season, respectively, driving these temporal changes are not yet completely resolved. Our study thus highlights the importance of species- and stand-specific assessments to identify the influencing environmental factors of tree growth at high elevation. Such assessments are important because they help predict treeline dynamics and carbon sequestration under an increasingly warmer and drier climate.

Author Contributions: W.O. and G.W. conceived the study and W.O. supervised the research activities. U.B., V.G., J.G., A.H., H.K., and B.W. collected and analyzed the dendrochronological data. W.O., G.W., and W.K. interpreted and discussed the data. W.O. wrote the manuscript, and G.W. and W.K. provided editorial advice. All authors have read and agreed to the published version of the manuscript.

Funding: This research received no external funding.

Conflicts of Interest: The authors declare no conflict of interest.

References

1.   Körner, C. A re-assessment of high elevation tree line positions and their explanation. *Oecologia* **1998**, *115*, 445–459. [PubMed]

2. Esper, J.; Cook, E.R.; Schweingruber, F.H. Low-frequency signals in long tree-ring chronologies for reconstructing past temperature variability. *Science* **2002**, *295*, 2250–2253. [CrossRef] [PubMed]
3. Wieser, G.; Tausz, M. Trees at Their Upper Limit: Treelife Limitation at the Alpine Timberline. In *Plant Ecophysiology*; Springer: Berlin/Heidelberg, Germany, 2007; Volume 5.
4. Körner, C. *Alpine Treelines. Funtional Ecology of the Global High Elevation Tree Limits*; Springer: Basel, Switzerland, 2012.
5. Körner, C.; Paulsen, J. A world-wide study of high altitude treeline temperatures. *J. Biogeogr.* **2004**, *31*, 713–732. [CrossRef]
6. Rossi, S.; Deslauriers, A.; Gričar, J.; Seo, J.-W.; Rathgeber, C.B.K.; Anfodillo, T.; Morin, H.; Levanic, T.; Oven, P.; Jalkanen, R. Critical temperatures for xylogenesis in conifers of cold climates. *Glob. Ecol. Biogeogr.* **2008**, *17*, 696–707. [CrossRef]
7. Carrer, M.; Urbinati, C. Age-dependent tree ring growth responses to climate of *Larix decidua* and *Pinus cembra* in the Italian Alps. *Ecology* **2004**, *85*, 730–740. [CrossRef]
8. Oberhuber, W. Influence of climate on radial growth of *Pinus cembra* within the alpine timberline ecotone. *Tree Physiol.* **2004**, *24*, 291–301. [CrossRef]
9. IPCC. Contribution of Working Groups I, II and III to the Fifth Assessment Report of the Intergovernmental Panel on Climate Change. In *Climate Change 2014: Synthesis Report*; Pachauri, R.K., Meyer, L.A., Eds.; IPCC: Geneva, Switzerland, 2014.
10. Auer, I.; Böhm, R.; Jurkovic, A.; Lipa, W.; Orlik, A.; Potzmann, R.; Schöner, W.; Ungersböck, M.; Matulla, C.; Briffa, K.; et al. HISTALP—Historical instrumental climatological surface time series of the greater Alpine region 1760–2003. *Int. J. Climatol.* **2007**, *27*, 17–46. [CrossRef]
11. Gobiet, A.; Kotlarski, S.; Beniston, M.; Heinrich, G.; Rajczak, J.; Stoffel, M. 21st century climate change in the European Alps–A review. *Sci. Total Environ.* **2014**, *493*, 1138–1151. [CrossRef]
12. Büntgen, U.; Frank, D.; Wilson, R.; Carrer, M.; Urbinati, C. Testing for tree-ring divergence in the European Alps. *Glob. Chang. Biol.* **2008**, *14*, 2443–2453. [CrossRef]
13. St George, S.; Esper, J. Concord and discord among Northern Hemisphere paleotemperature reconstructions from tree rings. *Quat. Sci. Rev.* **2019**, *203*, 278–281. [CrossRef]
14. D'Arrigo, R.; Wilson, R.; Liepert, B.; Cherubini, P. On the 'divergence problem' in northern forests: A review of the tree ring evidence and possible causes. *Glob. Planet. Chang.* **2008**, *60*, 289–305. [CrossRef]
15. Oberhuber, W.; Kofler, W.; Pfeifer, K.; Seeber, A.; Gruber, A.; Wieser, G. Long-term changes in tree-ring-climate relationships at Mt. Patscherkofel (Tyrol, Austria) since the mid-1980s. *Trees* **2008**, *22*, 31–40. [CrossRef] [PubMed]
16. Cazzolla Gatti, R.; Callaghan, T.; Velichevskaya, A.; Dudko, A.; Fabbio, L.; Battipaglia, G.; Liang, J. Accelerating upward treeline shift in the Altai Mountains under last-century climate change. *Sci. Rep.* **2019**, *9*, 7678. [CrossRef] [PubMed]
17. Wieser, G.; Oberhuber, W.; Gruber, A. Effects of climate change at treeline: Lessons from space-for-time studies, manipulative experiments, and long-term observational records in the Central Austrian Alps. *Forests* **2019**, *10*, 508. [CrossRef]
18. Babst, F.; Bouriaud, O.; Alexander, R.; Trouet, V.; Frank, D. Toward consistent measurements of carbon accumulation: A multi-site assessment of biomass and basal area increment across Europe. *Dendrochronologia* **2014**, *32*, 153–161. [CrossRef]
19. Salzer, M.W.; Larson, E.R.; Bunn, A.G.; Hughes, M.K. Changing climate response in near-treeline bristlecone pine with elevation and aspect. *Environ. Res. Lett.* **2014**, *9*, 114007. [CrossRef]
20. Sullivan, P.F.; Ellison, S.B.Z.; McNown, R.W.; Brownlee, A.H.; Sveinbjornsson, B. Evidence of soil nutrient availability as the proximate constraint on growth of treeline trees in northwest Alaska. *Ecology* **2015**, *96*, 716–727. [CrossRef]
21. Liang, E.; Leuschner, C.; Dulamsuren, C.; Wagner, B.; Hauck, M. Global warming related tree growth decline and mortality on the north-eastern Tibetan Plateau. *Clim. Chang.* **2016**, *134*, 163–176. [CrossRef]
22. Liu, B.; Wang, Y.; Zhu, H.; Liang, E.; Camarero, J.J. Topography and age mediate the growth responses of Smith fir to climate warming in the southeastern Tibetan Plateau. *Int. J. Biometeorol.* **2016**, *60*, 1577–1587. [CrossRef]

23. Harris, I.C.; Jones, P.D. *CRU TS4.01: Climatic Research Unit (CRU) Time-Series (TS) Version 4.01 of High-Resolution Gridded Data of Month-by-Month Variation in Climate (Jan. 1901–Dec. 2016)*; Centre for Environmental Data Analysis: Norwich, UK, 2017. [CrossRef]
24. Estrada, F.; Perron, P.; Martínez-López, B. Statistically derived contributions of diverse human influences to twentieth-century temperature changes. *Nat. Geosci.* **2013**, *6*, 1050–1055. [CrossRef]
25. Hegerl, G.C.; Brönnimann, S.; Schurer, A.; Cowan, T. The early 20th century warming: Anomalies, causes, and consequences. *WIREs Clim. Chang.* **2018**, *9*, e522. [CrossRef] [PubMed]
26. Leuschner, C.; Ellenberg, H. Ecology of Central European Forests. In *Vegetation Ecology of Central Europe*; Springer: Berlin/Heidelberg, Germany, 2017; Volume 1.
27. Tollmann, A. Geologie von Österreich Band 1. In *Die Zentralalpen*; Deuticke: Wien, Austria, 1977.
28. Holmes, R.L. Computer-assisted quality control in tree ring dating and measurement. *Tree Ring Bull.* **1983**, *43*, 69–78.
29. LeBlanc, D.C. Relationships between breast-height and whole-stem growth indices for red spruce on Whiteface mountain, New York. *Can. J. For. Res.* **1990**, *20*, 1399–1407. [CrossRef]
30. Biondi, F.; Qeadan, F. A theory-driven approach to tree-ring standardization: Defining the biological trend from expected basal area increment. *Tree Ring Res.* **2008**, *64*, 81–96. [CrossRef]
31. Peters, R.L.; Groenendijk, P.; Vlam, M.; Zuidema, P.A. Detecting long-term growth trends using tree rings: A critical evaluation of methods. *Glob. Chang. Biol.* **2015**, *21*, 2040–2054. [CrossRef]
32. Wieser, G.; Matyssek, R.; Luzian, R.; Zwerger, P.; Pindur, P.; Oberhuber, W.; Gruber, A. Effects of atmospheric and climate change at the timberline of the Central European Alps. *Ann. For. Sci.* **2009**, *66*, 402. [CrossRef]
33. Erb, K.H.; Kastner, T.; Luyssaert, S.; Houghton, R.A.; Kuemmerle, T.; Olofsson, P.; Haberl, H. Bias in the attribution of forest carbon sinks. *Nat. Clim. Chang.* **2013**, *3*, 854–856. [CrossRef]
34. Büntgen, U.; Frank, D.C.; Nievergelt, D.; Esper, J. Summer temperature variations in the European Alps, AD 755–2004. *J. Clim.* **2006**, *19*, 5606–5623. [CrossRef]
35. Carrer, M.; Nola, P.; Eduard, J.L.; Motta, R.; Urbinati, C. Regional variability of climate-growth relationships in *Pinus cembra* high elevation forests in the Alps. *J. Ecol.* **2007**, *95*, 1072–1083. [CrossRef]
36. Williams, A.P.; Allen, C.D.; Macalady, A.K.; Griffin, D.; Woodhouse, C.A.; Meko, D.M.; Swetnam, T.W.; Rauscher, S.A.; Seager, R.; Grissino-Mayer, H.D.; et al. Temperature as a potent driver of regional forest drought stress and tree mortality. *Nat. Clim. Chang.* **2013**, *3*, 292–297. [CrossRef]
37. Wieser, G.; Oberhuber, W.; Gruber, A.; Leo, M.; Matyssek, R.; Grams, T.E.E. Stable water use efficiency under climate change of three sympatric conifer species at the alpine treeline. *Front. Plant Sci.* **2016**, *7*, 799. [CrossRef] [PubMed]
38. Saulnier, M.; Edouard, J.L.; Corona, C.; Guibal, F. Climate/growth relationships in a *Pinus cembra* high-elevation network in the Southern French Alps. *Ann. For. Sci.* **2011**, *68*, 189–200. [CrossRef]
39. Babst, F.; Poulter, B.; Trouet, V.; Tan, K.; Neuwirth, B.; Wilson, R.J.S.; Carrer, M.; Grabner, M.; Tegel, W.; Levanič, T.; et al. Site- and species-specific responses of forest growth to climate across the European continent. *Glob. Ecol. Biogeogr.* **2013**, *22*, 706–717. [CrossRef]
40. Primicia, I.; Camarero, J.J.; Janda, P.; Čada, V.; Morrissey, R.C.; Trotsiuk, V.; Bače, R.; Teodosiu, M.; Svoboda, M. Age, competition, disturbance and elevation effects on tree and stand growth response of primary *Picea abies* forest to climate. *For. Ecol. Manag.* **2015**, *354*, 77–86. [CrossRef]
41. Szeicz, J.M.; MacDonald, G.M. Age dependent tree-ring growth response of subarctic white spruce to climate. *Can. J. For. Res.* **1994**, *24*, 120–132. [CrossRef]
42. Girardin, M.P.; Guo, X.J.; Bernier, P.Y.; Raulier, F.; Gauthier, S. Changes in growth of pristine boreal North American forests from 1950 to 2005 driven by landscape demographics and species traits. *Biogeosciences* **2012**, *9*, 2523–2536. [CrossRef]
43. Esper, J.; Niederer, R.; Bebi, P.; Frank, D. Climate signal age effects-Evidence from young and old trees in the Swiss Engadin. *For. Ecol. Manag.* **2008**, *255*, 3783–3789. [CrossRef]
44. Cherubini, P.; Dobbertin, M.; Innes, J.L. Potential sampling bias in long-term forest growth trends reconstructed from tree rings: A case study from the Italian Alps. *For. Ecol. Manag.* **1998**, *109*, 103–118. [CrossRef]
45. Brienen, R.J.W.; Gloor, E.; Zuidema, P.A. Detecting evidence for CO_2 fertilization from tree ring studies: The potential role of sampling biases. *Glob. Biogeochem. Cycles* **2012**, *26*. [CrossRef]

46. Nehrbass-Ahles, C.; Babst, F.; Klesse, S.; Nötzli, M.; Bouriaud, O.; Neukom, R.; Dobbertin, M.; Frank, D. The influence of sampling design on tree-ring-based quantification of forest growth. *Glob. Chang. Biol.* **2014**, *20*, 2867–2885. [CrossRef]

47. Bowman, D.M.J.S.; Brienen, R.J.W.; Gloor, E.; Phillips, O.L.; Prior, L.D. Detecting trends in tree growth: Not so simple. *Trends Plant Sci.* **2013**, *18*, 11–17. [CrossRef] [PubMed]

48. Duchesne, L.; Houle, D.; Ouimet, R.; Caldwell, L.; Gloor, M.; Brienen, R. Large apparent growth increase in boreal forests inferred from tree-rings are an artefact of sampling design. *Sci. Rep.* **2019**, *9*, 6832. [CrossRef] [PubMed]

49. Li, M.H.; Yang, J. Effects of microsite on growth of *Pinus cembra* in the subalpine zone of the Austrian Alps. *Ann. For. Sci.* **2004**, *61*, 319–325. [CrossRef]

50. Reich, P.B.; Sendall, K.M.; Stefanski, A.; Wei, X.; Rich, R.L.; Montgomery, R.A. Boreal and temperate trees show strong acclimation of respiration to warming. *Nature* **2016**, *531*, 633–636. [CrossRef]

51. Girardin, M.P.; Bouriaud, O.; Hogg, E.H.; Kurz, W.; Zimmermann, N.E.; Metsaranta, J.M.; Jong Rde Frank, D.C.; Esper, J.; Büntgen, U.; Guo, X.J.; et al. No growth stimulation of Canda's boreal forest under half-century of combined warming and CO_2 fertilization. *Proc. Natl. Acad. Sci. USA* **2016**, *113*, E8406–E8414. [CrossRef] [PubMed]

52. Liang, P.; Wang, X.; Sun, H.; Fan, Y.; Wu, Y.; Lin, X.; Chang, J. Forest type and height are important in shaping the altitudinal change of radial growth response to climate change. *Sci. Rep.* **2019**, *9*, 1336. [CrossRef]

53. Hoch, G.; Körner, C. The carbon charging of pines at the climatic treeline: A global comparison. *Oecologia* **2003**, *135*, 10–21. [CrossRef]

54. Hoch, G.; Körner, C. Global patterns of mobile carbon stores in trees at the high-elevation treeline. *Glob. Ecol. Biogeogr.* **2012**, *21*, 861–871. [CrossRef]

55. Gruber, A.; Pirkebner, D.; Oberhuber, W.; Wieser, G. Spatial and seasonal variations in mobile carbohydrates in *Pinus cembra* in the timberline ecotone of the Central Austrian Alps. *Eur. J. For. Res.* **2011**, *130*, 173–179. [CrossRef]

56. Handa, I.T.; Körner, C.; Hättenschwiler, S. A test of the treeline carbon limitation hypothesis by in situ CO_2 enrichment and defoliation. *Ecology* **2005**, *86*, 1288–1300. [CrossRef]

57. Dawes, M.A.; Hättenschwiler, S.; Bebi, P.; FHagedorn ITHanda CKörner, C. Rixen Species-specific tree growth responses to 9 years of CO_2 enrichment at the alpine treeline. *J. Ecol.* **2011**, *99*, 383–394.

58. Tranquillini, W. Physiological Ecology of the Alpine Timberline. In *Tree Existence in High Altitudes with Special Reference to the European Alps*; Ecological Studies 31; Springer: Berlin/Heidelberg, Germany, 1979.

59. Hadley, J.L.; Smith, W.K. Wind effects on needles of timberline conifers: Seasonal influence of mortality. *Ecology* **1986**, *67*, 12–19. [CrossRef]

Article

Artificial Top Soil Drought Hardly Affects Water Use of *Picea abies* and *Larix decidua* Saplings at the Treeline in the Austrian Alps

Gerhard Wieser [1,*], Walter Oberhuber [2], Andreas Gruber [3], Florian Oberleitner [4], Roland Hasibeder [4] and Michael Bahn [4]

[1] Department of Alpine Timberline Ecophysiology, Federal Research and Training Centre for Forests, Natural Hazards and Landscape (BFW), Rennweg 1, A-6020 Innsbruck, Austria
[2] Department of Botany, Leopold-Franzens-Universität Innsbruck, Sternwartestraße15, A-6020 Innsbruck, Austria
[3] Naturwerkstatt Tirol, itworks Personalservice & Beratung, gemeinnützige GmbH, 6511 Zams, Austria
[4] Department of Ecology, Leopold-Franzens-Universität Innsbruck, Sternwartestraße15, A-6020 Innsbruck, Austria
* Correspondence: gerhard.wieser@uibk.ac.at; Tel.: +43-512-573933-5120

Received: 2 August 2019; Accepted: 3 September 2019; Published: 6 September 2019

Abstract: This study quantified the effect of shallow soil water availability on sap flow density (Q_s) of 4.9 ± 1.5 m tall *Picea abies* and *Larix decidua* saplings at treeline in the Central Tyrolean Alps, Austria. We installed a transparent roof construction around three *P. abies* and three *L. decidua* saplings to prevent precipitation from reaching the soil surface without notably influencing the above ground microclimate. Three additional saplings from each species served as controls in the absence of any manipulation. Roofing significantly reduced soil water availability at a 5–10 cm soil depth, while soil temperature was not affected. Sap flow density (using Granier-type thermal dissipation probes) and environmental parameters were monitored throughout three growing seasons. In both species investigated, three years of rain exclusion did not considerably reduce Q_s. The lack of a significant Q_s-soil water content correlation in *P. abies* and *L. decidua* saplings indicates sufficient water supply, suggesting that whole plant water loss of saplings at treeline primarily depends on evaporative demand. Future work should test whether the observed drought resistance of saplings at the treeline also holds for adult trees.

Keywords: climate change; experimental rain exclusion; plant water availability; soil drought; treeline; sap flow; *Picea abies*; *Larix decidua*

1. Introduction

Concerns have been raised in regard to high altitude treelines (i.e., the ecotone between the upper limit of the closed continuous forest canopy and the treeless alpine zone above) [1,2], as they may undergo significant alterations due to climate change. During the last century, global surface temperature has increased by about 0.6 ± 0.2 °C, and in the Alps warming has been well above global average [3–5]. Global change models predict a further increase by 1.4–5.8 °C in upcoming decades and an increased occurrence of climate extremes, including more frequent and severe drought [6]. At the treeline in the Central Austrian Alps, the observed temperature increase was apparently most pronounced during spring and summer compared with autumn and winter [7]. Moreover, as shown by the Central Austrian Alps [8], higher temperatures coupled with a decline in relative humidity and thus a considerable increase in evaporative demand may reduce the water supply of adult trees [9]. Although drought effects on tree transpiration have been studied intensively in various climates worldwide [10] and references therein, the reasons for the intra-annual variability of transpiration and responses to extreme meteorological conditions such as those during the dry summer of 2003 [11–13] still await clarification for conifers at the treeline. Norway spruce (*Picea abies* (L.) Karst.) and European larch (*Larix decidua* Mill.) are the dominant tree species at high elevation sites in the Alps [14].

Reductions in water loss of adult *P. abies* and *L. decidua* trees upon rain exclusion were reported in the literature for a low elevation site in the Austrian Alps [15,16]. In an inner alpine dry valley, rain exclusion throughout three growing seasons considerably reduced the water loss of evergreen *P. abies*, while no effects of rain exclusion were detected in deciduous *L. decidua* [15,16]. At treeline, however, larch may benefit from the higher vulnerability of spruce to increased temperatures and drought [17–19]. The present study focuses on the effect of limited soil water availability in the absence of other soil disturbances on the tree transpiration of *P. abies* and *L. decidua* saplings at treeline. There is evidence that soil water deficits have a stronger impact on reducing tree water loss in adult *P. abies* than in adult *L. decidua* trees [20,21]. As seedling establishment at the treeline is an important issue and because saplings, due to the small size of their root systems, could be particularly vulnerable to reduced soil moisture in the top soil, we tested the hypothesis that (1) a decline in soil water availability in the absence of any other soil disturbance will result in a decline in the water loss of evergreen *P. abies*, while (2) deciduous *L. decidua* will not respond to soil water shortage. Experimental soil water shortage was incited by roofing the forest floor throughout three consecutive growing seasons while continuously monitoring sap flow density (Q_s) with thermal dissipation probes [20]. Findings are used to explore tree response in a future warmer environment within the treeline ecotone of the Central Austrian Alps.

2. Materials and Methods

2.1. Study Site and Experimental Design

The study was carried out in a south-exposed afforestation at the treeline above Neustift in the Stubai Valley, Tyrol, Austria (1980 m a.s.l.; 47°7′45″ N, 11°18′20″ E), adjacent to the LTER-Master site Stubai. The average annual temperature is 3 °C, the annual precipitation is 1097 mm, and the soil type is a dystric cambisol. The stand formed a sparsely open canopy permitting a dense understory of grass and herbaceous species. During the study period (2016–2018), the saplings were 17 ± 3 years old. The stem diameter 0.3 m above ground (= height of sensor installation; see below) was 7.9 ± 1.5 cm, and the average height was 4.9 ± 1.5 m. Saplings selected for the experiment were separated by a distance of at least 3–5 m.

Rain exclusion was achieved by roofing the forest floor according to the approach of [15,16]. A 1.2 mm thick rip-stop film fixed 30 cm above ground prevented precipitation fro reaching the soil surface by draining off rainfall downhill outside the roof without notably influencing the above ground microclimate. The sides were open for allowing air circulation, and the area covered around each individual tree was approximately 4 × 4 m (≈16 m²); (see Figure S1). The roofs were installed underneath the canopy around the stems of six individual saplings (three of *P. abies* and three of *L. decidua*; hereafter "rain exclusion" treatment). An identical number of *P. abies* and *L. decidua* saplings served as controls in the absence of any soil water manipulation ("control" treatment). Rain exclusion operated in 2016 from 16 August to 3 October and continued throughout the snow free period during 18 May–13 September 2017 and 23 May–13 September 2018.

2.2. Environmental Sap Flow Density Measurements and Stem Radial Increment

Air temperature (T_{air}) and relative humidity (RH; Vaisala HMP45AC, Helsinki, Finland), photosynthetic active radiation (PHAR; Delta T BF5H; Cambridge, UK), and precipitation (P; Young 52202, Traverse City, MI, USA) were monitored at 2 m above ground in an open area neighboring the saplings. These data were recorded with a CR1000 data logger (Campbell Scientific, Shepshed, UK) programmed to record 30 min averages of measurements taken by 1 min intervals. In order to determine seasonal differences in shallow soil water content (SWC) and soil temperature (T_s) between the control and the rain exclusion plots, three soil moisture sensors (Theta Probe Type ML2x, Delta-T, Cambridge, UK) and three soil temperature probes (HOBO Pendant; ONSET, Pocasset, MA, USA) were installed at 5–10 cm and 10 cm soil depths, respectively. Close to three saplings per treatment were used for sap flow measurements. The soil moisture sensors were connected to a DL6 data logger (Delta-T, Cambridge, UK), while the soil temperature probes were equipped with internal storages. The measuring interval for soil moisture and soil temperature was set to 30 min, and the mean SWC (vol %) and T_s of the control and the rain exclusion plots, respectively, were calculated by averaging all measurements per treatment.

Sap flow density (Q_s) through the trunks of the selected study saplings was monitored with thermal dissipation sensors [22] by battery-operated sap flow systems (M1 Sapflow System, PROSA-LOG; UP, Umweltanalytische Produkte GmbH, Cottbus, Germany). Each system consisted of a three-channel PROSA-LOG datalogger and a constant source for sensor heating. Each sensor consisted of a heated and an unheated pair of thermocouples, connected in opposite for measuring temperature difference. In each study tree, one 20 mm long sensor was installed into the outer xylem (0–20 mm from the cambium) 15 cm apart vertically on the north facing side of the stems, 0.3 m above ground. The upper probe of each sensor included a heater that was continuously supplied with a constant power of 0.2 W, whereas the lower probe was unheated, remaining at trunk temperature for reference. The sensors were shielded with a thick aluminium-faced foam cover to prevent exposure to rain and to avoid physical damage and thermal influences from radiation. The temperature difference between the upper heated probe and the lower reference probe was recorded every 30 min. Power for the sap-flow systems was provided by a car battery (12 VDC, 90 Ah), which was recharged by means of an 80 W solar panel and a charge controller.

For each sensor, Q_s (g m^{-2} s^{-1}) was calculated from the temperature difference between the two probes (ΔT) relative to the maximum temperature difference (ΔT_m), which occurred at times of zero flow according to the calibration equation determined by [23]

$$Q_s = 119 \times [(\Delta T_m - \Delta T)/\Delta T]^{1.231}. \tag{1}$$

Each night, ΔT_m was determined and used as a reference for the following day. This assumption of zero sap fluxes seems reasonable as night-time vapor pressure deficits were mostly low, and temperature courses of the sensors reached equilibrium most nights, suggesting that the refilling of internal reserves was complete.

After the 2018 growing season, two radii per tree were taken at the height of the sap-flow sensors using a borer with a 5 mm diameter increment. Cores were dried in the laboratory, mounted on a holder, and the surface was prepared with a razor blade. For contrast enhancement of tree-ring boundaries, white chalk powder was rubbed into the pores. Ring widths were measured to the nearest 1 μm using a reflecting microscope (Olympus SZ61, Tokyo, Japan) and the software package TSAP WIN Scientific (UGT-Müncheberg, Müncheberg, Germany).

2.3. Data Analysis

All the analyses were performed using the SPSS 16 software package (SPSS Inc. Chicago, IL, USA), and curve fits were performed using FigP for Windows (BIOSOFT, Cambridge, UK). Differences in the overall mean SWC and T_s between the control and the rain exclusion treatment before roof closure and during the periods of rain exclusion in 2016, 2017, and 2018, respectively, were analyzed by a one-way ANOVA. Due to the small sample sizes [24], differences in Q_s and stem radial growth between controls and the rain exclusion treatment were evaluated by a Mann–Whitney U-test using the exact probabilities for small sample sizes, and a probability level of $p < 0.1$ was considered statistically significant [24].

Because of the large variation in Q_s within saplings (spruce and larch), treatment (control vs. rain exclusion), and year (2016, 2017, and 2018), we used normalized Q_s data in this study. This was achieved by converting Q_s values of each tree to a ratio of the maximum daily mean value observed during each measurement period (Table 1). Subsequently, each tree had a normalized Q_s of 1, which allowed for a better comparison of Q_s to environmental variables between control and rain exclusion saplings.

Table 1. Maximum sap flow density (Q_{Smax}; g m^{-2} s^{-1}) of *Picea abies* and *Larix decidua* saplings in control and rain exclusion plots obtained during the study periods 2016, 2017, and 2018. Data are medians ± half total range of three saplings per treatment. No significant effects of rain exclusion were found.

Treatment	*Picea abies*			*Larix decidua*		
	2016	2017	2018	2016	2017	2018
control	3.8 ± 1.7	7.3 ± 1.0	7.5 ± 5.9	11.2 ± 4.8	9.5 ± 1.0	9.0 ± 1.3
rain exclusion	6.2 ± 1.6	10.6 ± 0.7	12.8 ± 1.9	11.8 ± 3.8	10.0 ± 1.9	9.3 ± 2.4

Normalized Q_s was also used to calculate relative canopy conductance. Canopy conductance related to ground surface area (g_c) can be calculated from whole tree transpiration rate per unit of ground surface area (T) and ambient vapor pressure deficit (VPD) according to [25]

$$g_c = \gamma \lambda \, T / \rho c_p \text{VPD}, \tag{2}$$

where γ is the psychometric constant (kPa K^{-1}), λ is the latent heat of water vaporisation (J kg^{-1}), ρ is the density of the air (kg m^{-3}), c_p is the specific heat of air at constant pressure (J kg^{-1} K^{-1}), and VPD is the vapor pressure deficit (kPa) calculated from T_{air} and RH. As we used normalized Q_s values in this study, we calculated a simplified relative g_c as the ratio of normalized Q_s and VPD.

Values of Q_s and of environmental variables were available at a 30 min resolution. In order to reduce the dimension of the data sets and to avoid the problem that stem capacitance may affect the analysis of transpiration responses to variation in environmental conditions [24], we averaged diurnal values of Q_s and environmental parameters to daily means. Finally, for each species (spruce and larch), the data set was pooled over all the saplings per treatment.

Regression analyses were performed to analyze the response of normalized Q_s values to soil water content (SWC), photosynthetic active radiation (PHAR), and vapor pressure deficit (VPD), as these environmental factors have often been found to be most closely related to Q_s in the conifers under study [18,26,27]. While correlation of normalized Q_s with SWC and PHAR was obtained by linear regression analysis, the relationship between normalized Q_s and VPD was analyzed using the following exponential saturation function:

$$Q_s = (1 - exp(-a \times \text{VPD})) \tag{3}$$

where a is a fitting parameter.

The response of g_c to VPD was examined using the Lohammer type equation:

$$g_c = g_{cmax} \times (1/(1 + (\text{VPD}/a))) \tag{4}$$

where g_{cmax} is the maximum canopy conductance observed during the study period, and a is a fitting parameter.

2.4. Environmental Conditions

Table 2 provides an overview of the microclimatic conditions during the growing seasons (1 May–31 October) of 2016, 2017, and 2018 at the study site. Daily mean photosynthetic active radiation (PHAR) averaged 247 µmol m^{-2} s^{-1} in 2016, 331 µmol m^{-2} s^{-1} in 2017, and 349 µmol m^{-2} s^{-1} in 2018.

Daily mean air temperature (T_{air}) was 9.4 °C in 2016 and in 2017, and 10.9 °C in 2018. The corresponding values for daily mean vapor pressure deficit (VPD) were 2.4, 2.7, and 2.8 hPa in 2016, 2017, and 2018, respectively. Precipitation (P) during the growing seasons 2016, 2017, and 2018 amounted to 830, 1171, and 563 mm, respectively. Roofing excluded 136–962 mm of the incoming P during the growing season (Table 2).

Table 2. Mean daily photosynthetic active radiation (PHAR), air temperature (T_{air}), vapor pressure deficit (VPD), and sum of precipitation (P) during the growing season (1 May–31 October) as well as the amount of precipitation excluded by roofing the forest floor in 2016, 2017, and 2018.

Year	PHAR (μmol m^{-2} s^{-1})	T_{air} (°C)	VPD (hPa)	P (mm)	Rainout Period	P Excluded (mm)	% P Excluded
2016	247	9.4	2.4	830	16 August–3 October	137	16
2017	331	9.4	2.7	1171	18 May–13 September	962	82
2018	349	10.9	2.8	563	23 May–23 September	512	91

Before rain exclusion (DOY 228 in 2016 and DOY 144 in 2017), daily mean soil water content (SWC) at a 5–10 cm soil depth did not differ significantly (p = 0.51) between control and rain exclusion plots (Figure 1). Due to frequent precipitation over the three investigation periods (Figure 1), daily mean SWC at a 5–10 cm soil depth of the control plots varied between 11 vol % (14 August 2018) and 43 vol % (29 August 2016) (Figure 1) averaging 35 vol % in 2016, 32 vol % in 2017 and, 30 vol % in 2018 (Table 3), indicating that control saplings did not suffer from shallow soil drought. As expected, rain exclusion caused SWC at a 5–10 cm soil depth to decline continuously throughout the end of rainout periods (Figure 1). Roofing caused SWC to be on average 21, 45, and 20% below the corresponding control levels in 2016, 2017, and 2018, respectively (all p < 0.001, Table 3).

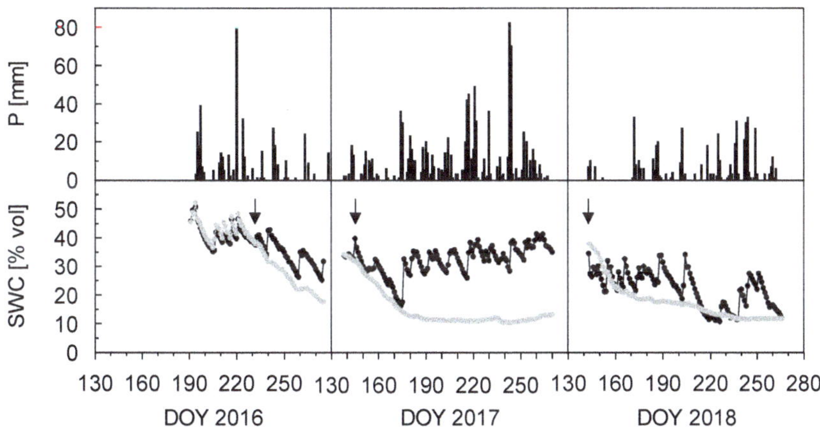

Figure 1. Seasonal course of daily sum of precipitation (P) and daily mean soil water content (SWC) at 5–10 cm soil depth in control (black) and rain exclusion plots (grey) from 10 July–2 October 2016, 18 May–27 September 2017, and 23 May–23 September 2018. SWC data are the mean of three sensors per treatment. Arrows indicate the start of rain exclusion.

Roofing, however, did not considerably influence soil temperature at a 10 cm soil depth. Average soil temperature at a 10 cm soil depth over 2016, 2017, and 2018 did not differ significantly (p > 0.48 each) between control (11.9 °C) and rain exclusion plots (11.8 °C; Table 3).

Table 3. Seasonal mean soil water content (vol %) at a 5–10 cm soil depth and soil temperature (°C) at a 10 cm soil depth in control and rain exclusion plots for the periods 16 August–3 October 2016, 18 May–13 September 2017, and 23 May–23 September 2018. Values are the mean ± SE of three sensors per treatment. Significant differences ($p < 0.1$) between control and rain exclusion plots are marked in bold and italics.

Treatment	Soil Water Content			Soil Temperature		
	2016	2017	2018	2016	2017	2018
control	*34.6 ± 2.8*	*32.3 ± 4.0*	*29.9 ± 3.8*	11.9 ± 0.7	11.6 ± 1.0	12.0 ± 0.9
rain exclusion	*27.9 ± 4.2*	*15.3 ± 3.0*	*15.7 ± 3.2*	11.9 ± 0.7	11.5 ± 1.0	12.0 ± 0.8

3. Results

3.1. Diameter Growth

In 2015, the year preceding the rain exclusion experiment, stem radial increment did not differ significantly between control and the rain exclusion *P. abies* and *L. decidua* saplings, respectively (Table 4). Rain exclusion had no effect on diameter growth in both species investigated (Table 4). Even after three years of treatment annual stem radial increment of *P. abies* and *L. decidua* saplings did not differ significantly between control and rain exclusion saplings (Table 4).

Table 4. Stem radial growth (mm) at the height of sensor installation of *Picea abies* and *Larix decidua* saplings in the year preceding the experiment (2015) and at the end of the study periods 2016, 2017, and 2018. Data are medians ± half total range of three saplings per treatment. No significant effects of rain exclusion were found.

Tree	Picea abies				Larix decidua			
	2015	2016	2017	2018	2015	2016	2017	2018
control	3.9 ± 1.0	3.9 ± 1.2	3.8 ± 1.1	5.3 ± 1.5	2.1 ± 0.2	1.8 ± 0.1	2.4 ± 0.2	2.1 ± 1.6
rain exclusion	3.3 ± 0.5	3.0 ± 0.1	2.6 ± 0.4	3.2 ± 0.4	2.8 ± 0.1	2.4 ± 1.1	2.9 ± 1.4	4.4 ± 18

3.2. Sap Flow Density and Influencing Factors

Before (DOY 228 in 2016 and DOY 144 in 2017) as well as during rain exclusion, normalized Q_s did not differ considerably between *P. abies* and *L. decidua* saplings in control and rain exclusion plots (Figure 2). In *P. abies* saplings exposed to rain exclusion, flow rates were 92% in 2016, 84% in 2017, and 103% in 2018 when compared to control *P. abies* saplings (all $p > 0.19$). The corresponding values for *L. decidua* saplings exposed to rain exclusion were 92, 107, and 101%, in 2016, 2017, and 2018, respectively (all $p > 0.32$).

Three years of rain exclusion did not significantly modify the response of Q_s to SWC at a 5–10 cm soil depth and evaporative demand in terms of PHAR and VPD. On a daily timescale Q_s and SWC were unrelated (Figure 3). Even when SWC at a 5–10 cm soil depth in the rain exclusion plots dropped below 15 vol %, normalized Q_s of spruce and larch saplings reached values up to 100% (Figure 3, exemplified for 2017) of their corresponding maximum values (see Table 1). With respect to PHAR and VPD, these results generally reflected positive correlations between Q_s and both environmental factors. We obtained linear correlations between normalized Q_s and PHAR (Figure 2, all $p < 0.001$). With respect to VPD, Q_s increased sharply at low VPD and tended to saturate at daily mean VPD values > 4 hPa in *P. abies* and > 7 hPa in *L. decidua* saplings growing in control and rain exclusion plots, respectively ($p < 0.001$ each). In contrast to normalized Q_s, normalized canopy conductance (g_c) declined with increasing VPD, a commonly observed relationship, which was not affected by the rainout treatment in *P. abies* and in *L. decidua* (Figure 4, $p < 0.001$ each).

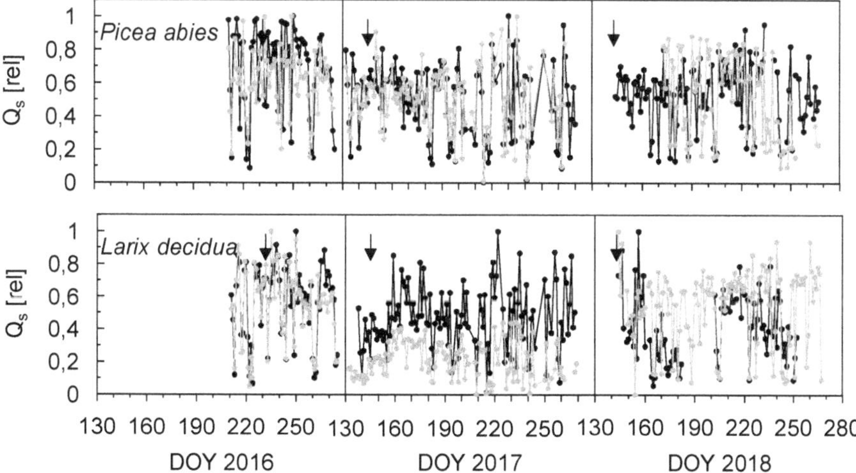

Figure 2. Seasonal course of daily mean normalized sap flow density (Q_s) of *Picea abies* and *Larix decidua* saplings in the control (black) and rain exclusion plots (grey) from 10 July–2 October 2016, 18 May–27 September 2017, and 23 May–23 September 2018. Data are the mean of three saplings per treatment. Arrows indicate the start of rain exclusion.

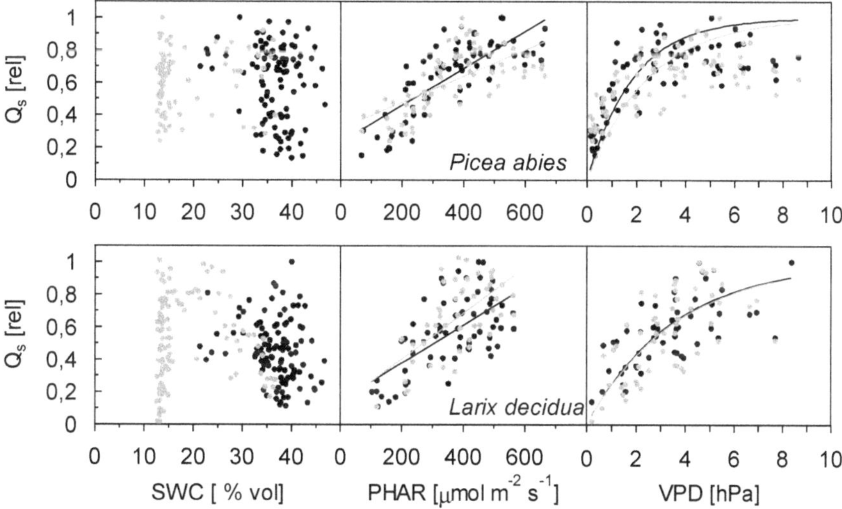

Figure 3. Daily mean normalized sap flow density (Q_s) of *Picea abies* and *Larix decidua* saplings in the control (black) and the rain exclusion treatment (grey) in relation to soil water content at a 5–10 cm soil depth (SWC, **left**), photosynthetic active radiation (PHAR, **middle**), and vapor pressure deficit (VPD, **right**) obtained in 2017. Points are mean values of three saplings per treatment. Points were fit by linear and exponential saturation functions for PHAR and VPD, respectively. *Picea abies*: PHAR: control: y = 0.001×PHAR + 0.23, R^2 = 0.61; rain exclusion: y = 0.0008×PHAR + 0.32, R^2 = 0.42; VPD: control: y = 1× (1−exp (−0.54×VPD)), R^2 = 0.69; rain exclusion: y = 1*(1−exp(−0.39×VPD)), R^2 = 0.54. *Larix decidua*: PHAR: control: y = 0.001×PHAR + 0.14, R^2 = 0.41; rain exclusion: y = 0.002×PHAR + 0.09, R^2 = 0.41; VPD: control: y = 1× (1−exp(−0.28×VPD)), R^2 = 0.48; rain exclusion: y = 1× (1−exp(−0.28×VPD)), R^2 = 0.46. All P < 0.001.

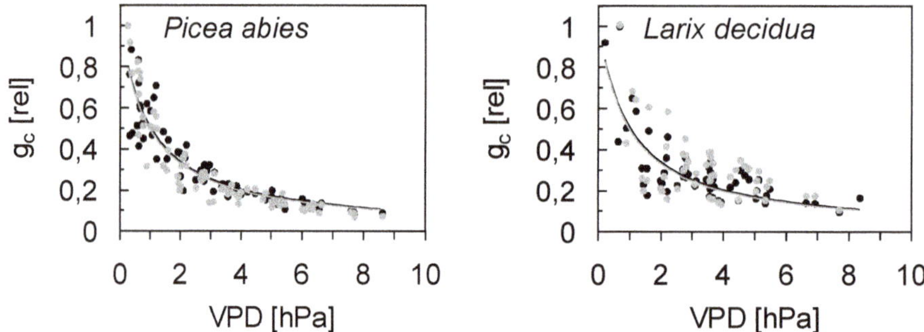

Figure 4. Daily mean normalized canopy conductance (g_c) of *Picea abies* and *Larix decidua* saplings in the control (black) and rain exclusion treatment (grey) in relation to vapor pressure deficit obtained in 2017. Points are mean values of three saplings per treatment. Points were fit by Lohammer type equation: *Picea abies*: control: y = 1/(1 + (VPD/0.93)), R^2 = 0.83; rain exclusion: y = 1/(1 + (VPD/0.98)), R^2 = 0.91 *Larix decidua*: control: y = 1/(1 + (VPD/0.94)), R^2 = 0.67; rain exclusion: y = 1/(1 + (VPD/0.94)), R^2 = 0.43. All $p < 0.001$.

4. Discussion

Our experimental approach was appropriate to manipulate shallow soil water availability, enabling the clarification of tree water loss in two co-occurring conifers in situ under a wide range of environmental conditions at the treeline. Moreover, differences in soil temperature between control and rain exclusion plots stayed within the typical variations at the study site, which confirmed the employed roofing system to prevent any change in soil temperature.

The lack of a significant Q_s–SWC correlation in *P. abies* and *L. decidua* saplings at our study site (Figure 3) indicates sufficient water supply as also observed in adult *P. abies* and *L. decidua* trees at treeline during periods of reduced soil water availability [9,20,28]. Even when SWC at a 5–10 cm soil depth in rain exclusion plots dropped to values below 15 vol %, Q_s of spruce and larch saplings reached values up to 100% of their corresponding maximum values. These findings suggest that, at treeline where VPD is generally considerably lower than at low elevation sites [29–32], whole tree water loss primarily depends on evaporative demand [9,20,27,33,34]. Furthermore, at treeline, small young trees are less coupled to the atmosphere and experience different environmental conditions as compared to adult trees [35], suggesting a smaller effect of increasing evaporative demand on Q_s in saplings as compared to tall trees, which are well coupled to the atmosphere.

Our findings suggest a great tolerance of saplings of both tree species to limited top soil SWC at the treeline. One possible assumption to explain this observation is that roots which extend below the subsoil can maintain a favorable water status, tap water from deeper pools when the upper layers are exhausted. It seems to be possible that, even if only a few fine roots are present in deeper and wet soil layers, these can contribute significantly to water uptake, as is known for forests in dry environments [36–40].

Our data also indicate that a decline in shallow soil water availability hardly affected water loss of *P. abies* and *L. decidua* saplings at treeline in the Central Austrian Alps, falsifying hypothesis (1) and confirming hypothesis (2). In this sense, *P. abies* saplings at treeline differ from adult *P. abies* trees that follow a water-saving strategy [20,21,27]. While reductions in water loss of *P. abies* and *L. decidua* upon rain exclusion were reported in the literature for a low elevation site in the Austrian Alps [15,16], we did not find reductions in Q_s of *P. abies* and *L. decidua* saplings, probably because high air humidity dampens evapotranspiration at high elevation [41]. Nevertheless, reduced soil water availability upon warming may endanger treeline saplings through water imbalances, in particular at their upper distribution limit, where tree habitat becomes stunted and deformed to "krummholz" in the "kampfzone" [32,33,42]. This latter zone already today is affected by temporal soil drought at wind-exposed ridges and leeward sun-exposed slopes with thin soil layers [14,27]. Finally, while there is ample evidence of upward shifts of treelines caused by climate warming [1,43–46] as well as non-climatic factors (REFs) [47], and further references therein], such as soil conditions [32,33,42] and couple effects of climate and land use change [2,21], it has been uncertain whether and to what extent such shifts can be limited by an increased occurrence of droughts. Our results indicate that, at the treeline, saplings of spruce and larch do not reduce transpirational water losses and growth under drought, suggesting that upward shifts of the treeline will likely not be limited by drier conditions in a warmer world.

5. Conclusions

Although our study suggests comparatively small effects of dry periods on Q_s at the treeline, it remains to be tested whether more severe drought events, likely to occur in the coming decades, have any effects on the Q_s of mature trees with higher water demand, especially during periods of high VPD. There is also evidence that, on a global scale, adult trees suffer from severe drought events [48]. Moreover, saplings differ in their response to drought from adult trees [49] due to ontogenetic changes in foliar gas exchange parameters [50]. It therefore remains to be investigated whether the surprisingly high drought resistance of saplings of the two conifer species at the treeline also holds for mature trees. Such studies, ideally based on long-term experiments manipulating both precipitation and evaporative demand, would also be important for understanding drought effects on forest regeneration under climate warming, as there is evidence that adult trees can enhance the survival and growth of saplings in a future warmer environment [51].

Supplementary Materials: The following Figure is available online at http://www.mdpi.com/1999-4907/10/9/777/s1. Figure S1A,B: Roof construction for experimental top-soil manipulation installed underneath the canopy around *P. cembra* saplings in a subalpine afforestation in the Sellrain Valley, Tyrol, Austria [7]. The same approach has been used in this study.

Author Contributions: M.B. conceived and co-ordinated the overall project. G.W., W.O., and A.G. designed the sapling experiment. G.W., W.O., A.G., F.O., and R.H. analyzed the data. G.W. and W.O. wrote the manuscript.

Funding: Part of this work was funded by Österreichische Akademie der Wissenschaften, Project "ClimLUC."

Acknowledgments: We would like to thank two anonymous reviewers for their comments on an earlier version of the manuscript.

Conflicts of Interest: The authors declare no conflict of interest.

References

1. Holtmeier, F.-K.; Broll, G.E. Treeline advance - driving processes and adverse factors. *Landsc. Online* **2007**, *1*, 1–33. [CrossRef]
2. Wieser, G.; Matyssek, R.; Luzian, R.; Zwerger, P.; Pindur, P.; Oberhuber, W.; Gruber, A. Effects of atmospheric and climate change at the timberline of the Central European Alps. *Ann. For. Sci.* **2009**, *66*, 402. [CrossRef] [PubMed]
3. Böhm, R.; Auer, I.; Brunetti, M.; Maugeri, M.; Nanni, T.; Schöner, W. Regional temperature variability in the European Alps. 1760–1998 from homogenized instrumental time series. *Int. J. Clim.* **2001**, *21*, 1779–1801. [CrossRef]
4. Rebetez, M.; Reinhard, M. Monthly air temperature trends in Switzerland 1901–2000 and 1975–2004. *Theor. Appl. Clim.* **2007**, *91*, 27–34. [CrossRef]
5. Ciccarelli, N.; Von Hardenberg, J.; Provenzale, A.; Ronchi, C.; Vargiu, A.; Pelosini, R. Climate variability in north-western Italy during the second half of the 20th century. *Glob. Planet. Chang.* **2008**, *63*, 185–195. [CrossRef]
6. Intergovernmental Panel on Climate Change (IPCC). Climate Change. In *The Physical Science Basis*; Cambridge University Press: Cambridge, UK, 2013.

7. Wieser, G.; Grams, T.E.; Matysssek, R.; Oberhuber, W.; Gruber, A. Soil warming increased whole-tree water use of Pinus cembra at the treeline in the Central Tyrolean Alps. *Tree Physiol.* **2015**, *35*, 279–288. [CrossRef] [PubMed]

8. Wieser, G.; Oberhuber, W.; Waldboth, B.; Gruber, A.; Siegwolf, R.T.; Grams, T.E.; Matyssek, R. Long-term trends in leaf level gas exchange mirror tree-ring derived intrinsic water-use efficiency of Pinus cembra at treeline during the last century. *Agric. For. Meteorol.* **2018**, *248*, 251–258. [CrossRef]

9. Obojes, N.; Meurer, A.; Newesely, C.; Tasser, E.; Oberhuber, W.; Mayr, S.; Tappeiner, U. Water stress limits transpiration and growth of European larch up to the lower subalpine belt in an inner-alpine dry valley. *New Phytol.* **2018**, *220*, 460–475. [CrossRef]

10. Hoover, D.L.; Wilcox, K.R.; Young, K.E. Experimental droughts with rainout shelters: A methodological review. *Ecosphere* **2018**, *9*, e02088. [CrossRef]

11. Reichstein, M.; Tenhunen, J.D.; Roupsard, O.; Ourcival, J.-M.; Rambal, S.; Miglietta, F.; Peressotti, A.; Pecchiari, M.; Tirone, G.; Valentini, R.; et al. Severe drought effects on ecosystem CO2 and H2O fluxes at three Mediterranean evergreen sites: Revision of current hypotheses? *Glob. Chang. Biol.* **2002**, *8*, 999–1017. [CrossRef]

12. Bréda, N.; Huc, R.; Granier, A.; Dreyer, E. Temperate forest trees and stands under severe drought: A review of ecophysiological responses, adaptation processes and long-term consequences. *Ann. For. Sci.* **2006**, *63*, 625–644. [CrossRef]

13. Granier, A.; Reichstein, M.; Breda, N.; Janssens, I.; Falge, E.; Ciais, P.; Grunwald, T.; Aubinet, M.; Berbigier, P.; Bernhofer, C.; et al. Evidence for soil water control on carbon and water dynamics in European forests during the extremely dry year: 2003. *Agric. For. Meteorol.* **2007**, *143*, 123–145. [CrossRef]

14. Aulitzky, H. Grundlagen und Anwendungen des vorläufigen Wind-Schnee-Ökogramms. *Mitt. Forstl. Bundesversuchsanstalt Mariabrunn* **1963**, *60*, 763–834.

15. Leo, M.; Oberhuber, W.; Schuster, R.; Grams, T.E.E.; Matyssek, R.; Wieser, G. Evaluating the effect of plant water availability on inner alpine coniferous trees based on sap flow measurements. *Eur. J. For. Res.* **2014**, *133*, 691–698. [CrossRef]

16. Schuster, R.; Oberhuber, W.; Gruber, A.; Wieser, G. Soil drought decreases water-use of pine and spruce but not of larch in a dry inner alpine valley. *Austrian J. For. Res.* **2016**, *133*, 1–17.

17. Schmidt, O. *Fichtenwälder im Klimawandel*; Bayerische Landesanstalt für Wald und Forstwirtschaft: Freising, Germany, 2009.

18. Wolfslehner, G.; Koeck, R.; Hochbichler, E.; Steiner, H.; Frank, G.; Formayer, H.; Arbeiter, F. *Ökologische und waldbauliche Eigenschaften der Lärche (Larix decidua MILL.)—Folgerungen für die Waldbewirtschaftung in Österreich unter Berücksichtigung des Klimawandels*; Endbericht Start-Clim. 2010: Anpassung an den Klimawandel: Weitere Beiträge zur Erstellung einer Anpassungsstrategie für Österreich Auftraggeber: BMLFUW, BMWF, BMWFJ, ÖBF; Austrian Academy of Science: Vienna, Austria, 2011.

19. Hartl-Meier, C.; Dittmar, C.; Zang, C.; Rothe, A. Mountain forest growth response to climate change in the Northern Limestone Alps. *Trees* **2014**, *28*, 819–829. [CrossRef]

20. Anfodillo, T.; Rento, S.; Carraro, V.; Furlanetto, L.; Urbinati, C.; Carrer, M. Tree water relations and climatic variations at the alpine timberline: Seasonal changes of sap flux and xylem water potential *in Larix decidua* Miller, *Picea abies* (L.) Karst. and *Pinus cembra* L. *Ann. For. Sci.* **1998**, *55*, 159–172. [CrossRef]

21. Wieser, G. Lessons from the timberline ecotone in the Central Tyrolean Alps: A review. *Plant Ecol. Divers.* **2012**, *5*, 127–139. [CrossRef]

22. Granier, A. Une nouvelle méthode pour la mesure du flux de sève brute dans le tronc des arbres. *Ann. For. Sci.* **1985**, *42*, 193–200. [CrossRef]

23. Granier, A. Evaluation of transpiration in a Douglas-fir stand by means of sap flow measurements. *Tree Physiol.* **1987**, *3*, 309–320. [CrossRef]

24. Bortz, J.; Lienert, G.A.; Boenke, K. *Verteilungsfreie Methoden in der Biostatistik*; Springer: Berlin, Germany, 2008.

25. Köstner, B.M.M.; Schulze, E.-D.; Kelliher, F.M.; Hollinger, D.Y.; Byers, J.N.; Hunt, J.E.; McSeveny, T.M.; Meserth, R.; Weir, P.L. Transpiration and canopy conductance in a pristine broad-leaved forest of Nothofagus: An analysis of xylem sap flow and eddy correlation measurements. *Oecologia* **1992**, *91*, 350–359. [CrossRef] [PubMed]

26. Zimmermann, R.; Vygodskaya, N.N.; Ziegler, W.; Schulze, E.D.; Wirth, C.; McDonald, K.C. Canopy transpiration in a chronosequence of Central Siberian pine forests. *Glob. Chang. Biol.* **2000**, *6*, 25–37. [CrossRef]

27. Matyssek, R.; Wieser, G.; Patzner, K.; Blaschke, H.; Häberle, K.-H. Transpiration of forest trees and stands at different altitude: Consistencies rather than contrasts? *Eur. J. For. Res.* **2009**, *128*, 579–596. [CrossRef]

28. Badalotti, A.; Anfodillo, T.; Grace, J. Evidence of osmoregulation in Larix decidua at Alpine treeline and comparative responses to water availability of two co-occurring evergreen species. *Ann. For. Sci.* **2000**, *57*, 623–633. [CrossRef]

29. Baumgartner, A. Mountain climates from a perspective of forest growth. In *Mountain Environments and Subalpine Forest Growth*; Benecke, U., Davis, M.R., Eds.; New Zealand Forest Service: Wellington, New Zealand, 1980; pp. 27–39.

30. Fliri, F. *Das Klima der Alpen im Raume von Tirol*; Monographien zur Landeskunde Tirols I Universitätsverlag Wagner: Innsbruck, Austria; Munchen, Germany, 1975.

31. Veit, H. *Die Alpen—Geoökologie und Landschaftsentwicklung*; Ulmer: Stuttgart, Germany, 2002.

32. Wieser, G.; Tausz, M. *Trees at Their Upper Limit: Treelife Limitation at the Alpine Timberline*; Springer: Berlin, Germany, 2007; Volume 5.

33. Tranquillini, W. Physiological ecology of the alpine timberline. In *Tree Existence at High Altitudes with Special Reference to the European Alps*; Springer: Berlin, Germany, 1979; Volume 31.

34. Wieser, G.; Gruber, A.; Oberhuber, W. Sap flow characteristics and whole-tree water use of *Pinus cembra* across the treeline ecotone of the central Tyrolean Alps. *Eur. J. For. Res.* **2012**, *133*, 287–295. [CrossRef]

35. Gruber, A.; Zimmenmann, J.; Wieser, G.; Oberhuber, W. Effects of climate variables on intra-annual stem radial increment in *Pinus cembra* (L.) along the timberline ecotone. *Ann. For. Sci.* **2009**, *66*, 503. [CrossRef]

36. Dawson, D.E. Determining water use by trees and forests from isotopic, energy balance, and transpirational analyses: The role of tree site and hydraulic lift. *Tree Physiol.* **1994**, *18*, 177–184.

37. Villar-Salvador, P.; Castro-Diez, P.; Perez-Rontome, C.; Montserrat-Martí, G. Stem xylem features in three *Quercus* (Fagaceae) species along a climatic gradient in NE Spain. *Trees* **1997**, *12*, 90–96. [CrossRef]

38. Sarris, D.; Siegwolf, R.; Körner, C. Inter- and intra-annual stable carbon and oxygen isotope signals in response to drought in Mediterranean pines. *Agric. For. Meteorol.* **2013**, *168*, 59–68. [CrossRef]

39. Vincke, C.; Thiry, Y. Water table is a relevant source for water uptake by a Scots pine (*Pinus sylvestris* L.) stand: Evidences from continuous evapotranspiration and water table monitoring. *Agric. For. Meteorol.* **2008**, *148*, 1419–1432. [CrossRef]

40. Brito, P.; Lorenzo, J.R.; González-Rodríguez, Á.M.; Morales, D.; Wieser, G.; Jiménez, M.S.; González-Rodríguez, Á.M. Canopy transpiration of a semi arid Pinus canariensis forest at a treeline ecotone in two hydrologically contrasting years. *Agric. For. Meteorol.* **2015**, *201*, 120–127. [CrossRef]

41. Ellenberg, H. *Vegetation Mitteleuropas mit den Alpen in Okologischer, Dynamischer und Historischer Sicht*; Verlag Eugen Ulmer: Stuttgart, Germany, 1996.

42. Holtmeier, F.-K. *Mountain Timberlines. Ecology, Patchiness, and Dynamics*; KLuwer Academic Publishers: Dortrecht, The Netherlands; Boston, MA, USA; London, UK, 2000; Volume 14.

43. Boisvert-Marsh, L.; Perie, C.; de Blois, S. Shifting with climate? Evidence for recent changes in tree species distribution at high latitudes. *Ecosphere* **2014**, *5*, 1–33. [CrossRef]

44. Dyderski, M.K.; Paz, S.; Frelich, L.E.; Jagodzinski, A.M. How much does climate change threaten European forest tree species distributions? *Glob. Chang. Biol.* **2018**, *24*, 1150–1163. [CrossRef] [PubMed]

45. Hanewinkel, M.; Cullmann, D.A.; Schelhaas, M.-J. Climate change may cause severe loss in the economic value of European forest land. *Nat. Clim. Chang.* **2013**, *3*, 203–207. [CrossRef]

46. Sykes, M.; Prentice, I. Climate change, tree species distributions and forest dynamics: A case study in the mixed conifer/northern hardwoods zone of northern Europe. *Clim. Chang.* **1996**, *34*, 161–177. [CrossRef]

47. Tognetti, R.; Alados, C.; Bebi, P.; Grunewald, K.; Zhiyanski, M.; Andonowski, V.; Hofgaard, A.; La Porta, N.; Bratanova-Doncheva, S.; Edwards-Jonášová, M.; et al. Drivers of treeline shift in different European mountains. *Clim. Res.* **2017**, *73*, 135–150.

48. Bennett, A.C.; McDowell, N.G.; Allen, C.D.; Anderson-Teixeira, K.J. Larger trees suffer most during drought in forests worldwide. *Nat. Plants* **2015**, *1*, 15139. [CrossRef]

49. He, J.-S.; Zhang, Q.-B.; Bazzaz, F.A. Differential drought responses between saplings and adult trees in four co-occurring species of New England. *Trees* **2005**, *19*, 442–450. [CrossRef]

50. Thomas, S.C.; Winner, W.E. Photosynthetic differences between saplings and adult trees: An integration of field results by meta-analysis. *Tree Physiol.* **2002**, *22*, 117–127. [CrossRef]
51. Andivia, E.; Madrigal-González, J.; Villar-Salvador, P.; Zavala, M.A. Do adult trees increase conspecific juvenile resilience to recurrent droughts? Implications for forest regeneration. *Ecosphere* **2018**, *9*, e02282. [CrossRef]

 *forests*

Article

Winter Embolism and Recovery in the Conifer Shrub *Pinus mugo* L.

Stefan Mayr [1], Peter Schmid [1] and Sabine Rosner [2,*

[1] Department of Botany, University of Innsbruck, Sternwartestr. 15, 6020 Innsbruck, Austria;
    stefan.mayr@uibk.ac.at (S.M.); peter25schmid@gmail.com (P.S.)
[2] Institute of Botany, University of Natural Resources and Life Sciences, BOKU Vienna,
    Gregor-Mendel-Straße 33, 1180 Vienna, Austria
* Correspondence: sabine.rosner@boku.ac.at; Tel.:+43-1-47654-83117

Received: 30 September 2019; Accepted: 22 October 2019; Published: 24 October 2019

Abstract: Research Highlights: Pronounced winter embolism and recovery were observed in the Alpine conifer shrub *Pinus mugo* L. Data indicated that the hydraulic courses and underlying mechanism were similar to timberline trees. Background and Objectives: At high elevation, plants above the snow cover are exposed to frost drought and temperature stress during winter. Previous studies demonstrated winter stress to induce low water potentials (Ψ) and significant xylem embolism (loss of conductivity, or LC) in evergreen conifer trees, and recovery from embolism in late winter. Here, we analyzed xylem hydraulics and related structural and cellular changes in a conifer shrub species. Materials and Methods: The uppermost branches of Pinus mugo shrubs growing at the Alpine timberline were harvested over one year, and the Ψ, water content, LC, proportion of aspirated pits, and carbohydrate contents were analyzed. Results: Minimum Ψ (-1.82 ± 0.04 MPa) and maximum LC ($39.9\% \pm 14.5\%$) values were observed in mid and late winter, followed by a recovery phase. The proportion of aspirated pits was also highest in winter ($64.7\% \pm 6.9\%$ in earlywood, $27.0\% \pm 1.4\%$ in latewood), and decreased in parallel with hydraulic recovery in late winter and spring. Glucose and fructose contents gradually decreased over the year, while starch contents (also microscopically visible as starch grains in needle and stem tissues) increased from May to July. Conclusions: The formation and recovery of embolism in *Pinus mugo* were similar to those of timberline trees, as were the underlying mechanisms, with pit aspiration enabling the isolation of embolized tracheids, and changes in carbohydrate contents indicating adjustments of osmotic driving forces for water re-distribution. The effects of future changes in snow cover regimes may have pronounced and complex effects on shrub-like growth forms, because a reduced snow cover may shorten the duration of frost drought, but expose the plants to increased temperature stress and impair recovery processes.

Keywords: alpine timberline; conifer shrub; pit aspiration; refilling; winter stress; xylem embolism

---

1. Introduction

At high elevation, plants, unless protected by the snow cover, are exposed to high stress intensities during winter. They have to cope with temperature stress (i.e., low minimum temperatures and high temperature fluctuations) as well as winter drought. Especially evergreen trees were demonstrated to experience frost drought, because water is lost via (cuticular) transpiration over the crown, while the frozen soil blocks water uptake for months [1]. Consequently, low water potentials (Ψ) develop, which, in combination with frequent freeze–thaw events, can lead to the formation of xylem embolism.

Embolism blocks the water transport and thus the water supply of distal tissues. On drought, low Ψ causes air entry into xylem conduits via pits ("air seeding"; [2]), and disruption of the metastable water columns. The Ψ thresholds for embolism are species-specific and depend on xylem structures [3–5].

Embolism can also be caused by freeze–thaw events [6] when bubbles, which are formed in conduits during freezing, expand on thawing. This occurs when bubbles entrapped in the ice are large and Ψ of the melting sap is low. At high elevation, tree crowns are exposed to numerous freeze–thaw events, and more than 100 frost cycles per winter were reported at the European Alp´s timberline [7,8], causing embolism even in conifers with small tracheids. Accordingly, winter embolism was found in several conifer trees growing at high elevation: At the Rocky Mountains´ treeline, ca. 35% loss of conductivity (LC) was observed in *Pinus albicaulis* Engelm. and more than 25% LC was observed in *Larix lyallii* Parl. [9]. At the European alpine timberline, conifers (*Picea abies* L. Karst., *Pinus mugo* L., and *Juniperus communis* L.) can exhibit up to 100% LC, which is induced by the combination of drought and freeze–thaw stress [10,11].

However, how can heavily embolized trees survive, considering that more than 50% LC [12,13] was supposed to be critical or even lethal for conifers? Interestingly, studies reported a stepwise decrease in the LC of several conifers (*Abies*, *Juniperus*, *Larix*, *Picea*, *Pinus*, *Thuja* and *Tsuga*) during the end of winter and spring [7,9–11,14]. Ref. [15] found recovery from 60% LC in *Pseudotsuga menziesii* (Mirb.) Franco, and [1] reported the recovery process of *Picea abies* to be based on active, cellular processes. The decrease in LC corresponded to changes in aquaporin density in the needle mesophyll as well as to changes in starch contents in mesophyll, xylem, and phloem. Aquaporins enable water transport between cells [16–18], and may therefore be responsible for water shifts towards embolized xylem. Carbohydrates might be important to create the osmotic driving force for water shifts [19–22]. Ref. [1] also assumed that recovery was based on water uptake over branches, similar to the needle uptake of water demonstrated by [23] on *Picea glauca* (Moench) Voss and according to the findings of other authors e.g., [24–27]. Thereby, snow, covering branches in late winter and melting upon increasing temperatures, may be an important source for water uptake and recovery. According to [21], hydraulic recovery under negative Ψ also requires an isolation of embolized conduits from functional xylem areas. In conifers, this isolation might be based on the special pit architecture [28–30], as the proportion of aspirated pits corresponded well to winter embolism in *Picea abies* [1].

Thus, the snow cover plays a dual role for tree hydraulics: It impairs tree water relations when ice formed in roots and the stem base under/in the snow layer blocks the uptake of soil water for months. In contrast, snow melting from branches may help trees recover from winter embolism. While these processes and respective dynamics (i.e., increasing frost drought until late winter followed by recovery) are relatively well understood in tall, adult trees e.g., [1,9,15], the situation of small trees or shrubs is more complex. As demonstrated in [31], snow cover protection has a major influence on the development of Ψ and embolism in small trees. The same is probably true for shrubs e.g., [32,33] but respective knowledge is scarce.

In the present study, we analyzed the formation and recovery of embolism over one year in *Pinus mugo*, which is a coniferous shrub of the Alpine "Krummholz" zone. This species is well adapted to snow loads (as well as avalanches) as the branches are very flexible and, consequently, often temporarily buried under the snow cover. The complex and unknown snow cover "history" makes the interpretation of winter measurements unavoidably difficult, but interesting from a hydraulic point of view, as embolism formation and recovery periods might alternate. We measured Ψ, water content, LC, and the proportion of aspirated pits in branch and needle xylem as well as carbohydrate contents of needles, stem wood (secondary xylem), and secondary phloem (Figure 1). We hypothesized an overall weaker annual course of embolism formation and recovery compared to trees as well as recovery periods already in winter. Recovery was expected to correspond, as in trees, to changes in carbohydrate contents and the proportion of aspirated pits.

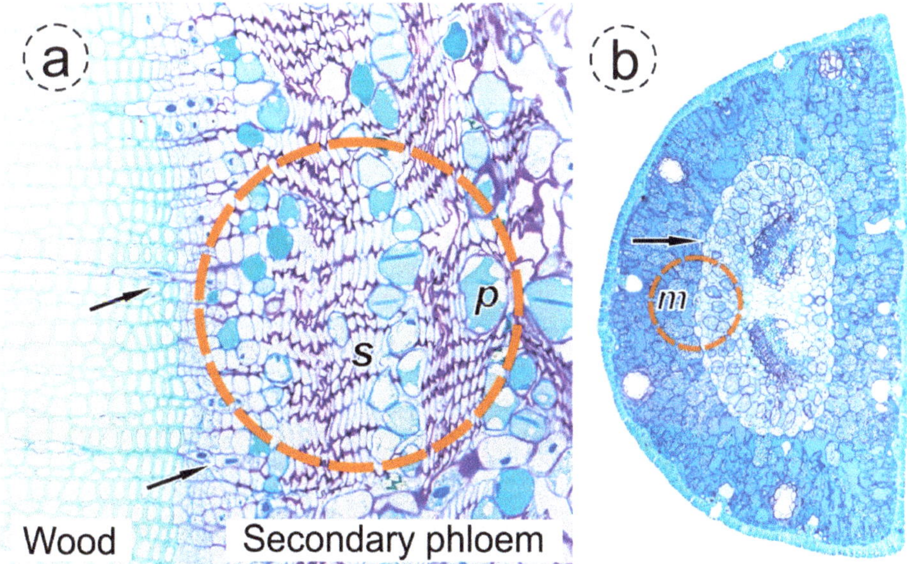

Figure 1. Information on histological anatomical investigations in *Pinus mugo* (**a**) branches and (**b**) needles. Regions of interest for histological staining are indicated with orange circles, the arrows in (**a**) indicate parenchymatic ray cells, the arrow in (**b**) indicates the endodermis, m, mesophyll; p, parenchyma cell; s, sieve cell.

2. Materials and Methods

2.1. Plant Material

This study was performed on knee pine (*Pinus mugo* Turra) growing at Birgitz Köpfl, Tyrol, Austria (2035 m, 47°11′N/11°19′E). For measurements, the uppermost, sun-exposed branches of shrubs (height up to 2.5 m) growing in a northwest exposed slope were used. For logistic reasons, samples harvested at the timberline were sometimes stored at −20 °C until taking measurements of sugar/starch contents and anatomy.

Measurements during winter and spring were performed in parallel with a study on *Picea abies* [1]. Climate data (daily minimum, mean and maximum air temperatures, daily sum of precipitation) of a nearby weather station (Mt. Patscherkofel, 2.252 m) were provided by ZAMG Zentralanstalt für Meteorologie und Geodynamik, Austria.

2.2. Seasonal Course

Samples were harvested from November 2010 to November 2011 in ca. monthly intervals between 11:00 and 14:00, respectively. Three shrubs were randomly selected and one branch per shrub was harvested, whereby in winter, branches above the snow cover were used. First, end twigs (ca. 10 cm) of branches were sampled and enclosed in plastic bags for transport to the laboratory and analysis of water potential (Ψ; see Section 2.3) and needle sugar/starch content. Second, branches (up to 1 m in length, basal diameter up to 1.5 cm, age up to 10 years) were cut and enclosed in plastic bags. Main stems of these branches were used for analysis of the loss of conductivity (LC), proportion of aspirated pits, and xylem and phloem starch content. In addition, needles were sampled for microscopic analyses (sugar/starch, proportion of aspirated pits).

2.3. Water Potential (Ψ)

Ψ measurements were made directly after field trips. For the analysis of Ψ, terminal segments of three side twigs (length up to 10 cm) per selected branch were cut, tightly wrapped in plastic bags, and transported to the laboratory within 1–2 h. Then, the Ψ of samples was measured with a Scholander apparatus (Model 1000, PMS, USA), whereby the entire end twig was sealed into the pressure chamber before the pressure was slowly increased until the water front was visible at the cut stem surface. The discrimination of water and resin was possible due to different viscosities and bubble sizes. Ψ values were averaged per branch.

2.4. Loss of Conductivity (LC) and Water Content (WC)

Samples for LC measurements were stored and wrapped in plastic bags in a fridge at ca. 10 °C until the following day at a maximum. LC was measured with a Xyl'em system (Bronkhorst, France) on three subsamples and averaged per branch. Segments (length ~ 4 cm, xylem diameter ~ 0.8 cm) of the main stem were cut under water, and LC was determined by measuring the increase in hydraulic conductivity after the removal of xylem embolism by repeated high-pressure flushes [34]. The sample preparation and measurement procedure followed a previously described protocol [11]: Flushing (at 80 kPa) and conductivity measurements (at 4 kPa) were done with distilled, filtered (0.22 μm), and degassed water containing 0.005% (v/v) "Micropur" (Katadyn Products Inc., Wallisellen, Switzerland) to prevent microbial growth. Flushing was repeated until measurements showed no further increase in conductivity, and the loss of conductivity was calculated as $LC = (1 - K_{min}/K_{max}) \times 100$, where K_{min} is the initial conductivity, and K_{max} is the maximal conductivity.

WC was measured gravimetrically on additional segments (length ca. 2 cm) cut from the main stem of samples (ME2355, Sartorius, Götingen, Germany). Bark and wood were separated, and the fresh weight was determined before samples were dried at 80 °C in and oven (48 hours) and re-weighted.

2.5. Anatomical Analysis

Fresh branch segments, and needles were placed directly into 100% ethanol in order to avoid pit aspiration [35,36]. Samples remained in 100% ethanol for 48 h, and were thereafter embedded in Technovit 7100 (Heraeus Kulzer, Germany). Semi-thin transverse sections (1–2 μm) were made on a rotary microtome (RM2235 Leica, Germany). Sections were thereafter stained for 3 min with Toluidine blue (Merck, Germany) solution (0.03 g/100 mL tap water (pH 7)), washed in distilled water, dehydrated in 99% alcohol, and mounted in Euparal (Carl Roth, Germany). The staining of starch grains was performed with Lugol´s solution (Fluka, Sigma-Aldrich, USA). Lugol´s solution is a solution of 2 g of potassium iodide with 1 g of iodine in distilled water. Samples were stained for 2min, washed with distilled water, and embedded in glycerol.

Starch grains and bordered pits were observed with a Leica CTR6000 microscope equipped with a digital camera (DMC2900 Leica, Germany). Pits in the xylem of needles and in the sapwood were classified as "closed" if the pit membrane was aspirated (Figure 2a,b) or as "open" if the pit membrane was positioned in the middle of the pit chamber (Figure 2c,d). Toluidine blue stains the torus of the pit in dark purple-dark blue, which gives a strong contrast to the turquoise-stained lignified pit chamber walls and secondary cell walls of the tracheids [1]. The classification of pit aspiration in order to assess the proportion of aspirated pits in sapwood was done for the earlywood and latewood of the latest three annual rings within a tangential window of 2–3 mm on 2–3 specimens from each tree. Regarding latewood, only pits in the last formed five cell rows close to the ring border were analyzed (Figure 2a). The screening of pit aspiration in earlywood was done mainly in the first formed 10 cell rows. In total, 7163 bordered pits were analyzed: on average, 512 pits per tree.

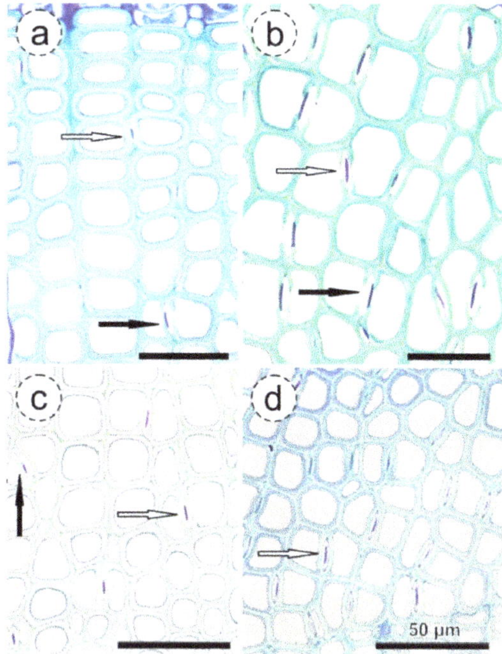

Figure 2. Examples of open and aspirated bordered pits in sapwood of *Pinus mugo*. (**a**) 24 November 2010, (**b**) 22 February 2011, (**c**) 15 June 2011, and (**d**) 6 July 2011, open arrows point at open bordered pits, closed arrows on aspirated bordered pits, reference bars = 50 µm.

2.6. Glucose, Fructose, and Starch Content

Analyses were performed for all samples harvested between November 2010 and August 2011. The needles were cut from branches, and from the main stem, and xylem and phloem samples were separated. Dried and finely ground plant material was incubated two times with 80% ethanol (v/v) at 75 °C with polyvinylpyrrolidon (PVP 40, Sigma Chemicals, USA) added. The supernatants containing the low molecular weight sugars were collected, the solvent was evaporated, and the residue was dissolved in citrate buffer pH 4.0. After dissociating the sucrose with invertase (β-fructosidase), the content of glucose was determined photometrically [37]. Fructose was converted to glucose with isomerase (phosphoglucose isomerase) and quantified using a standard test (Enzytec E1247, r-biopharm, Germany) measuring the absorption of NADPH [38] at 340 nm. The precipitate containing the starch was incubated with sodium hydroxide (0.5 M, 1 h at 60 °C), neutralized with hydrochloric acid, treated with amylase (amyloglucosidase) to break down the starch to glucose, and measured as described (Enzytec E1210, r-biopharm, Germany).

2.7. Statistics

Values were averaged per branch, which is the statistical unit throughout the study, and are given as the mean ± SE. Differences were tested with Student's t test and correlations with Pearson's coefficient were determined at $p = 0.05$.

3. Results

3.1. Water Potential and Embolism

During summer months (May–September; Figure 3b), Ψ varied between −1 and −1.5 MPa. Note that these measurements were taken around midday on sunny days and thus reflect values during transpiration. In winter, a decrease to −1.82 ± 0.04 MPa (22 February), followed by a steep increase to −0.34 ± 0.04 MPa (4 April), was observed. Values before (22 February) and after (19 May) this phase of intense winter drought significantly differed from peak Ψ in midwinter (22 March). The course of Ψ followed the course of the water content of the bark and (less) to that of the xylem, although no significant correlation was observed. LC was 15% at a maximum from November 2010 to the end of 2010 and from April 2011 to November 2011 (Figure 3c). In winter months, a maximum of 39.9% ± 14.5% was reached at 22 March, and significant differences between midwinter (22 February) and late winter values (19 May) were found.

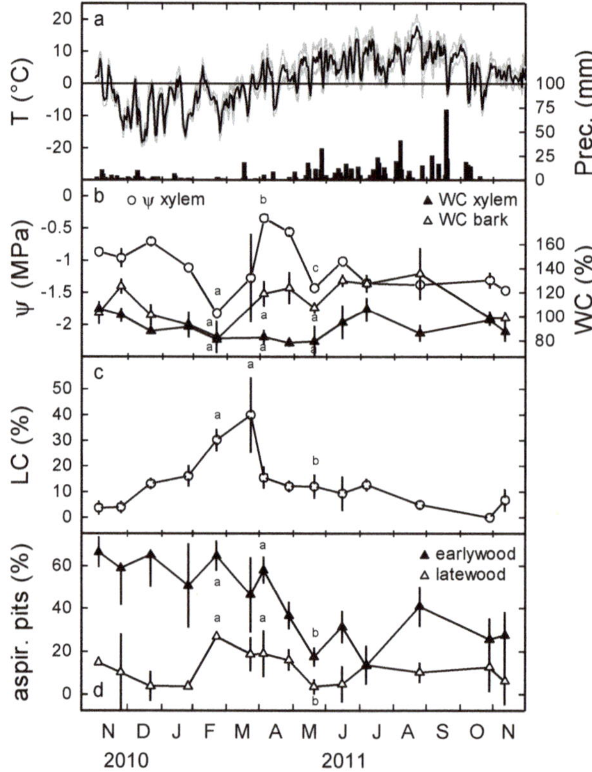

Figure 3. Climate, course of water potential and content, loss of conductivity, and pit aspiration. (**a**) Air temperatures (T; daily minimum, mean, and maximum) and daily sum of precipitation (prec.) from Mt. Patscherkofel (2.252 m provided by ZAMG Zentralanstalt für Meteorologie und Geodynamik, Austria). (**b**) Water potential (Ψ) and water content in stem bark and xylem. (**c**) Loss of conductivity in branch xylem (LC). (**d**) Proportion of aspirated pits (asp. pits) in earlywood and latewood. Mean ± SE. Different letters indicate significant differences measurements on 22 February, 19 May, and the sampling date of the highest values within this period.

3.2. Proportion of Aspirated Pits

The proportion of aspirated pits was overall higher in earlywood than in latewood (Figure 3d). The highest proportion of aspirated pits was observed on 22 February (64.7% ± 6.9% in earlywood (Figure 2b), 27.0% ± 1.4% in latewood), after which pit aspiration gradually decreased (17.5% ± 4.1% in earlywood, 3.6% ± 3.2% in latewood on 19 May). The strongest decrease was observed between 4 April and 19 May (significant differences in earlywood). A high proportion of latewood pits did not aspirate, even during winter time (Figure 2a). The lowest proportion of aspirated pits could be found from May to July (Figure 2c,d, Figure 3d). Although Figure 3 indicates a parallel decrease in LC and proportion of aspirated pits, due to the limited number of data points, no significant correlation was observed.

In needles, no screening for the proportion of aspirated pits could be made, because in each transverse section of the vascular bundles, only a few number of bordered pits could be found. However, we have hints that the bordered pits of the needle xylem are as well closed in winter (Figure 4a,b) and open in summer (Figure 4c,d) until—at least—early September (Figure 4e).

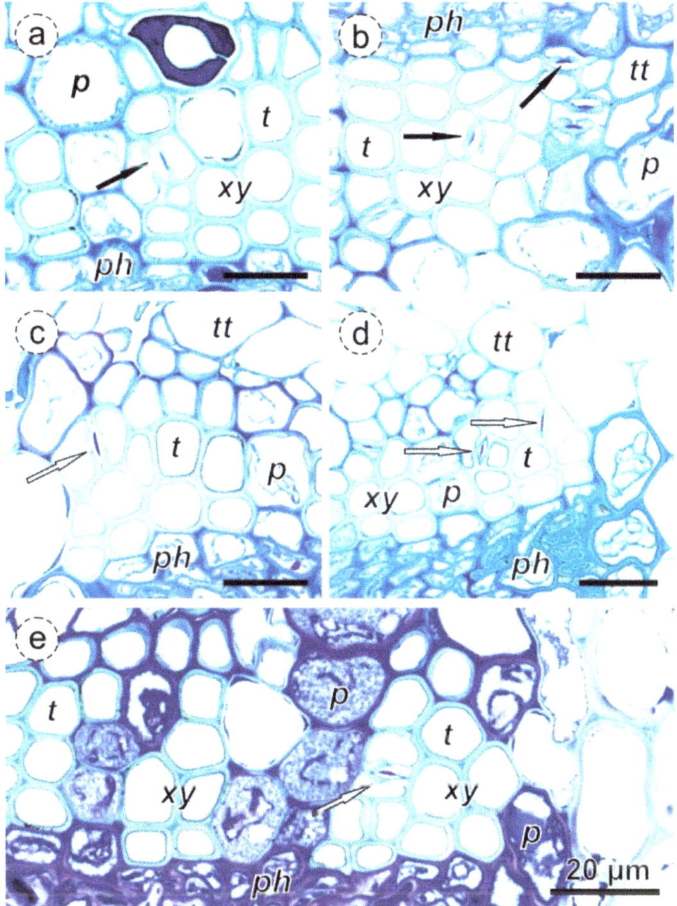

Figure 4. Pit aspiration in the xylem of the needle vascular bundles at different dates. (**a,b**) 24 November 2010, (**c,d**) 15 June 2011, and (**e**) 10 September 2011; p, parenchyma cell; ph, phloem; t, tracheid; tt, transfusion tracheid; xy, xylem; open arrows point at open pits, closed arrows at aspirated pits, reference bars = 20 µm.

3.3. Glucose, Fructose and Starch Content

Most pronounced courses in carbohydrate contents were found in needles (Figure 5a): While glucose and fructose concentrations gradually decreased from November 2010 (3.3% ± 0.5% DW glucose, 2.9% ± 0.4% DW fructose) to August 2011 (1.5% ± 0.3% DW glucose, 1.1% ± 0.2% DW fructose), starch contents were low until May 2010, peaked in July 2011 (9.3% ± 1.1% DW), and then decreased to 1.4% ± 0.8% DW (August 2011). July values were significantly higher than those in March and August. Annual course patterns were similar but less clear in stem sapwood and secondary phloem, with the highest starch contents reached in July (Figure 5b,c).

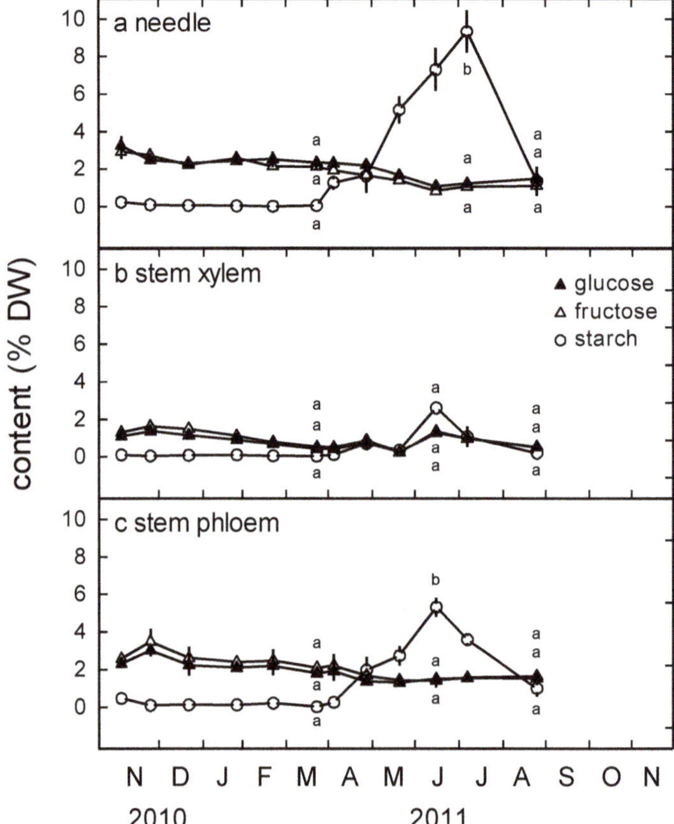

Figure 5. Course of glucose, fructose, and starch content. Carbohydrate contents (% of dry weight) in (**a**) needles, (**b**) stem xylem, and (**c**) stem phloem. Mean ± SE. Different letters indicate significant differences between measurements on 22 March, 24 August, and the sampling date of the highest values within this period.

Histochemical observation of the starch grains (Figure 6) in needles and secondary phloem confirmed the results of the chemical analyses. Firstly, a lower number of starch grains were found in the living cells of the secondary phloem, i.e., parenchyma cells of the rays and of the tangential parenchyma bands between the sieve cells (Figure 6a), compared to the needles (Figure 6b,c) and secondly, the maximum amount of starch was present in July 2011 in both organs. Starch grains were detectable until November 2011, but not in the period from 21 December 2010 until 22 March 2011. In midwinter and early spring, when starch was absent in the needles (Figure 7a), several white-colored vesicles were found in the mesophyll cells, and the cells of the endodermis contained small translucent (hydrophobic) droplets. During late spring and summer, no such droplets were found in the endodermis (Figure 7b); instead, the cells were filled with quite big starch grains. Starch grains in the endodermis were much bigger than those in the parenchyma cells of the mesophyll or near the vascular bundles (Figure 6b). The Casparian strip of the endodermis is impregnated with lignin, as indicated by its turquoise-greenish color when stained with toluidine blue.

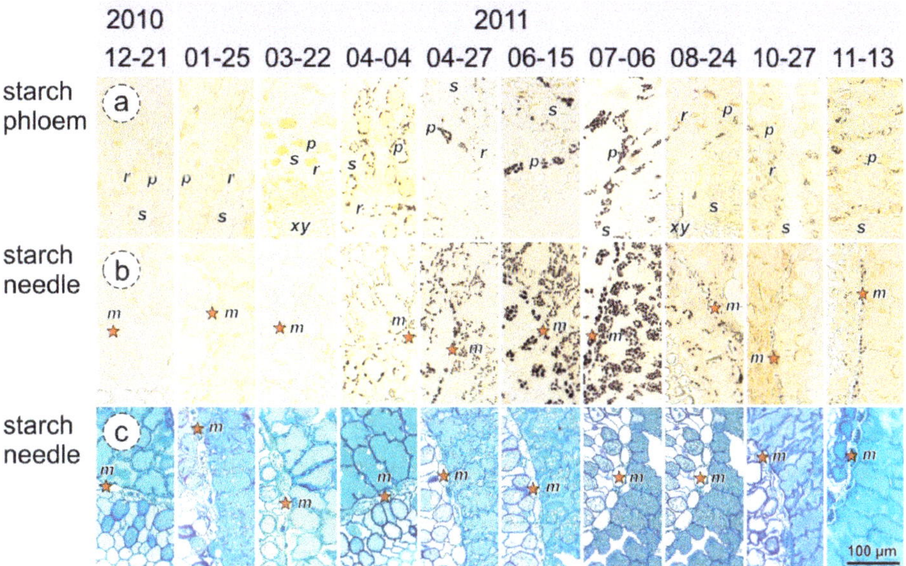

Figure 6. Histochemical detection of starch in secondary phloem and needles. (**a**) Secondary phloem (and parts of sapwood) stained with Lugol´s solution, (**b**) areas next to the needle endodermis stained with Lugol´s solution, and (**c**) areas next to the needle endodermis stained with toluidine blue at different times of the season. m, mesophyll; p, parenchyma cell in the secondary phloem; r, ray cell in the secondary phloem; s, sieve cells; xy, secondary xylem; stars indicate the locations of the endodermis, bar − 100 μm (the bar is representative for all photos). For regions of interest, see Figure 1.

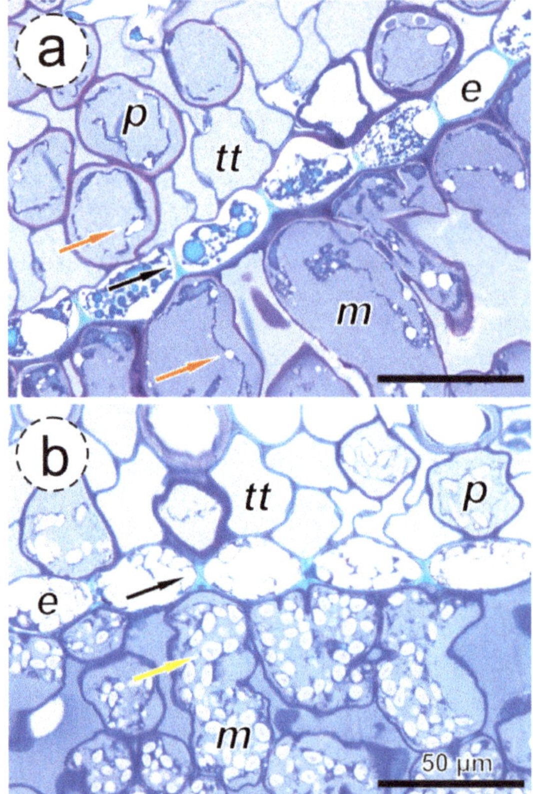

Figure 7. Endodermis and the adjacent tissues of *Pinus mugo* needles in winter and summer. (**a**) 15 January 2011 and (**b**) 6 July 2011. e, endodermis; m, mesophyll; p, parenchyma cell; tt, transfusion tracheid; black arrows point at the Casparian strips, orange arrows point at translucent droplets, and the yellow arrow points to a starch grain, bar = 50 µm. For orientation, see Figure 1.

4. Discussion

Winter stress induced embolism in *Pinus mugo*, as observed in other conifers growing at high elevation. *Pinus mugo* also showed recovery during late winter and spring, and findings indicate that underlying mechanisms were similar to those found in timberline trees: pit aspiration probably enabled the isolation of embolized tracheids until their functionality was re-established (Figure 3d), and changes in carbohydrate contents demonstrate cellular activity during late winter, which might enable osmotic adjustment to create driving forces for water re-distribution. However, in contrast to our hypothesis, the annual course in water potential (Ψ), loss of conductivity (LC), and related structural parameters were not more complex than that in trees, despite the shrub habit of *Pinus mugo* and respective more snow coverage.

The maximum LC in *Pinus mugo* reached during the study period was about 40% (Figure 3c), which is remarkably similar to the LC in *Picea abies* (43.4% ± 1.6%; [1]) observed in the same winter season and at the same study site. Accordingly, the minimum Ψ was about −1.8 MPa in *Pinus mugo* (Figure 3b) and −1.69 ± 0.11 MPa in *Picea abies*. This indicates that conifer shrubs and trees were exposed to identical winter drought stress, and that the more frequent cover by snow did not protect shrubs from dehydration during winter months. This may be due to the following two explanations: First, note that uppermost branches were harvested during winter. These were probably the branches experiencing the most stress and being protected by snow for relatively short periods only. [33] already demonstrated that branches permanently covered by snow develop only moderate Ψ and LC. Second, trees can store a lot of water in the trunk, which may buffer water losses induced by frost drought. [31] found water storage to increase and drought effects to decrease with tree size. The small branches of *Pinus mugo* cannot store larger amounts of water and thus Ψ decreased, even when drought stress was lower than in neighboring trees. In the branches harvested for this study, we could also show that even the bordered pits of the xylem from the needle vascular bundles can be aspirated during winter time (Figure 4a,b) in order to prevent the further embolism of adjacent tracheids. Relatively short-term snow covers on upper branches may also explain why we did not observe complex sequences of embolism formation and recovery but, as in trees [7,9–11,14,15], a period of embolism formation during midwinter and of recovery during late winter and spring. Similar to findings in *Picea abies* [1], embolism formation and repair thereby corresponded to changes in pit aspiration and carbohydrate contents.

The course of the proportion of aspirated pits followed the course of LC during the recovery period with decreasing LC and the opening of pits from late winter onwards (Figure 3c,d). In contrast, the proportion of aspirated pits was already high from the beginning of winter, while LC progressively increased. This indicates that some pits closed independently of embolism formation, maybe due to early frost events. Accordingly, the correlation (earlywood pits $r^2 = 0.25$) of the proportion of aspirated pits versus LC was weaker than that observed in *Picea abies* [1]. Overall, earlywood tracheids showed a higher proportion of aspirated pits, while latewood pits remained open more frequently. It remains to be studied whether this is due to lower embolism or less efficient pit aspiration in latewood [35]. As earlywood tracheids and their pits are hydraulically much more relevant than latewood tracheids, the findings again underline that the special conifer pit architecture [28,39] plays a central role for conifer hydraulic safety and pit aspiration, as well as probably for embolism recovery in secondary xylem (Figure 2), and even in the xylem of the vascular bundles of the needles (Figure 4).

Observed recovery also corresponded to changes in carbohydrate contents (Figure 5). Similar to *Picea abies* [1], a first peak in starch concentrations was observed in late winter (4 April), indicating relevant redistributions of carbohydrates within the plants. However, starch was not present year-round in *Pinus mugo* secondary phloem and needles. In the period from December until March, lipid droplets were found, especially in the endodermis cells (Figure 7a). According to [40], fats or lipid droplets are present the whole year; in some species, they are the only visible winter food reserve. Such species, including mainly diffuse porous angiosperm, but also the genus *Pinus*, are termed as "fat trees" [41,42]. In our study, starch grains were not present until the end of March (Figure 6). It has to be considered that the first changes in starch contents were observed when the snow cover was still present and the plants did not have access to soil water. However, the link between changes in carbohydrate contents and xylem refilling has yet to be demonstrated. According to [22], it is likely that carbohydrates are required to create an osmotic force, which enables pulling water into water films on the cell walls of embolized conduits also; see [43–45]. However, we did not observe relevant changes in glucose or fructose during recovery, probably because the formation of osmotic active sugars occurs locally and in comparably low overall concentrations. Unfortunately, a separate analysis of tracheid contents, as in the much bigger *Populus* [44] vessels, is hardly possible in conifers. The remarkable drop in starch content in late summer (Figures 5 and 6) indicates the beginning of transition to lipid pools and the displacement of storage pools to stem and roots.

Overall, embolism and refilling in *Pinus mugo* were surprisingly similar to the seasonal patterns observed in other conifers. The relevance of snow cover protection for both embolism formation and repair complicates predictions of the winter stress situation under future climate: in the Alps, the temperature increase was above the global average during the past decades [46,47], so that reductions in snow cover heights and durations have to be expected in future. An estimation of future precipitation changes is difficult, as the mountain terrain is complex. However, some studies for instance indicate an increase of spring droughts [48–50]. A reduced snow cover may reduce the time of blocked access to soil water and thus reduce frost drought. At the same time, the lack of snow protection may expose the plants to increased temperature amplitudes and freeze–thaw cycles, which can also induce embolism (see introduction). Furthermore, a lack of snow cover in late winter may hinder refilling processes, when water cannot be taken up over the branch surface. Shrub-like growth forms, such as *Pinus mugo*, might be especially affected by changes in snow cover regimes, but more studies are required to unravel the significance and dynamics of processes during embolism formation and recovery.

Author Contributions: Conceptualization, S.M. and S.R.; methodology, S.M., P.S., and S.R.; formal analysis, S.M., P.S., and S.R.; investigation, S.M., P.S., and S.R.; resources, S.M. and S.R.; data curation, S.M., P.S., and S.R.; writing—original draft preparation, S.M. and S.R.; writing—review and editing, S.M. and S.R.; visualization, S.M. and S.R.; supervision, S.M.; project administration, S.M. and S.R.; funding acquisition, S.M. and S.R.

Funding: This research was funded by the Austrian Science Fund, grant number I826–B25 and P32203_B.

Acknowledgments: We thank Birgit Dämon for his excellent assistance. Climate data were provided by ZAMG Zentralanstalt für Meteorologie und Geodynamik, Austria.

Conflicts of Interest: The authors declare no conflict of interest.

References

1. Mayr, S.; Schmid, P.; Laur, J.; Rosner, S.; Charra-Vaskou, K.; Dämon, B.; Hacke, U.G. Uptake of water via branches helps timberline conifers refill embolized xylem in late winter. *Plant Physiol.* **2014**, *164*, 1731–1740. [CrossRef] [PubMed]
2. Tyree, M.T.; Zimmermann, M.H. *Xylem Structure and the Ascent of Sap*, 2nd ed.; Springer: Berlin, Germany, 2002; p. 284. [CrossRef]
3. Hacke, U.G.; Sperry, J.S. Functional and ecological xylem anatomy. *Perspect. Plant Ecol. Evol. Syst.* **2001**, *4*, 97–115. [CrossRef]
4. Sperry, J.S. Evolution of water transport and xylem structure. *Int. J. Plant Sci.* **2003**, *164*, S115–S127. [CrossRef]
5. Choat, B.; Jansen, S.; Brodribb, T.J.; Cochard, H.; Delzon, S.; Bhaskar, R.; Bucci, S.J.; Field, T.S.; Gleason, S.M.; Hacke, U.G.; et al. Global convergence in the vulnerability of forests to drought. *Nature* **2012**, *491*, 752–755. [CrossRef]
6. Pittermann, J.; Sperry, J.S. Analysis of freeze-thaw embolism in conifers. The interaction between cavitation pressure and tracheid size. *Plant Physiol.* **2006**, *140*, 374–382. [CrossRef]
7. Mayr, S.; Schwienbacher, F.; Bauer, H. Winter at the alpine timberline. Why does embolism occur in Norway spruce but not in stone pine? *Plant Physiol.* **2003**, *131*, 780–792. [CrossRef]
8. Mayr, S.; Wieser, G.; Bauer, H. Xylem temperatures during winter in conifers at the alpine timberline. *Agric. For. Meteorol.* **2006**, *137*, 81–88. [CrossRef]
9. Sparks, J.P.; Black, R.A. Winter hydraulic conductivity end xylem cavitation in coniferous trees from upper and lower treeline. *Arct. Antarct. Alp. Res.* **2000**, *32*, 397–403. [CrossRef]
10. Mayr, S.; Wolfschwenger, M.; Bauer, H. Winter-drought induced embolism in Norway spruce (*Picea abies*) at the Alpine timberline. *Physiol. Plant.* **2002**, *115*, 74–80. [CrossRef]
11. Mayr, S.; Hacke, U.; Schmid, P.; Schwienbacher, F.; Gruber, A. Frost drought in conifers at the alpine timberline: Xylem dysfunction and adaptations. *Ecology* **2006**, *87*, 3175–3185. [CrossRef]
12. Brodribb, T.; Cochard, H. Hydraulic failure defines the recovery and point of no return in water-stressed conifers. *Plant Physiol.* **2009**, *149*, 575–584. [CrossRef] [PubMed]
13. Hammond, W.M.; Yu, K.L.; Wilson, L.A.; Will, R.E.; Anderegg, W.R.; Adams, H.D. Dead or dying? Quantifying the point of no return from hydraulic failure in drought-induced tree mortality. *New Phytol.* **2019**, *223*, 1834–1843. [CrossRef] [PubMed]
14. Sperry, J.S.; Nichols, K.L.; Sullivan, J.E.M.; Eastlack, S.E. Xylem embolism in ring-porous, diffuse-porous, and coniferous trees of northern Utah and interior Alaska. *Ecology* **1994**, *75*, 1736–1752. [CrossRef]
15. McCulloh, K.A.; Johnson, D.M.; Meinzer, F.C.; Lachenbruch, B. An annual pattern of native embolism in upper branches of four tall conifer species. *Am. J. Bot.* **2011**, *98*, 1007–1015. [CrossRef]
16. Maurel, C.; Verdoucq, L.; Luu, D.T.; Santoni, V. Plant aquaporins: Membrane channels with multiple integrated functions. *Annu. Rev. Plant Biol.* **2008**, *59*, 595–624. [CrossRef]

17. Gomes, D.; Agasse, A.; Thiébaud, P.; Delrot, S.; Gerós, H.; Chaumont, F. Aquaporins are multifunctional water and solute transporters highly divergent in living organisms. *Biochim. Biophys. Acta* **2009**, *1788*, 1213–1228. [CrossRef]
18. Heinen, R.B.; Ye, Q.; Chaumont, F. Role of aquaporins in leaf physiology. *J. Exp. Bot.* **2009**, *60*, 2971–2985. [CrossRef]
19. Holbrook, N.M.; Zwieniecki, M.A. Embolism repair and xylem tension: Do we need a miracle? *Plant Physiol.* **1999**, *120*, 7–10. [CrossRef]
20. Hacke, U.G.; Sperry, J.S. Limits to xylem refilling under negative pressure in *Laurus nobilis* and Acer negundo. *Plant Cell Environ.* **2003**, *26*, 303–311. [CrossRef]
21. Zwieniecki, M.A.; Holbrook, N.M. Confronting Maxwell's demon: Biophysics of xylem embolism repair. *Trends Plant Sci.* **2009**, *14*, 530–534. [CrossRef]
22. Nardini, A.; Lo Gullo, M.A.; Salleo, S. Refilling embolized xylem conduits: Is it a matter of phloem unloading? *Plant Sci.* **2011**, *180*, 604–611. [CrossRef] [PubMed]
23. Laur, J.; Hacke, U.G. Exploring *Picea glauca* aquaporins in the context of needle water uptake and xylem refilling. *New Phytol.* **2014**, *203*, 388–400. [CrossRef] [PubMed]
24. Katz, C.; Oren, R.; Schulze, E.-D.; Milburn, J.A. Uptake of water and solutes through twigs of *Picea abies* (L.) Karst. *Trees* **1989**, *3*, 33–37. [CrossRef]
25. Sparks, J.P.; Campbell, G.S.; Black, R.A. Water content, hydraulic conductivity, and ice formation in winter stems of *Pinus contorta*: A TDR case study. *Oecologia* **2001**, *127*, 468–475. [CrossRef]
26. Burgess, S.S.O.; Dawson, T.E. The contribution of fog to the water relations of *Sequoia sempervirens* (D. Don): Foliar uptake and prevention of dehydration. *Plant Cell Environ.* **2004**, *27*, 1023–1034. [CrossRef]
27. Limm, E.B.; Simonin, K.A.; Bothman, A.G.; Dawson, T.E. Foliar water uptake: A common water acquisition strategy for plants of the redwood forest. *Oecologia* **2009**, *161*, 449–459. [CrossRef]
28. Pittermann, J.; Sperry, J.S.; Hacke, U.G.; Wheeler, J.K.; Sikkema, E.H. Torus-margo pits help conifers compete with angiosperms. *Science* **2005**, *310*, 1924. [CrossRef]
29. Hacke, U.G.; Jansen, S. Embolism resistance of three boreal conifer species varies with pit structure. *New Phytol.* **2009**, *182*, 675–686. [CrossRef]
30. Plavcová, L.; Jansen, S.; Klepsch, M.; Hacke, U.G. Nobody's perfect: Can irregularities in pit structure influence vulnerability to cavitation? *Front. Plant Sci.* **2013**, *12*, 453. [CrossRef]
31. Mayr, S.; Charra-Vaskou, K. Winter at the alpine timberline causes complex within-tree patterns of water potential and embolism in *Picea abies*. *Physiol. Plant.* **2007**, *131*, 131–139. [CrossRef]
32. Mayr, S.; Gruber, A.; Bauer, H. Repeated freeze-thaw cycles induce embolism in drought stressed conifers (Norway spruce, stone pine). *Planta* **2003**, *217*, 436–441. [CrossRef] [PubMed]
33. Mayr, S.; Gruber, A.; Schweinbacher, F.; Dämon, B. Winter-embolism in a "Krummholz"-Shrub (*Pinus mugo*) growing at the alpine timberline. *Austrian J. For. Sci.* **2003**, *1*, 29–38.
34. Sperry, J.S.; Donnelly, J.R.; Tyree, M.T. A method for measuring hydraulic conductivity and embolism in xylem. *Plant Cell Environ.* **1988**, *11*, 35–40. [CrossRef]
35. Liese, W.; Bauch, J. On the closure of bordered pits in conifers. *Wood Sci. Technol.* **1967**, *1*, 1–13. [CrossRef]
36. Rosner, S.; Gierlinger, N.; Klepsch, M.; Karlsson, B.; Evans, R.; Lundqvist, S.-O.; Světlík, J.; Børja, I.; Dalsgaard, L.; Andreassen, K.; et al. Hydraulic and mechanical dysfunction of Norway spruce sapwood due to extreme summer drought in Scandinavia. *For. Ecol. Manag.* **2018**, *409*, 527–540. [CrossRef]
37. Bauer, H.; Plattner, K.; Volgger, W. Photosynthesis in Norway spruce seedlings infected by the needle rust *Chrysomyxa rhododendri*. *Tree Physiol.* **2000**, *20*, 211–216. [CrossRef]
38. Bergmeyer, H.U.; Bernt, E. Sucrose. In *Methods of Enzymatic Analysis*, Bergmeyer, H.U., Ed., Academic Press: New York, NY, USA, 1974; Volume 3, pp. 1176–1179.
39. Hacke, U.G.; Sperry, J.S.; Pittermann, J. Analysis of circular bordered pit function–II. Gymnosperm tracheids with torus-margo pit membranes. *Am. J. Bot.* **2004**, *91*, 386–400. [CrossRef]
40. Murmanis, L.; Evert, R.F. Parenchyma cells of secondary phloem in *Pinus strobus*. *Planta* **1967**, *73*, 301–318. [CrossRef]
41. Ziegler, H. Storage, Mobilization and Distribution of Reserve Material in Trees. In *The Formation of Wood in Forest Trees*; Zimmermann, M.H., Ed.; Academic Press: New York, NY, USA, 1964; pp. 303–320.

42. Höll, W. Storage and mobilization of carbohydrates and lipids. In *Trees-Contributions to Modern Tree Physiology*; Rennenberg, H., Eschrich, W., Ziegler, H., Eds.; Blackhuhys Publishers: Leiden, The Netherlands, 1997; pp. 197–211.
43. Bucci, S.J.; Scholz, F.G.; Goldstein, G.; Meinzer, F.C.; Sternberg, L.D.L. Dynamic changes in hydraulic conductivity in petioles of two savanna tree species: Factors and mechanisms contributing to the refilling of embolized vessels. *Plant Cell Environ.* **2003**, *26*, 1633–1645. [CrossRef]
44. Secchi, F.; Zwieniecki, M.A. Analysis of xylem sap from functional (nonembolized) and nonfunctional (embolized) vessels of *Populus nigra*: Chemistry of refilling. *Plant Physiol.* **2012**, *160*, 955–964. [CrossRef]
45. Barnard, D.M.; Lachenbruch, B.; McCulloh, K.A.; Kitin, P.; Meinzer, F.C. Do ray cells provide a pathway for radial water movement in the stems of conifer trees? *Am. J. Bot.* **2013**, *100*, 322–331. [CrossRef] [PubMed]
46. Böhm, R.; Auer, I.; Brunetti, M.; Maugeri, M.; Nanni, T.; Schöner, W. Regional Temperature variability in the European Alps: 1760-1998 from homogenized instrumental time. *Int. J. Climatol.* **2001**, *1801*, 1779–1801. [CrossRef]
47. Rebetez, M.; Reinhard, M. Monthly air temperature trends in Switzerland 1901–2000 and 1975–2004. *Theor. Appl. Climatol.* **2008**, *91*, 27–34. [CrossRef]
48. Ciccarelli, N.; von Hardenberg, J.; Provenzale, A.; Ronchi, C.; Vargiu, A.; Pelosini, R. Climate variability in north-western Italy during the second half of the 20th century. *Glob. Planet. Chang.* **2008**, *63*, 185–195. [CrossRef]
49. Beniston, M. Mountain climates and climatic change: An overview of processes focusing on the European Alps. *Pure Appl. Geophys.* **2005**, *162*, 1587–1606. [CrossRef]
50. Elkin, C.; Gutiérrez, A.G.; Leuzinger, S.; Manusch, C.; Temperli, C.; Rasche, L.; Bugmann, H. A 2 °C warmer world is not safe for ecosystem services in the European Alps. *Glob. Chang. Biol.* **2013**, *19*, 1827–1840. [CrossRef] [PubMed]

Article

Weak Apical Control of Swiss Stone Pine (*Pinus cembra* L.) May Serve as a Protection against Environmental Stress above Treeline in the Central European Alps

Walter Oberhuber [1], Theresa Andrea Geisler [1], Fabio Bernich [1] and Gerhard Wieser [2],*

[1] Department of Botany, Leopold-Franzens-Universität Innsbruck, Sternwartestraße15, A-6020 Innsbruck, Austria
[2] Department of Alpine Timberline Ecophysiology, Federal Research and Training Centre for Forests, Natural Hazards and Landscape (BFW), Rennweg 1, A-6020 Innsbruck, Austria
* Correspondence: gerhard.wieser@uibk.ac.at; Tel.: +43-512-573-933-5120

Received: 31 July 2019; Accepted: 27 August 2019; Published: 28 August 2019

Abstract: At the treeline in the Central European Alps, adverse climate conditions impair tree growth and cause krummholz formation of Swiss stone pine (*Pinus cembra* L.). Multi-stemmed trees (tree clusters) are frequently found in the treeline ecotone and are generally thought to originate from seed caches (multiple genets) of the European nutcracker (*N. caryocatactes*) or due to repeated damage of the leader shoot by browsing or mechanical stress (single genet). Additionally, lack of apical control can lead to upward bending of lateral branches, which may obscure single-genet origin if the lower branching points are overgrown by vegetation and the humus layer. The multi-stemmed growth form may serve as a means of protection against extreme environmental stress during winter, especially at wind-exposed sites, because leeward shoots are protected from, e.g., ice particle abrasion and winter desiccation. The aims of this study therefore were to analyze in an extensive field survey: (i) whether weak apical control may serve as a protection against winter stress; and (ii) to what extent the multi-stemmed growth form of *P. cembra* in the krummholz zone is originating from a single genet or multiple genets. To accomplish this, the growth habit of *P. cembra* saplings was determined in areas showing extensive needle damage caused by winter stress. Multi-stemmed saplings were assigned to single and multiple genets based on determination of existing branching points below the soil surface. The findings revealed that upward bending of lateral branches could protect saplings against winter stress factors, and, although multi-stemmed *P. cembra* trees were primarily found to originate from multiple genets (most likely seed caches), about 38% of tree clusters originated from upward bending of (partially) buried branches. The results suggest that weak apical control of *P. cembra* in the sapling stage might be an adaptation to increase survival rate under severe climate conditions prevailing above treeline during winter.

Keywords: apical control; multi-stemmed growth form; *Pinus cembra*; treeline

---

1. Introduction

The alpine treeline is a conspicuous climate-driven ecological boundary, which designates the upper elevational limit of tree growth [1–3]. There is extensive evidence that at high altitude cold temperatures during the growing season, which directly limit cell division and differentiation in meristematic tissues ("carbon-sink-limitation hypothesis"), are a major cause of treeline formation [4–7]. Above the treeline, adverse climatic conditions (e.g., frost drought, wind abrasion, and late frosts) frequently occur such that height growth of trees is strongly suppressed and tree stature is dominated by stunted multi-stemmed architecture (krummholz).

Swiss stone pine (*Pinus cembra* L.) is a key species forming the alpine treeline in the Central European Alps [8], which is reached at about 2400 m–2500 m above sea level (a.s.l.) under natural conditions, i.e., without human interference [3]. The heavy wingless seeds of *P. cembra* are almost exclusively dispersed by the European nutcracker (*Nucifraga caryocatactes* L.). At sheltered sites, krummholz forms of this species may be found up to ca. 2800 m a.s.l. (see [9]). Three growth forms of *P. cembra* can be distinguished: (1) single-trunk; (2) single genet multi-trunk caused by repeated damage of the leader shoot; and (3) multi-genet tree cluster attributable to multiple germination of seed caches [10]. The latter two forms are morphologically similar and generally thought to be distinguishable only by genetic analysis [11]. Although several coniferous tree species, e.g., spruces, firs, and larches, are able to form clonal tree groups by layering, i.e., formation of adventitious roots on buried lateral branches [12–14], cluster formation by vegetative sprouting has not been reported to occur in *P. cembra* [2,3]. Most pine species have a dominant main stem and distinct lateral branches which grow shorter and more horizontal than the vertical leader shoot [15], i.e., strong apical control is exerted by the leader shoot. Lack of apical (hormonal) control, e.g., due to mechanical or biotic damage of the leader, leads to upward bending of branches [16]. Shoot architecture of *P. cembra* is generally characterized by a vertical main trunk and upward bending of lateral branches.

This study aimed to increase our understanding of tree adaptation to extreme environmental conditions prevailing in the treeline ecotone. There are no reports on the influence of apical control on resistance against environmental stress in the treeline ecotone. Therefore, in an extensive field survey, we evaluated whether weak apical control may protect *P. cembra* against winter stress (frost desiccation, and snow mold). Furthermore, determining a morphological feature that consistently distinguishes single genet multi-trunk trees from multi-genet tree clusters originating from seed caches was a second aim of this study.

2. Materials and Methods

The study area is situated in the treeline ecotone on Mt. Patscherkofel (2246 m a.s.l.; 47°12′ 33″ N; 11°27′40″ E), which is located in the Central European Alps in western Austria. Mean annual temperature and precipitation (1967–2015) at the top of Mt. Patscherkofel are 0.2 °C and 883 mm with minimum in winter (December–February: 139 mm), respectively. The geology is dominated by gneisses and shist. Three sites (based on information gathered in Table 1) were selected on a ridge facing W to WNW (slope 15°–20°) above the current treeline at ca. 2100 m–2150 m a.s.l. Selected sites are frequently exposed to extreme winds (foehn), frequently reaching >100 km/h. Exposed mineral soil is common and dominating plants (*Loiseleuria procumbens* (L.) Desv., *Calluna vulgaris* (L.) Hull., *Juncus trifidus* L. and lichens, e.g., *Thamnolia vermicularis* (Sw.) Ach.ex Schaer and *Alectoria ochroleuca* (Hoffm.) A. Massal) are known to be resistant against wind, frost and winter desiccation [8].

Table 1. Number of *P. cembra* individuals at a wind-exposed ridge above treeline belonging to single- or multi-stemmed growth form. Multi-stemmed growth form is divided into multi-genet individuals developing from seed caches, and single-genet individuals, which developed upright growing lower branches. Percentages are given in parenthesis. For multi-stemmed growth form the mean number of shoots ± standard deviation per site are given. Significant differences ($p < 0.05$) in the mean values of the three plots (± SD) between single- and multi-stemmed and between multi-genet and single genet, respectively, are marked in bold and italics.

		Single Stemmed		Multi-Stemmed		
site	n			*n* shoots (mean ± SD)	Multi-genet	Single genet
1	31	11 (35)	20 (65)	2.9 ± 1.3	9 (45)	11 (55)
2	30	6 (20)	24 (80)	3.5 ± 1.6	20 (83)	4 (17)
3	31	9 (29)	22 (71)	2.6 ± 0.6	13 (59)	9 (41)
Mean		*9 ± 3 (28 ± 8)*	*22 ± 2 (72 ± 8)*	3.0 ± 0.5	*14 ± 6 (62 ± 19)*	*8 ± 5 (38 ± 19)*

The study was conducted on *P. cembra*, which is the dominant and widespread conifer along the treeline ecotone. European larch (*Larix decidua* Mill.) is scattered at some locations. The importance of apical control in *P. cembra* saplings as a mean to protect against winter stress was studied at wind exposed ridges and depressions with a long duration of snow cover showing pronounced needle damage after winter. In each of the three sites, one 100 m^2 plot was established where all *P. cembra* saplings (single stemmed and clusters, i.e., multi-stemmed trunks) without any visible damage to the leader shoot were counted. Multi-stemmed trunks were assigned to single and two or more genets by digging out all leaders until able to be traced back to the main stem and determining whether below the soil surface upward bending of lateral branches (i.e., single genet) or vertical leader shoots originating from seed caches (i.e., multi-genets) occurred.

Differences in the overall mean number between single stemmed and multi-stemmed *P. cembra* individuals were analyzed by one-way analysis of variance (ANOVA). One-way ANOVA was also used to test for differences between multi-genet and single genet saplings. A probability level of $p < 0.05$ was considered as statistically significant. Statistical analyses were made with the SPSS 16 software package for windows (SPSS, Inc., Chicago, IL, USA).

3. Results

Stem height and diameter of selected saplings ($n = 92$) were 39.2 ± 22.4 cm and 1.9 ± 1.2 cm, respectively, and annual shoot growth amounted to 4.4 ± 1.4 cm (mean values ± standard deviation). These growth variables were not significantly different among single-stemmed and multi-stemmed saplings. A potentially protective function of weak apical control in *P. cembra* against severe winter drought is depicted in Figures 1 and 2. Several examples of upward bending lower branches with connection to the main stem below the soil surface, whose vertical growing lateral branches are obscured by the humus layer and intensively growing dwarf-shrubs (*L. procumbens*, *C. vulgaris*), are depicted in Figure 3.

Figure 1. *Pinus cembra* sapling with upward-bending lateral branches (**left**) and *P. cembra* showing strong needle damage on the wind-exposed side caused by ice particle abrasion and winter desiccation, while sheltered leeward side remained largely undamaged (**right**).

Figure 2. Multi-stemmed appearance of *P. cembra* individuals (single genet) showing strong needle damage on the wind-exposed side caused by ice particle abrasion and winter desiccation, while sheltered leeward side remained largely undamaged (**left**). Upright growing lower branch after vegetation and raw humus layer was removed (**right**).

Figure 3. Examples of weak apical control of excavated basal branches of *P. cembra* saplings showing no damage to the leader shoot.

Above the treeline, the number of single stemmed *P. cembra* individuals (9 ± 3) was significantly lower ($p = 0.02$) when compared to the number (22 ± 2) of multi-stemmed *P. cembra* saplings, which corresponds to 28 ± 8% and 72 ± 8% single- and multi-stemmed *P. cembra* saplings, respectively (Table 1). The multi-stemmed growth form originates from 14 ± 6 (62 ± 19%) seed caches from multiple genet individuals and 8 ± 5 (38 ± 19%) seed caches from upright growing branches, i.e., single genet individuals (Table 1), which however was not statistically significant ($p = 0.19$). Upward-bending branches of a *P. cembra* sapling growing in a depression with long-lasting snow cover and frequent occurrence of snow mold is shown in Figure 4.

Figure 4. Branches of *P. cembra* in a hollow with long-lasting snow cover, where *Rhododendron ferrugineum* L. is the dominating shrub (**left**). Predominantly dead lower branches caused by snow mold. Note shoot tips with undamaged green needles (**right**).

4. Discussion

4.1. Multi-Stemmed Growth Form

Multi-stemmed trunks of *P. cembra* are frequently occurring in the alpine treeline ecotone [11,17,18]. This growth form is frequently reported to result from seed caches [11,19] or damage of the leader shoot caused by browsing or mechanical disturbance (e.g., winter desiccation, wind abrasion, frost, and avalanche [2,20]). In the case of destroyed apical meristems, apical control is lost and lateral branches bend upwards and form vertical shoots [16]. Here, we report that, in about 38% of multi-stemmed *P. cembra* saplings, lower branches at the base of the stem tend to grow upright without any visible damage to the leader. Hence, weak apical control may exist in *P. cembra* at the sapling stage, which can be explained by the competitive-sink hypothesis developed in [21]. This hypothesis states that branches compete with the subjacent stem for branch-produced carbohydrates, i.e., when the subjacent stem sink for carbohydrates is small, as is to be expected at the sapling stage, the branch is largely released from apical control and develops vertical growth.

At wind-exposed ridges, tree cluster formation favors growth of dwarf-shrubs (especially *L. procumbens* and *C. vulgaris*) by locally trapped litter and wind-blown organic and soil particles, which may increase nutrient content and water storage capacity of the soil [2]. Because the branching point close to the soil surface may be overgrown by vegetation and development of a humus layer, the origin from a single genet can be obscured. By carefully removing vegetation and the humus layer, multi-stemmed trunks of saplings and small trees can unequivocally be allocated to lateral branches of a main stem (single genet) or genetic different individuals (multi genet) originating from seed caches. Multi-stemmed trunks merge with time, which survive and grow better than single-stemmed trees, most likely due to better structural stability against wind [11,22,23]. Upright branches will ultimately merge with increasing tree age and radial growth, forming multi-stemmed trees, which—without genetic analysis—can mistakenly be assigned to originate from multi genet seed caches.

Analyzing the genetic diversity in multi-stemmed *P. cembra*, the authors of [11,23] found that 7 out of 22 and 1 out of 4 *P. cembra* tree clusters, i.e., 30% and 25%, respectively, were single genet multi-stemmed trees. Their findings are quite similar to our morphological approach, which was based on determination of the branching pattern below the soil surface. Hence, the assumption of a genetically controlled delay in dormancy release from varying stem diameter in multi-stemmed clusters may be false, because multi-genet shoots can be hidden by upright growing branches belonging to the same single genet tree individual.

4.2. Protective Function of Cluster Formation and Weak Apical Control During Tree Establishment

In the alpine treeline ecotone, natural regeneration of *P. cembra* is concentrated on locally higher terrain, such as ridges, shoulders and rock buttresses where wind exposure is high, snow depth is low, snow cover duration is short and there is sparse vegetation. Tree seedling establishment is controlled by micro-relief and/or shelter from low stature alpine vegetation, e.g., dwarf shrubs [24,25]. Once seedlings emerge from the ground vegetation and are exposed to the convective conditions of the atmosphere, wind effects on exposed ridges are important for tree survival during winter [3,12,17]. Damage of needles and buds is common at wind-exposed sites with little snow cover due to injuries from abrasive blowing snow, ice and mineral particles (for review see [26]), which reduces cuticle resistance leading to winter desiccation. Winter desiccation results from gradual water loss through transpiration, which cannot be compensated due to deep and long-lasting soil frost at snow-free ridges [27–29]. Saplings growing in clusters or single genet trees with multi-stemmed trunks caused by weak apical control create a wind-barrier effect that reduces mechanical damage at the sheltered leeward side especially during winter. This is corroborated by the finding that afforestation of *P. cembra* in groups in the subalpine zone yielded greater success than planting solitary trees ([30]; for a review, see [31]). Tree groups produce a snowdrift at their leeward side, which prevents deep soil frost and provides soil moisture at the beginning of the growing season [19]. On the other hand, weak apical control is also favorable for tree establishment in snow-rich concave topography and leeward slopes. Here, lower branches are less likely to be at risk for occurrence of snow mold under a long-lasting snow cover, if lower branches grow upright in the early sapling stage. Age- and/or size-related changes in apical control are to be expected, however, as a tree faces different challenges to survival with increasing tree age and/or size (cf. [32]). Hence, findings of this study suggest that upward bending of basal branches, i.e., a weak apical control in the sapling stage, might be an adaptation to extreme environmental conditions prevailing in the alpine treeline ecotone in cool temperate and boreal mountains, especially during winter. Our interpretation is corroborated by strong apical control found in co-occurring European larch (*L. decidua* [33]), which, due to its deciduous behavior, is less prone to winter drought and snow mold.

5. Conclusions

The results of this field survey suggest that cluster formation due to weak apical control of lateral branches at the stem base might be an adaptation to extreme environmental conditions prevailing above treeline within the study area. About one third of multi-stemmed *P. cembra* clusters in the krummholz zone belonged to a single genet. Multi-stemmed trees serve as protection against winter stress factors (snow and ice particle abrasion, and winter desiccation), have a better structural stability against wind and possibly improve resource acquisition due to trapping of litter and wind-blown soil. However, as this study was carried out within the inner Alpine dry zone of the Central European Alps, where the local climate is strongly influenced by southern "Foehn" type winds [34], further research needs to clarify the above demonstrated growth patterns of *P. cembra* saplings at sites differing in environmental conditions, especially winter stress.

Author Contributions: W.O. designed the study. W.O., T.A.G. and F.B. contributed to data acquisition. W.O., T.A.G. and G.W. interpreted and discussed the data. W.O. and G.W. wrote the manuscript.

Funding: This research received no external funding.

Acknowledgments: The authors thank two anonymous reviewers for their comments on earlier versions of the manuscript.

Conflicts of Interest: The authors declare no conflict of interest.

References

1. Wieser, G.; Tausz, M. *Trees at Their Upper Limit: Treelife Limitations at the Alpine Timberline*; Springer: Dordrecht, The Netherlands, 2007.
2. Holtmeier, F.K. Mountain Timberlines: Ecology, Patchiness, and Dynamics. In *Advances in Global Change Research*; Springer: Berlin, Germany, 2009; Volume 36, p. 437.
3. Körner, C. *Alpine Treelines: Functional Ecology of the Global High Elevation Tree Limits*; Springer: Basel, Switzerland, 2012; p. 220.
4. Körner, C. A re-assessment of high elevation treeline positions and their explanation. *Oecologia* **1998**, *115*, 445–459. [CrossRef] [PubMed]
5. Grace, J.; Berninger, F.; Nagy, L. Impacts of climate change on the tree line. *Ann. Bot.* **2002**, *90*, 537–544. [CrossRef] [PubMed]
6. Oberhuber, W. Influence of climate on radial growth of *Pinus cembra* within the alpine timberline ecotone. *Tree Physiol.* **2004**, *24*, 291–301. [CrossRef] [PubMed]
7. Körner, C.; Paulsen, J. A world-wide study of high altitude treeline temperatures. *J. Biogeogr.* **2004**, *31*, 713–732. [CrossRef]
8. Ellenberg, H.; Leuschner, C. *Vegetation Mitteleuropas mit den Alpen: In Ökologischer, Dynamischer und Historischer Sicht*; Verlag Eugen Ulmer: Stuttgart, Germany, 2010.
9. Available online: https://www.lfi.ch/resultate/meldungen/logbuch.php (accessed on 12 July 2019).
10. Linhart, Y.B.; Tomback, D.F. Seed dispersal by nutcrackers causes multi-trunk growth form in pines. *Oecologia* **1985**, *67*, 107–110. [CrossRef] [PubMed]
11. Tomback, D.F.; Holtmeier, F.K.; Mattes, H.; Carsey, K.S.; Powell, M.L. Tree clusters and growth form distribution in *Pinus cembra*, a bird-dispersed pine. *Arct. Alp. Res.* **1993**, *25*, 374–381. [CrossRef]
12. Jeník, J. Clonal growth in woody plants: A review. *Folia Geobot.* **1994**, *29*, 291–306. [CrossRef]
13. Bellingham, P.J.; Sparrow, A.D. Resprouting as a life history strategy in woody plant communities. *Oikos* **2000**, *89*, 409–416. [CrossRef]
14. Del Tredici, P. Sprouting in temperate trees: A morphological and ecological review. *Bot. Rev.* **2001**, *67*, 121–140. [CrossRef]
15. Turnbull, C.G. Shoot architecture II: Control of branching. In *Plant Architecture and Its Manipulation*; Turnbull, C.G., Ed.; Blackwell Publishing: Oxford, UK, 2005; Volume 17, pp. 92–120.
16. Wilson, B.F. Apical control of branch growth and angle in woody plants. *Am. J. Bot.* **2000**, *87*, 601–607. [CrossRef]
17. Tranquillini, W. *Physiological Ecology of the Alpine Timberline: Tree Existence at High Altitudes with Special Reference to the European Alps*; Springer: Berlin, Germany, 1979.
18. Kratochwil, A.; Schwabe, A. Wuchsformen der Arve (*Pinus cembra* L.) in Abhängigkeit von der ornithochoren Ausbreitung-im Vergleich mit Weidbuchen (*Fagus sylvatica* L.). *Diss. Bot.* **1993**, *196*, 107–134.
19. Holtmeier, F.K.; Broll, G. Broll Feedback effects of clonal groups and tree clusters on site conditions at the treeline: Implications for treeline dynamics. *Clim. Res.* **2017**, *73*, 85–96. [CrossRef]
20. Schuster, W.S.; Mitton, J.B. Relatedness within clusters of a bird-dispersed pine and the potential for kin interactions. *Heredity* **1991**, *67*, 41–48. [CrossRef]
21. Wilson, B.F.; Gartner, B.L. Effects of phloem girdling in conifers on apical control of branches, growth allocation and air in wood. *Tree Physiol.* **2002**, *22*, 347–353. [CrossRef] [PubMed]
22. Till-Bottraud, I.; Fajardo, A.; Rioux, D. Multi-stemmed trees of *Nothofagus pumilio* second growth forest in Patagonia are formed by highly related individuals. *Ann. Bot.* **2012**, *110*, 905–913. [CrossRef] [PubMed]
23. Höhn, M.; Ábrán, P.; Vendramin, G.G. Genetic analysis of Swiss stone pine populations (*Pinus cembra* L. subsp. *cembra*) from the Carpathians using chloroplast microsatellites. . *Acta Silv. Lign. Hung* **2005**, *1*, 39–47.
24. Smith, W.K.; Germino, M.J.; Hancock, T.E.; Johnson, D.M. Another perspective on altitudinal limits of alpine timberlines. *Tree Physiol.* **2003**, *23*, 1101–1112. [CrossRef]
25. Renard, S.M.; McIntire, E.J.; Fajardo, A. Winter conditions—Not summer temperature—Influence establishment of seedlings at white spruce alpine treeline In Eastern Quebec. *J. Veg. Sci.* **2016**, *27*, 29–39. [CrossRef]
26. Holtmeier, F.K.; Broll, G. Wind as an ecological agent at treelines in North America, the Alps, and the European Subarctic. *Phys. Geogr.* **2010**, *31*, 203–233. [CrossRef]

27. Larcher, W. Frosttrocknis an der Waldgrenze und in der alpinen Zwergstrauchheide auf dem Patscherkofel bei Innsbruck. *Veröffentlichungen Mus. Ferdinandeum Innsbr.* **1957**, *37*, 49–81.

28. Holzer, K. Winterliche Schäden an Zirben nahe der alpinen Baumgrenze. *Cent. Gesamte Forstwes.* **1959**, *76*, 232–244.

29. Hadley, J.L.; Smith, W.K. Influence of krummholz mat microclimate on needle physiology and survival. *Oecologia* **1987**, *73*, 82–90. [CrossRef] [PubMed]

30. Aulitzky, H.; Turner, H.; Mayer, H. Bioklimatische Grundlagen einer standortsgemäßen Bewirtschaftung des subalpinen Lärchen-Arvenwaldes. *Mitt. Eidg. Anst. Forstl. Vers.* **1982**, *58*, 327–577.

31. Schönenberger, W. Cluster afforestation for creating diverse mountain forest structes—A review. *For. Ecol. Manag.* **2001**, *145*, 121–128. [CrossRef]

32. Day, M.E.; Greenwood, M.S.; Diaz-Sala, C. Age- and size-related trends in woody plant shoot development: Regulatory pathways and evidence for genetic control. *Tree Physiol.* **2002**, *22*, 507–513. [CrossRef] [PubMed]

33. Prendin, A.L.; Petit, G.; Fonti, P.; Rixen, C.; Dawes, M.A.; von Arx, G. Axial xylem architecture of Larix decidua exposed to CO_2 enrichment and soil warming at the tree line. *Funct. Ecol.* **2018**, *32*, 273–287. [CrossRef]

34. Fliri, F. *Das Klima der Alpen im Raume von Tirol*; Universitätsverlag Wagner: Innsbruck, Austria, 1975.

Communication

High Resolution Maps of Climatological Parameters for Analyzing the Impacts of Climatic Changes on Swiss Forests

Andreas Paul Zischg [1,*], **Päivi Gubelmann** [2], **Monika Frehner** [3,4] and **Barbara Huber** [2]

[1] Institute of Geography, Oeschger Centre for Climate Change Research, University of Bern, 3012 Bern, Switzerland
[2] Abenis AG, 7000 Chur, Switzerland
[3] Forest Engineering Frehner, 7320 Sargans, Switzerland
[4] Department of Environmental Systems Science, Institute of Terrestrial Ecosystems, Swiss Federal Institute of Technology Zurich, 8092 Zurich, Switzerland
*   Correspondence: andreas.zischg@giub.unibe.ch; Tel.: +41-31-631-88-39

Received: 17 June 2019; Accepted: 22 July 2019; Published: 25 July 2019

Abstract: Assessing the impacts of climatic changes on forests requires the analysis of actual climatology within the forested area. In mountainous areas, climatological indices vary markedly with the micro-relief, i.e., with altitude, slope, and aspect. Consequently, when modelling potential shifts of altitudinal belts in mountainous areas due to climatic changes, maps with a high spatial resolution of the underlying climatological indices are fundamental. Here we present a set of maps of climatological indices with a spatial resolution of 25 by 25 m. The presented dataset consists of maps of the following parameters: average daily temperature high and low in January, April, July, and October as well as of the year; seasonal and annual thermal continentality; first and last freezing day; frost-free vegetation period; relative air humidity; solar radiation; and foehn conditions. The parameters represented in the maps have been selected in a knowledge engineering approach. The maps show the climatology of the periods 1961–1990 and 1981–2010. The data can be used for statistical analyses of forest climatology, for developing tree distribution models, and for assessing the impacts of climatic changes on Swiss forests.

Keywords: forest climatology; Switzerland; temperature; relative air humidity; thermal continentality; foehn winds; expert elicitation; knowledge engineering

---

1. Introduction

Forests in mountainous areas are providing important ecosystem services [1]. Most importantly, mountain forests protect settlements, roads and other infrastructure from natural hazards [2]. This protection function may be altered by the effects of climatic changes, invasive species, disturbances, or unsustainable forest management practices [3–6]. Moreover, forest disturbances can lead to an increase in flood hazards by large wood in rivers [7]. Hence, assessing the long-term sustainability and stability of mountain forests is a key requirement for assessing natural hazard risks in the Alps [8,9]. Forest managers today need to adapt to the effects of climatic changes. This requires accurate, reliable and localized information about potential future dynamics of forests [10]. As empirical knowledge on localized effects of climatic changes is yet rare, the assessment of future dynamics is merely been done with the help of simulation models (e.g., References [11–14]). However, as the vulnerability of forests to climate changes is not always known, these approaches require local expertise and expert knowledge to be integrated into the procedures for climate impact assessment [15]. Thus, for confronting these challenges and for closing the science-practice gap in forest management [16,17], transdisciplinary approaches or even post-normal science approaches [18] are needed.

The Federal Office for the Environment in Switzerland is aiming at developing an expert system for supporting forest managers in their decisions regarding the adaptation of their forests to the effects of climatic changes in the long-term. As a first and basic step to achieve this goal, a fundamental dataset of climatological indices had to be elaborated. These data aim at providing fundamental geodata for statistical analyses of the spatial distribution of the different forest typologies and their altitudinal belts, for developing tree distribution models, and for assessing the impacts of climatic changes on Swiss forests. The parameters of forest climatology that are represented in maps have been selected by a combination of a literature review [19] and an expert elicitation or knowledge engineering approach [20–23]. In an iterative process, we screened first the literature for environmental parameters that influence the spatial and altitudinal distribution of tree species and forest typologies. In a second step, we discussed this selection of parameters with local experts. Consequently, we added further parameters or deleted parameters from the list not being assessed as regionally relevant.

The final set of maps show these climatological indices: average daily temperature high and low in January, April, July, and October as well as of the year; seasonal and annual thermal continentality; first and last freezing day; frost-free vegetation period; relative air humidity; solar radiation; and foehn conditions. In the following, we present the data and methods used to elaborate these maps. The overall dataset was made available in the data repository ZENODO [24]

2. Maps of Climatological Forest Parameters

All of the elaborated maps have a spatial resolution of 25 by 25 m. The map extent is aligned to the digital elevation model DHM25 of the Federal Office of Topography swisstopo [25]. The raster data are georeferenced in the Swiss coordinate system CH1903/LV03 (EPSG 21781) and archived in an ASCII-grid format.

2.1. Air Temperature

Air temperature is one of the most important climate parameters for the altitudinal belts of forest typologies. In addition, daily temperature fluctuations are important for forest climatology. In continental, inner-alpine areas at the same altitude, the average daily temperature fluctuations are remarkably greater than on the northern edge of the Alps or even on an isolated summit in the Jura mountains or in the Prealps. In analyzing temperature fluctuations for forest climatology purposes one has to distinguish between the temperature range (i.e., the difference between the mean value of all daily temperature maxima and the mean value of all daily temperature minima of a considered period) and the daily temperature variation (i.e., the difference between the mean high and low of the averaged daily course of temperature of a considered period). The reason is that both the temperature maximum and the temperature minimum can occur at any time of day. The mean daily variation of air temperature is the difference between the mean daily temperature high, which usually occurs at a station at 2:00 p.m., and the mean daily temperature low, which usually occurs at sunrise. To determine the mean high and low, the mean annual daily cycle between 1961 and 1990 is to be calculated from 0:00 a.m. to 11:50 p.m. The temperature curve of a certain period (e.g., month) can then be determined as a function of the time of day. This average temperature curve is similar to a sinusoidal curve. The highest value of this temperature curve is now referred to as the mean daily high of the temperature, the lowest value of this curve correspondingly as the mean daily low of the temperature. Therefore, the mean daily maximum temperature must be higher or at least equal to the mean daily temperature high, and the mean daily minimum temperature must always be lower or equal to the mean daily temperature low. The mean daily temperature low is occurring around 7.00 a.m.; the mean daily high is occurring at around 2:00 p.m. The difference between the mean daily high and the mean daily low, i.e., the mean daily variation of the air temperature at a certain location, is a measure of the thermal continentality of a location because it represents the mean radiation-related temperature change of a location. In contrast, the mean daily temperature range is always greater than the mean (radiation-related) daily variation. The difference between the mean daily range and the mean daily variation describes the aperiodic component. The aperiodic component is a measure of the frequency with which air masses with different temperatures are replaced. The reason why the mean daily variation was taken into account is that the effect of the lowest value is quite different when it occurs at night than when it occurs during the day. If frost occurs during the night, it is much more harmful to the plant, as in many cases it is accompanied by radiation which causes even lower minima on the surface of the plant. During the day, on the other hand, an air temperature that falls below 0 °C due to a brief cold advection hardly has a negative effect on the plant surface, as the radiation from the plant surface causes significantly higher temperatures. Conversely, a brief nocturnal foehn event at night, which drives the temperature to high values, has hardly any effect on a forest tree, since the tree is photosynthetically inactive at night. If, on the other hand, a maximum temperature value is coupled with high irradiation, this has a great influence on the plants.

In Switzerland, there are several places with a high daily variation of temperature. These are inner-Alpine valleys in the Valais, in Graubünden or in northern Ticino, where average daily variations of 10 °C and more are possible. Conversely, these areas are relatively unaffected by air mass changes, so that the aperiodic component there is relatively small at 1.5 to 2 °C. On the other hand, exposed peaks in the Pre-Alps show only a very low daily temperature variation, which barely exceeds 2 °C. However, the aperiodic component can reach up to 3 °C in these layers, because air mass changes take place very frequently in such layers. Generally, it can be said that the mean daily temperature cycle throughout Switzerland with a range from 2 to over 10 °C shows clearly greater regional differences than the aperiodic component, which only fluctuates between about 2 and 4 °C.

The mean temperature was derived by averaging the temperature lows and highs of the periods. In addition, we elaborated a map of the mean temperature during the vegetation period (April–September).

2.1.1. Interpolation

Since temperatures in Switzerland have only been measured every 10 min since the 1980s, the daily variations and aperiodic components could only be determined in the period 1981–2010. In order to obtain daily variations for the desired time period 1961–1990, we calculated the mean minimum and mean maximum at all stations for this time period and corrected them for the aperiodic components of the years 1981–2010. With this method, the mean daily low and the mean daily high could be determined at all stations of MeteoSwiss. This method was applied to the entire year as well as to the months of January, April, July and October, which are representative for the four seasons. The basis for the now calculated maps were thus the mean daily lows and the mean daily highs of the period 1961–1990 in the whole year, as well as in the months January, April, July and October. Since the absolute maximum and minimum temperatures can be significant for the flora, maps of the absolute temperature minima (1864 to 1990) and the absolute temperature maxima (1864 to 1990) were also produced.

The interpolation of the measured values to the entire space in the targeted high spatial resolution bases on a gradient approach. First, we elaborated temperature maps at a selected altitude. Second, these temperature maps at the different altitudes were overlaid. Finally, the gradients between neighboring temperature layers were used to interpolate and project temperatures on topography by using the digital terrain model.

For this, we calculated the values of the stations at altitudes of 500, 1000, 1500, 2000 and 3500 m with the aid of seasonal gradients derived by neighboring station data. These projected values were interpolated over the entire area at the altitudes mentioned. The temperature maps at an altitude of 3500 m were prepared in order to model the gradients between 2000 m and 3500 m over an area. The values at an altitude of 3500 m a.s.l. were based on the measured value of the Jungfraujoch station only. We used a slight horizontal gradient from north to south (+1 °C in the south at Chiasso and −1 °C in the north at Schaffhausen) to correct for the overall north-south temperature gradient as suggested by Reference [26].

2.1.2. Local Temperature Corrections

In a further step, we corrected the resulting maps with selected microclimatological biases, i.e., we slightly corrected the interpolated temperatures at remarkably sunny and shadowy hillslopes, and we added temperature inversion effects of cold air layers in topographic depressions. In a mountainous country like Switzerland with its high variability in topography, considerable temperature differences often occur within short distances. This is particularly the case when a strong temperature inversion close to the ground can form in an area during clear nights due to radiation.

The temperature on northern slopes was lowered during the day but increased on southern slopes as suggested by [27]. These corrections were made for the maps of January, October, and the year. In April and July, the exposure differences are negligibly small due to the strong valley winds and the unstable air stratification. To apply these corrections, we computed the solar radiation. The temperature at maximally shadowed slopes was corrected by −1 °C and at maximally sunny slopes by +1 °C. Temperature corrections between these extremes have been interpolated linearly and usually reached small amounts of less than +/−0.5 °C, since in most cases the stations are located in only slightly inclined positions.

More important were the corrections at stations with cold air layers in deeper valleys and topographic depressions. Here, the temperature lows could reach up to −10 °C at individual stations, for example in January at the Samedan station in the Upper Engadine, which lies in an extremely cold air layer. Especially in closed valley basins of the Alps and the Jura, such inversions often form, which are also called cold air lakes. Cold air layers are of great importance for the flora, as frost occurs in cold air layers much later in spring than on a slope at the same altitude. It can happen that tree species such as beech or, to a lesser extent, fir, which are sensitive to late frost or extremely low winter temperatures, no longer thrive in areas with strong cold air layers. Therefore, it is important to map the cold air layers in topographic depressions as accurately as possible, at least those that influence entire valleys. For some areas, such as Goms, Upper Engadine or La Brévine in the Neuchâtel Jura, the strength of the cold air layers could be estimated well, as there are measuring stations at the bottom of the valley. In other places, the strength of the cold air layers had to be estimated on the basis of extensive thermal surveys and data from older climate stations. With these methods, it was possible to determine the temperature deviation from the free atmosphere in the deepest part of the valley bottom. In a second step, the thickness of a cold air layer had to be determined. It was generally assumed that no cold air layer existed above a damming obstacle, e.g., a valley bar or a valley narrowing. In most cases, these obstacles are not more than 150 m thick. However, there are areas with a height of the damming obstacle of up to 600 m. In topographical depressions without a drain, such as those found in the Jura mountains, the procedure was basically the same as in a valley. Here, too, the lowest air temperature was assumed for the lowest point of the basin. It was also assumed that no cold air layer existed above the deepest surrounding edge of the basin. For each cold air layer, the height of the temperature inversion layer and the height of the valley bottom above sea level is derived. Moreover, a temperature at the bottom of each cold air layer was assumed on the basis of neighboring temperature measurements. The temperature of the original maps was corrected in the mapped cold air layers by means of the altitudinal temperature gradients. Example maps are shown in Figure 1. The resulting maps are listed and described in Table 1.

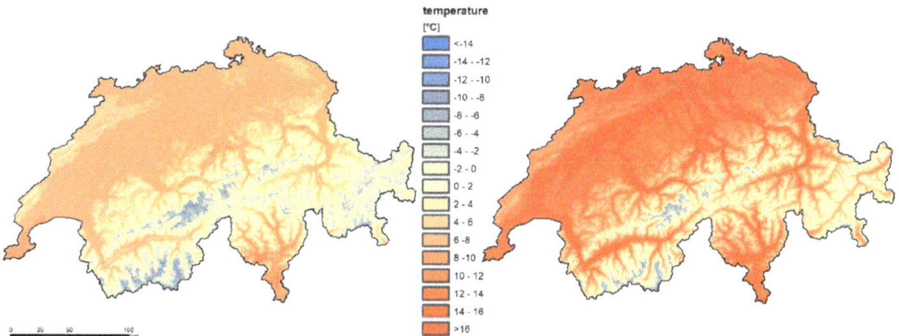

Figure 1. Mean daily temperature low (left) and (high) during the year in the period 1981–2010.

Table 1. Resulting temperature maps and intermediary files.

Map Abbreviation	Parameter	Unit	Period
TJANMIN6190	mean temperature low in January	°C	1961–1990
TJANMAX6190	mean temperature high in January	°C	1961–1990
TJANMEAN6190	mean temperature in January	°C	1961–1990
TAPRMIN6190	mean temperature low in April	°C	1961–1990
TAPRMAX6190	mean temperature high in April	°C	1961–1990
TAPRMEAN6190	mean temperature in April	°C	1961–1990
TJULMIN6190	mean temperature low in July	°C	1961–1990
TJULMAX6190	mean temperature high in July	°C	1961–1990
TJULMEAN6190	mean temperature in July	°C	1961–1990
TOCTMIN6190	mean temperature low in October	°C	1961–1990
TOCTMAX6190	mean temperature high in October	°C	1961–1990
TOCTMEAN6190	mean temperature in October	°C	1961–1990
TYYMIN6190	mean temperature low in the year	°C	1961–1990
TYYMAX6190	mean temperature high in the year	°C	1961–1990
TYYMEAN6190	mean temperature in the year	°C	1961–1990
TAMJJASMEAN6190	mean temperature April–September	°C	1961–1990
TABSMIN	absolute minimum temperature of the period	°C	1894–1990
TABSMAX	absolute maximum temperature of the period	°C	1894–1990
TJANMIN9110	mean temperature low in January	°C	1981–2010
TJANMAX8110	mean temperature high in January	°C	1981–2010
TJANMEAN8110	mean temperature in January	°C	1981–2010
TAPRMIN8110	mean temperature low in April	°C	1981–2010
TAPRMAX8110	mean temperature high in April	°C	1981–2010
TAPRMEAN8110	mean temperature in April	°C	1981–2010
TJULMIN8110	mean temperature low in July	°C	1981–2010
TJULMAX8110	mean temperature high in July	°C	1981–2010
TJULMEAN9110	mean temperature in July	°C	1981–2010
TOCTMIN8110	mean temperature low in October	°C	1981–2010
TOCTMAX9110	mean temperature high in October	°C	1981–2010
TOCTMEAN8110	mean temperature in October	°C	1981–2010
TYYMIN8110	mean temperature low in the year	°C	1981–2010
TYYMAX8110	mean temperature high in the year	°C	1981–2010
TYYMEAN8110	mean temperature in the year	°C	1981–2010
COLDAIRLAYER	temperature difference between the upper and lower boundary of the cold air layer, the altitude of the lower and upper boundary of the cold air layer	°C m a.s.l.	1981–2010

2.2. Daily Temperature Variation

The temperature maps provided the basis for the elaboration of maps showing the daily temperature variation. These temperature continentality maps result from the difference between the temperature low and the temperature high of the respective time period (month or year). The resulting maps are listed and described in Table 2.

Table 2. Resulting temperature continentality maps and intermediary files.

Map Abbreviation	Parameter	Unit	Period
CONTJAN	mean daily temperature variation in January	°C	1981–2010
CONTAPR	mean daily temperature variation in April	°C	1981–2010
CONTJUL	mean daily temperature variation in July	°C	1981–2010
CONTOCT	mean daily temperature variation in October	°C	1981–2010
CONTYY	mean daily temperature variation in the year	°C	1981–2010
CONTABS	absolute daily temperature variation in the period	°C	1894–1990

2.3. Average First and Last Freezing Day of the Year

The last frost day in spring is decisive for the occurrence of many tree species (frost tolerance and phenology). The first frost in autumn has less of an influence, as it limits the growth period of the trees. The maps showing the average first and average last frost day of the year were both calculated with the same procedure. The maps are based on measurement data for the period 1961–1990. A day was defined as a "frost day" if the minimum temperature of that day is below 0 °C. In a first step, we calculated the mean last frost day of the year for the period 1961–1990 for each meteorological station. Each station thus received a value between 1 and 365 (1 = 1 January). Based on this analysis, maps of the last frost have been elaborated for the heights of 500 m, 1000 m, 1500 m and 2000 m a.sl. For the calculation of the mean last frost day on the real heights of the digital terrain model with a target resolution of 25 m, the gradients from the frost maps on 500, 1000, 1500 and 2000 m a.sl. on each grid point were calculated between the different height levels. Below 500 m a.s.l. the gradient between 500 and 1000 m a.s.l. was used, above 2000 m a.s.l. the gradient between 1500 and 2000 m was used. The map for the mean first frost day was calculated according to a similar procedure. Both maps were subsequently corrected for cold air layers. For each cold air layer, the correction (in days) for the lowest point of the lake was determined. The correction value decreases linearly towards the upper boundary of the cold air layer reaching zero at the upper boundary. Finally, fixed limits were set for the occurrence of the last and first frost day. For the average last frost day, this is day 196 (July 15). For the average first frost day, this is day 225 (August 13). From these maps, the duration of frost-free vegetation period has deviated. The duration of the frost-free vegetation period is particularly decisive for frost-sensitive tree species such as beech and fir. The resulting maps are listed and described in Table 3.

Table 3. Resulting map of first and last freezing day of the year.

Map Abbreviation	Parameter	Unit	Period
LFD	average last frost day in a year	no. of day	1981–2010
FFD	average first frost day in a year	no. of day	1981–2010
VEGPER	durations of frost-free vegetation period	days	1981–2010

2.4. Relative Air Humidity

Air humidity is particularly important for vegetation during the day because the plant is photosynthetically active and at low air humidity either evaporates a lot or has to close its stomata, which prevents it from assimilating carbon. Air humidity during the night, in contrast, is of much less important, as the plant hardly evaporates during the night as a result of the stoma closure.

In accordance with the procedure for the production of temperature maps with altitude gradients and area interpolation, we elaborated maps of the monthly average relative air humidity at 1:30 p.m. for the months of January, April, July, and October as well as for the annual average in the period 1981–2010. In a second step, we elaborated maps of the monthly and annual average relative air humidity at 6:00 a.m. for the months of January, April, July and October for calculating the mean in the period 1981–2010. Figure 2 shows the average annual relative air humidity at 1:30 p.m. in Switzerland. The resulting maps are listed and described in Table 4.

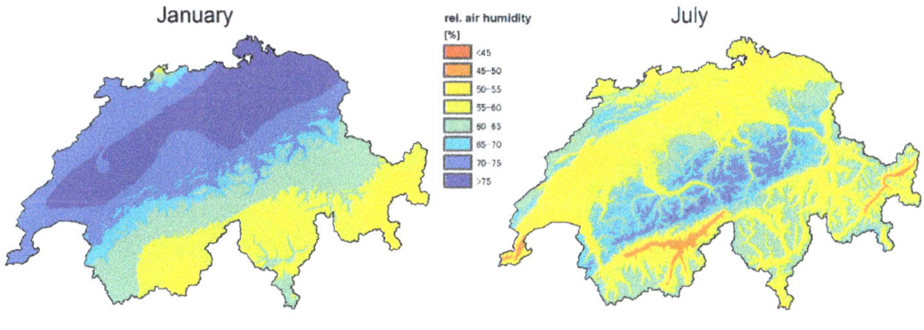

Figure 2. Mean daily relative air humidity at 1:30 p.m. in January (left) and July (high) in the period 1981–2010.

Table 4. Resulting maps of relative air humidity.

Map Abbreviation	Parameter	Unit	Period
MLFJAN8110	mean daily relative air humidity in January	%	1981–2010
MLFAPR8110	mean daily relative air humidity in April	%	1981–2010
MLFJUL8110	mean daily relative air humidity in July	%	1981–2010
MLFOCT8110	mean daily relative air humidity in October	%	1981–2010
MLFYY8110	mean relative air humidity in the year	%	1981–2010
LFJAN8110	mean relative air humidity at 1:30 p.m. in January	%	1981–2010
LFAPR8110	mean relative air humidity at 1:30 p.m. in April	%	1981–2010
LFJUL8110	mean relative air humidity at 1:30 p.m. in July	%	1981–2010
LFOCT8110	mean relative air humidity at 1:30 p.m. in October	%	1981–2010
LFYY8110	mean relative air humidity at 1:30 p.m. in the year	%	1981–2010

2.5. Solar Radiation

For elaborating maps of the global radiation, we used the data and followed the procedure of Reference [26]. The maps were recalculated with the resolution of the digital terrain model with a spatial resolution of 25 m. The monthly mean of the global radiation of the months January, April, July and October, as well as the annual mean of the global radiation, were elaborated. The time period was 1984–1993. The ratio between direct radiation and global radiation was estimated according to [26]. The resulting maps are listed and described in Table 5.

Table 5. Resulting maps of solar radiation.

Map Abbreviation	Parameter	Unit	Period
GLOBRADJAN	average global radiation in January	Wh/m^2	1984–1993
GLOBRADAPR	average global radiation in April	Wh/m^2	1984–1993
GLOBRADJUL	average global radiation in July	Wh/m^2	1984–1993
GLOBRADOCT	average global radiation in October	Wh/m^2	1984–1993
GLOBRADYY	average annual global	Wh/m^2	1984–1993

2.6. Foehn Conditions

The foehn wind is of great importance for vegetation, as it has a decisive influence on plants due to its high wind speed, low air humidity and relatively high temperature. In the northern Alpine valleys, for example, the foehn prolongs the vegetation period, but also has a very drying effect, so that some hygrophilic plants are severely weakened in their competitive power. In Switzerland, there are essentially two typical foehn phenomena. One is the well-known southern foehn, which is generally called the foehn. This phenomenon usually occurs during southern to south-western currents towards the Alps and affects the areas north of the main Alpine ridge. The second, pronounced foehn phenomenon is the northern foehn, which occurs in the valleys south of the Alpine ridge. The northern foehn can occur in almost any valley. We analyzed the meteorological data in terms of foehn frequency, relative air humidity, and foehn temperature. For this we distinguished foehn conditions from other conditions with the following combination of criteria:

- Relative air humidity: during the day $< = 50\%$, at night $< = 55\%$.
- Wind speed: $> = 5$ km/h
- Wind direction range: typical wind direction in ° with foehn $+/- 60°$

Each criterion has to be met. The monitoring network in Switzerland has two types of stations that can be used for foehn examinations. Firstly, these are the automatic measuring stations at which certain climate parameters are measured every 10 min. The foehn hours can be counted directly at such stations. However, since there are only a few such stations in the foehn areas, those climate stations had to be taken into consideration which are only observed three times a day, namely in the morning at 7.30 a.m., in the afternoon at 1.30 p.m. and in the evening at 7.30 p.m.. In order to be able to estimate the foehn hours of a day, a simple procedure was used: If a particular station has a foehn condition at a date, the foehn duration of this day was set to 8 h. If a station shows a foehn condition on two dates on a certain day, this results in a foehn duration of 16 h and if all three dates finally show a foehn condition, the foehn duration of this day is set to 24 h. According to the procedure described above, the number of hours with foehn, the temperature during foehn and the relative air humidity during foehn were evaluated for the year as well as for the months of January, April, July and October. The information obtained at the stations was subsequently used to draw isolines for the number of foehn hours, the temperature and the relative humidity in the southern foehn areas for the year and the months of January, April, July and October at heights of 500 m, 1000 m and 2000 m a.s.l. The isolines were drawn manually by the experts because usual kriging algorithms do not consider special topographic effects of foehn in valleys. With the help of these isolines, the number of foehn hours, the temperature at foehn and the relative humidity at foehn could be calculated for each grid point, analogously to the procedure for the temperature values at the heights mentioned above. Between the heights 1000 and 2000 m as well as between the heights 500 and 1000 m a.s.l. the height gradients of the number of foehn hours, the temperature at foehn and the relative air humidity at foehn were calculated. With the help of the calculated values at 1000 m and the gradient between 1000 and 2000 m or the gradient between 500 and 1000 m, the number of foehn hours, the foehn temperature and the relative air humidity could then be determined at each grid point of the digital terrain model at the true altitude at altitudes. The following data were then used for the calculation of the temperature at foehn and the relative humidity at foehn. For grid points with a height of more than 2000 m a.s.l., we assumed that the frequency of the foehn and the relative humidity during foehn do not change, for the temperature during foehn a uniform height gradient of $-0.7\,°C/100$ m increase was assumed. With these assumptions, it was also possible to obtain the foehn parameters above 2000 m a.s.l. on each grid point. Below 500 m a.s.l. we assumed that the gradient, which was calculated between the altitudes of 500 m and 1000 m a.s.l., was also valid at altitudes below 500 m a.s.l. With the area-wide data of the number of foehn hours, the temperature during foehn and the relative humidity during foehn at 500 m and the gradient between 500 m and 1000 m, the foehn parameters for the grid points, which have a height of less than 500 m, were finally determined. For the areas with northern foehn, the same procedure was applied as for the areas with southern foehn. The foehn duration in hours was subsequently divided by the duration of the period to determine the relative frequency as a share of the period. Figure 3 shows the annual frequency of foehn conditions and average relative air humidity during foehn conditions. The resulting maps are listed and described in Table 6.

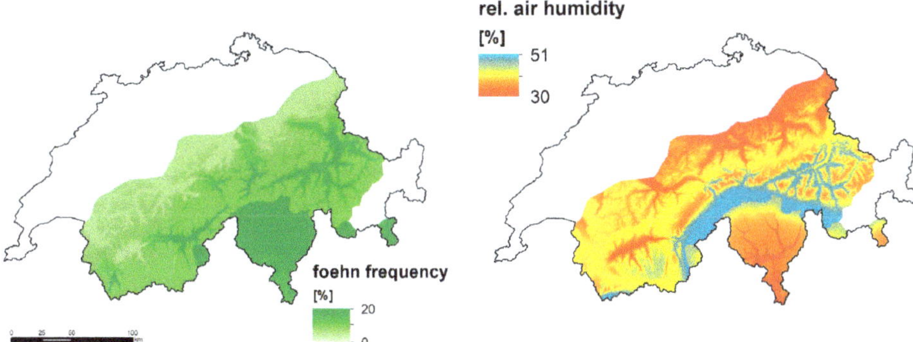

Figure 3. Annual frequency of foehn condition (left) and mean annual relative air humidity during foehn conditions (right) in the period 1981–2010.

Table 6. Resulting maps of foehn conditions.

Map Abbreviation	Parameter	Unit	Period
FOEHNHJAN	mean frequency of foehn conditions in January (time with foehn/time)	-	1981–2010
FOEHNHAPR	mean frequency of foehn conditions in April (time with foehn/time)	-	1981–2010
FOEHNHJUL	mean frequency of foehn conditions in July (time with foehn/time)	-	1981–2010
FOEHNHOCT	mean frequency of foehn conditions in October (time with foehn/time)	-	1981–2010
FOEHNHYY	mean frequency of foehn conditions in the year (time with foehn/time)	-	1981–2010
FOEHNTJAN8110	mean temperature during foehn conditions in January	°C	1981–2010
FOEHNTAPR8110	mean temperature during foehn conditions in April	°C	1981–2010
FOEHNTJUL8110	mean temperature during foehn conditions in July	°C	1981–2010
FOEHNTOCT8110	mean temperature during foehn conditions in October	°C	1981–2010
FOEHNTYY8110	mean temperature during foehn conditions in the year	°C	1981–2010
FOEHNFJAN	mean relative air humidity during foehn conditions in January	%	1981–2010
FOEHNFAPR	mean relative air humidity during foehn conditions in April	%	1981–2010
FOEHNFJUL	mean relative air humidity during foehn conditions in July	%	1981–2010
FOEHNFOCT	mean relative air humidity during foehn conditions in October	%	1981–2010
FOEHNFYY	mean relative air humidity during foehn conditions in the year	%	1981–2010

3. Discussion and Conclusions

The presented maps provide a fundamental dataset for the analyses of climate change impacts on mountain forests in Switzerland. The datasets complement other relevant data (e.g., precipitation, wind, geology, duration of heatwaves and droughts) and can be used for statistical analyses of forest climatology or as a baseline for adding the regionalized temperature increase given by climate models. The high-resolution maps can be used to train regression models for the location of treelines and vegetation belts in mountainous areas and tree distribution models aimed at analyzing potential future shifts due to climatic changes. The maps cover the whole of Switzerland. Thus, the presented data allow us to analyze the regional variability of altitude levels of tree lines and vegetation belts. Existing data can be downscaled at the same spatial resolution if needed for extended analyses. This is the case for precipitation data. The maps have been validated by the pool of experts. The interpolation and correction methods allowed the consideration of the effects of the micro-relief on temperature. This is on the one side novel; however these corrections base on assumptions and may introduce uncertainties. The interpolation error of the temperature maps is 0.1 °C. The presented datasets can furthermore be used to derive other maps such as for example potential evapotranspiration if combined with wind data. In future analyses, it is to be evaluated if the duration and intensity of heatwaves and droughts are relevant climatological indices. In principle, the data can be used also for other purposes such as for example agriculture. However, since only very rare meteorological stations are located above 2500 m a.s.l., the presented maps have only limited validity above this altitude.

Author Contributions: Conceptualization, all; methodology and analyses, all; project administration, B.H.; writing, A.Z.

Funding: This research was funded by the Swiss Federal Office for the Environment and the Swiss Federal Institute of Forest, Snow and Landscape Research WSL within the research program "Forests and climate change".

Acknowledgments: We thank the experts involved in the research project for making their expertise explicit. Moreover, we thank MeteoSwiss and SwissTopo for providing their data.

Conflicts of Interest: The authors declare no conflict of interest.

Data Availability: The described data is available at ZENODO, doi:10.5281/zenodo.3245891.

References

1. Grêt-Regamey, A.; Brunner, S.H.; Kienast, F. Mountain Ecosystem Services: Who Cares? *Mt. Res. Dev.* **2012**, *32*, S23–S34. [CrossRef]

2. Klein, J.A.; Tucker, C.M.; Steger, C.E.; Nolin, A.; Reid, R.; Hopping, K.A.; Yeh, E.T.; Pradhan, M.S.; Taber, A.; Molden, D.; et al. An integrated community and ecosystem-based approach to disaster risk reduction in mountain systems. *Environ. Sci. Policy* **2019**, *94*, 143–152. [CrossRef]

3. Kulakowski, D.; Bebi, P.; Rixen, C. The interacting effects of land use change, climate change and suppression of natural disturbances on landscape forest structure in the Swiss Alps. *Oikos* **2011**, *120*, 216–225. [CrossRef]

4. Moos, C.; Fehlmann, M.; Trappmann, D.; Stoffel, M.; Dorren, L. Integrating the mitigating effect of forests into quantitative rockfall risk analysis—Two case studies in Switzerland. *Int. J. Disaster Risk Reduct.* **2017**, *32*, 55–74. [CrossRef]

5. Moos, C.; Toe, D.; Bourrier, F.; Knüsel, S.; Stoffel, M.; Dorren, L. Assessing the effect of invasive tree species on rockfall risk—The case of Ailanthus altissima. *Ecol. Eng.* **2019**, *131*, 63–72. [CrossRef]

6. Reyer, C.P.O.; Bathgate, S.; Blennow, K.; Borges, J.G.; Bugmann, H.; Delzon, S.; Faias, S.P.; Garcia-Gonzalo, J.; Gardiner, B.; Gonzalez-Olabarria, J.R.; et al. Are forest disturbances amplifying or canceling out climate change-induced productivity changes in European forests? *Environ. Res. Lett.* **2017**, *12*, 034027. [CrossRef] [PubMed]

7. Zischg, A.; Galatioto, N.; Deplazes, S.; Weingartner, R.; Mazzorana, B. Modelling Spatiotemporal Dynamics of Large Wood Recruitment, Transport, and Deposition at the River Reach Scale during Extreme Floods. *Water* **2018**, *10*, 1134. [CrossRef]

8. Grêt-Regamey, A.; Bebi, P.; Bishop, I.D.; Schmid, W.A. Linking GIS-based models to value ecosystem services in an Alpine region. *J. Environ. Manag.* **2008**, *89*, 197–208. [CrossRef] [PubMed]

9. Kräuchi, N.; Brang, P.; Schönenberger, W. Forests of mountainous regions: Gaps in knowledge and research needs. *For. Ecol. Manag.* **2000**, *132*, 73–82. [CrossRef]

10. Lindner, M.; Maroschek, M.; Netherer, S.; Kremer, A.; Barbati, A.; Garcia-Gonzalo, J.; Seidl, R.; Delzon, S.; Corona, P.; Kolström, M.; et al. Climate change impacts, adaptive capacity, and vulnerability of European forest ecosystems. *For. Ecol. Manag.* **2010**, *259*, 698–709. [CrossRef]

11. Rutherford, G.N.; Bebi, P.; Edwards, P.J.; Zimmermann, N.E. Assessing land-use statistics to model land cover change in a mountainous landscape in the European Alps. *Ecol. Model.* **2008**, *212*, 460–471. [CrossRef]

12. Huber, N.; Bugmann, H.; Lafond, V. Global sensitivity analysis of a dynamic vegetation model: Model sensitivity depends on successional time, climate and competitive interactions. *Ecol. Model.* **2018**, *368*, 377–390. [CrossRef]

13. Seidl, R.; Fernandes, P.M.; Fonseca, T.F.; Gillet, F.; Jönsson, A.M.; Merganičová, K.; Netherer, S.; Arpaci, A.; Bontemps, J.-D.; Bugmann, H.; et al. Modelling natural disturbances in forest ecosystems: A review. *Ecol. Model.* **2011**, *222*, 903–924. [CrossRef]

14. Zischg, A. Floodplains and Complex Adaptive Systems—Perspectives on Connecting the Dots in Flood Risk Assessment with Coupled Component Models. *Systems* **2018**, *6*, 9. [CrossRef]

15. Grêt-Regamey, A.; Brunner, S.H.; Altwegg, J.; Christen, M.; Bebi, P. Integrating Expert Knowledge into Mapping Ecosystem Services Trade-offs for Sustainable Forest Management. *Ecol. Soc.* **2013**, *18*, 34. [CrossRef]

16. Pluess, A.R.; Augustin, S.; Brang, P. *Wald im Klimawandel. Grundlagen für Adaptationsstrategien*; 1. Auflage; Haupt Verlag: Bern, Switzerland, 2016; ISBN 3258079951.

17. Fabian, Y.; Bollmann, K.; Brang, P.; Heiri, C.; Olschewski, R.; Rigling, A.; Stofer, S.; Holderegger, R. How to close the science-practice gap in nature conservation? Information sources used by practitioners. *Biol. Conserv.* **2019**, *235*, 93–101. [CrossRef]

18. Ravetz, J. The post-normal science of precaution. *Futures* **2004**, *36*, 347–357. [CrossRef]

19. Huber, B.; Zischg, A.; Burnand, J.; Frehner, M.; Carraro, G. *Mit welchen Klimaparametern kann man Grenzen plausibel erklären, die in NaiS (Nachhaltigkeit und Erfolgskontrolle im Schutzwald) verwendet werden um Ökogramme auszuwählen? Schlussbericht des Projektes im Forschungsprogramm "Wald und Klimawandel" des Bundesamtes für Umwelt BAFU, Bern und der Eidg. Forschungsanstalt WSL*; ETH Zurich: Birmensdorf, Switzerland, 2015.

20. Gharari, S.; Hrachowitz, M.; Fenicia, F.; Gao, H.; Savenije, H.H.G. Using expert knowledge to increase realism in environmental system models can dramatically reduce the need for calibration. *Hydrol. Earth Syst. Sci.* **2014**, *18*, 4839–4859. [CrossRef]

21. Staffler, H.; Pollinger, R.; Zischg, A.; Mani, P. Spatial variability and potential impacts of climate change on flood and debris flow hazard zone mapping and implications for risk management. *Nat. Hazards Earth Syst. Sci.* **2008**, *8*, 539–558. [CrossRef]

22. Zischg, A.; Fuchs, S.; Keiler, M.; Meißl, G. Modelling the system behaviour of wet snow avalanches using an expert system approach for risk management on high alpine traffic roads. *Nat. Hazards Earth Syst. Sci.* **2005**, *5*, 821–832. [CrossRef]

23. Zischg, A.; Schober, S.; Sereinig, N.; Rauter, M.; Seymann, C.; Goldschmidt, F.; Bäk, R.; Schleicher, E. Monitoring the temporal development of natural hazard risks as a basis indicator for climate change adaptation. *Nat. Hazards* **2013**, *67*, 1045–1058. [CrossRef]

24. Zischg, A. High Resolution Maps of Climatological Parameters for Analyzing the Impacts of Climatic Changes on Swiss Forests. Zenodo. 2019. Available online: https://doi.org/10.5281/zenodo.3245891 (accessed on 24 July 2019). [CrossRef]

25. *dhm25*; swisstopo: Knitz, Switzerland, 2012.

26. Z'Graggen, L. Strahlungsbilanz der Schweiz. Ph.D. Thesis, ETH Zurich, Zurich, Switzerland, 2001.

27. Volken, D. Mesoklimatische Temperaturverteilung im Rhone-und Vispertal. Ph.D. Thesis, ETH Zurich, Zurich, Switzerland, 2008.

 *forests* 

Article

Latitude and Weather Influences on Sun Light Quality and the Relationship to Tree Growth

Camilo Chiang [1,2,*], Jorunn E. Olsen [3], David Basler [4], Daniel Bånkestad [2] and Günter Hoch [1]

1   Department of Enviromental Siences – Botany, Schönbeinstrasse 6, 4056 Basel, Switzerland
2   Department of Research and Development, Heliospectra, Box 5401,414 58 Göteborg, Sweden
3   Department of Plant Sciences, Faculty of Biosciences, Norwegian University of Life Sciences, P.O. Box 5003, N-1432 Ås, Norway
4   Department of Organismic and Evolutionary Biology, Harvard University, Cambridge, MA 02138, USA
*   Correspondence: Camilo.chiang@unibas.ch; Tel.: +46-727-764-880

Received: 24 June 2019; Accepted: 22 July 2019; Published: 24 July 2019

Abstract: Natural changes in photoperiod, light quantity, and quality play a key role in plant signaling, enabling daily and seasonal adjustment of growth and development. Growing concern about the global climate crisis together with scattered reports about the interactive effects of temperature and light parameters on plants necessitates more detailed information about these effects. Furthermore, the actual light emitting diode (LED) lighting technology allows mimicking of light climate scenarios more similar to natural conditions, but to fully exploit this in plant cultivation, easy-to-apply knowledge about the natural variation in light quantity and spectral distribution is required. Here, we aimed to provide detailed information about short and long-term variation in the natural light climate, by recording the light quantity and quality at an open site in Switzerland every minute for a whole year, and to analyze its relationship to a set of previous tree seedling growth experiments. Changes in the spectral composition as a function of solar elevation angle and weather conditions were analyzed. At a solar elevation angle lower than 20°, the weather conditions have a significant effect on the proportions of blue (B) and red (R) light, whereas the proportion of green (G) light is almost constant. At a low solar elevation, the red to far red (R:FR) ratio fluctuates between 0.8 in cloudy conditions and 1.3 on sunny days. As the duration of periods with low solar angles increases with increasing latitude, an analysis of previous experiments on tree seedlings shows that the effect of the R:FR ratio correlates with the responses of plants from different latitudes to light quality. We suggest an evolutionary adaptation where growth in seedlings of selected tree species from high latitudes is more dependent on detection of light quantity of specific light qualities than in such seedlings originating from lower latitudes.

Keywords: light quantity; light quality; spectrometer; shoot elongation; tree seedlings

1. Introduction

Light is one of the main environmental signals affecting plant biology, with multiple physiological responses being controlled by changes in light quantity, quality, and photoperiod [1–3]. Although the effects of the natural daily variation in light quality on plants have not been quantified in situ, it could be shown experimentally by using artificial lighting of defined wavelength ranges, that specific developmental processes in plants are differently affected by different fractions of the sunlight spectrum [4–7]. Punctual measurements comparing sun light spectral composition at different solar elevation angles showed a lower fraction of blue (B, 400–500 nm) and red (R, 600–700 nm) light and a higher fraction of green (G, 500–600 nm) light in the middle of the day than at sunset [1]. Smith et al. [1] quantified the effect of the weather conditions on light quality at high solar elevation angles and showed that clouds and dust cover have a small effect on the light spectra, mainly affecting the light in the B and R ranges, not unlike the changes in the spectral composition of sun light that occur when passing through a plant canopy. Yet, detailed information about the dynamic changes in these light qualities, especially with respect to their potential impact on plant biology, have so far not been reported, although there is substantial knowledge about static light quality effects on gas-exchange and other plant physiological processes [8–10].

At the short wavelength end of the sun spectra, effects of ultraviolet (UV) light on plants (e.g., shoot elongation, production of UV-protecting secondary compounds) and the associated UV-B receptors (UVR8) and UV-A-blue light receptors, have been well described in plants [4]. Interestingly, the signaling effects of UV light on plants is reduced at higher radiation, implying that UV as a plant signal may be most important during twilight. The next section of the light spectrum, the B light, which is mainly sensed by cryptochromes, phototropins, and other blue light-UV-A receptors, affects, for example, stomatal opening and plant phototropism. High percentages of B light have been shown to affect plant morphology [5]. Although chlorophyll, as the central plant pigment of the photosynthetic light reaction, absorbs mainly B and R light, G light has also been shown to contribute to photosynthesis and to be especially important at lower canopy levels (i.e., the so-called 'green shade') and at deeper levels of the leaves [6]. At the longer wavelength end of the visible sun spectra, R and far-red (FR) light have important signal functions for plants. The ratio between red to far red (R:FR) is sensed by the phytochrome system and changes in the R:FR ratio can influence important physiological processes like growth, germination, and flowering. In addition, FR has an important role in optimizing photosynthesis upon combined action of the PSII and PSI, increasing the photosynthetic efficacy [7].

Recent developments in light emitting diode (LED) lighting systems potentially enable the mimicking of more natural light quality changes during plant cultivation in indoor growth facilities [11]. Due to the high degree of absorption of B and R light by photosynthesis-related pigments and higher electric efficiency [8], these two wavelength ranges tend to be dominating in commercial LED lamp systems. However, the knowledge about the changes in light quality related to the solar elevation angle, latitude, time of the day, and the day of year, as well as the weather in general [1,12] has been so far reported mainly from an atmosphere-physical approach, and has not been transferred to actual lighting systems used for plant culture in greenhouses or growth chambers.

Changes of light quality in the morning and evening hours may be an especially important plant signal at higher latitudes where twilight conditions persist for a substantial period. Several studies have shown how different ecotypes of tree species react differently to R or FR light treatments as day extension, and it has been hypothesized that this could be due to adaptations to the light quality at the end of the day at their site of origin [13–15]. Additionally, it has been shown that light quality can interact with other environmental factors, like temperature, where higher temperatures have shown to reduce the promoting effect of FR light on growth [16]. Understanding the role of the light quality variation in plants is a crucial factor to predict the effect of the currently rising temperatures, especially in marginal areas such as those close to the latitudinal range limits of trees.

In the current study, we present detailed, easy-to-apply, and continuous field measurements of the natural changes of the spectral composition of sunlight over a full year at a mid-latitudinal site (47° N). This data was then combined with an analysis of studies investigating the effect of light quality on growth of tree seedlings of different latitudinal origin. Our study thereby focuses on the comprehensive effects of sun light quality changes due to weather conditions and time of the year, excluding further, smaller-scale modifications of the light spectra due to the presence of 'green shade' below a canopy. Here, we investigated the correlation between wavelength-specific light quantity requirements of tree seedlings from different latitude origins and the natural availability of these wavelengths due to geographical, annual, and diurnal changes, at their respective origin. Such a correlation would indicate ecotypic adaptations of tree populations to the specific spectral light quality and dynamics at their original site.

2. Materials and Methods

2.1. Light Spectra Recordings

A USB2000+XR1-ES spectrometer (25 μm entrance slit, range 200–1000 nm, 1.5 nm resolution, Ocean Optics Inc., Largo, Florida, USA) was installed twelve meters above ground level at the Botanical garden of the University of Basel (257 m AMSL, 47° 33′ 30.3″ N, 7° 34′ 52.4″ E, Basel, Switzerland) to acquire the light spectrum during a chronological year under a ensured shadow-free environment with minimalized light reflection from buildings, surface water bodies, or vegetation in the surrounding areas. Light spectra from 200–1000 nm were recorded every minute from 21 February 2018 to 21 February 2019 using a single board computer (Raspberry pi 2, Cambridge, UK) allowing dynamic change in the integration time, reducing the electric noise in the measurements through the use of 75–85% of the saturation point of the equipment. The optical fiber was installed at a 90° angle relative to the horizon. A cosine corrector made of Spectralon was used to capture environmental light coming from 180° (CC-3-UV-S, Ocean Optics Inc.). The cosine corrector was replaced every three months. The spectrometer was additionally equipped with a fan to avoid heat accumulation on hot days and was calibrated with a calibration lamp (HL-3 plus, Ocean Optics Inc.), once before mounting, and then every 3 months during the measurement year. The calibration lamp was warmed up for 15 min before the calibrations that were performed using a boxcar width of 2 wavelengths every 6 nm in both directions and the average of 5 measurements for each curve.

2.2. Light Energy Calculations

To acquire an initial dark library, the spectrometer was set in darkness at 20 °C, and a dark spectrum was recorded for integration times between 100 ms and 10 s every 100 ms. For each light measurement, the corresponding or interpolated dark spectrum was removed from the raw measurement in the corresponding integration time. The remaining count was multiplied with the corresponding calibration file of the calibration lamp. The resultant count was then divided by the area (m^2) of the cosine corrector (1.19×10^{-5} m^2) and then divided by the integration time of each measurement (s). Additionally, the energy in each particular wavelength was calculated multiplying each specific frequency by the Planck constant. To obtain the photon flux of each wavelength as μmol photons m^{-2} s^{-1}, the resultant was divided by the Avogadro number and the pre-calculated energy of each wavelength [17].

2.3. Light Quantity and Quality Proportions

To simplify the results from a biological and practical point of view, from each measurement three proportions were separately calculated from the visible light spectra: The percentage of blue (B), green (G), and red (R). For the calculation of B, G, and R, the light spectrum as μmol photons m^{-2} s^{-1} was integrated from 400 to 700 nm to obtain the total photosynthetic photon flux density (PPFD). Furthermore, every 100 nm between 400 and 700 nm, the proportions of B, G, and R where calculated. The B proportion corresponded to the percentage of photons from 400 to 500 nm compared with the total PPFD, G from 500 to 600 nm and R from 600 to 700 nm, respectively. Additionally, the red to far red (R:FR) ratio was also calculated. This was done through the division of the sum of photons (as μmol m^{-2} s^{-1}) between 655 and 665 nm and the sum of photons between 725 and 735 nm, respectively [18]. For the analysis of the weather conditions on the light spectra through the day, the sunniest and the most cloud-covered day of each month were selected from the recorded data ($n = 12$). After this, a locally estimated scatterplot smoothing (LOESS) regression was fitted for both weather conditions (i.e., clear sky and overcast).

2.4. Solar Elevation Angle Calculation

To quantify the average effect of the weather and remove the effect of the time of the day and day length through the year, the collected sun spectral data were analyzed as a function of the solar elevation angle. For each measurement throughout the observation year, the solar elevation angle was calculated based on the geographic position and time of the day and the day of year using the solar position calculator available online [19] and confirmed through the OCE R package based on the NASA-provided Fortran program, using equations from "The Astronomical Almanac".

2.5. Literature Review

To relate our light quality measurements to potential effects on growth of tree seedlings from the boreal/temperate zone, we conducted a literature search and performed an analysis on a set of published experiments that investigated the effect of light quality on seedling growth of selected tree species from different latitudes: The conifers *Pinus sylvestris* L., *Picea abies* (L.) H.Karst, and *Abies lasiocarpa* (Hook.) Nuttall, as well as the deciduous *Betula pendula* Roth. (more information in Supplementary Table S1). We exclusively choose studies that (1) were conducted under similar controlled conditions, (2) treated tree seedlings with different R:FR ratio light, (3) made quantitative growth measurements on potted seedlings of trees, (4) ran the experiments for at least one month (i.e., between 35 and 50 days) and/or (5) used different tree populations of different latitudinal origins [14–16,20–23]. The treatments corresponded to day extensions with different R:FR ratios and main light periods of 9–12 h with similar light quantities (W m^{-2}) during the day and day extension/night treatment. Growth was measured at the end of the experiments as the distance between the soil and apical bud (shoot elongation) or the elongation of the needles, depending on the study. To quantitatively compare the results among the experiments, the measured growth parameters (i.e., either needle elongation or shoot elongation), were analyzed by considering only the effect size, i.e., growth relative to the average growth under pure R light day extensions and, if the experiment included more than one ecotype, the average growth of the most southern ecotype investigated under pure R light day extensions. For the analysis, the effect of the different light quality treatments and the population origin on the measured growth variables was analyzed through forward selection and backward elimination on a single dataset, where both variables were included in a two-way analysis of variance (ANOVA) with light quality and latitudinal origin as fixed factors. All analyses were performed using R 3.6 [24].

3. Results

3.1. Light Quality Changes Throughout the Day

Our field radiation measurements under different weather conditions and time of the day showed a reduction of the blue (B) light proportion and an increase of the green (G), red (R), and far red (FR) light proportion from sunrise and sunset to the middle of the day (Figure 1A–D). The analysis of multiple days with either clear or overcast conditions throughout the year revealed quality changes induced by the weather conditions that can be of similar magnitude as the diurnal effects of the solar elevation angle on the B and R fraction of the spectrum (Figure 2). In the middle of the day, the presence of clouds increased the B fraction, depending on the cloud cover density and height, with a simultaneous reduction of the R fraction. Weather conditions had no significant effect on values on the R:FR ratio in the middle of the day.

3.2. Effect of Weather on Light Quality Changes at Low Solar Elevations Angles

The effect of the weather conditions on the light quality was significantly stronger at lower solar elevation angles. At solar angels below 20°, overcast conditions led to a significantly lower proportion of B light and a higher proportion of R light, while the effect was much weaker for G light (Figure 2). At solar elevation angles below 1°, close to 37 % of the incoming PPFD consisted of B light, while G light and R light accounted for 31 and 30% of the PPFD, respectively, independently of the weather conditions. During a clear sky after sunrise and before sunset (sun angles between 5 and 8°), an average of 40, 33, and 28% of the light was coming from B, G, and R light, whereas under cloudy conditions at the same solar elevation angles, the values for these light qualities were 34, 32, and 34%, respectively

(Figure 2). A strong effect of low solar elevation angles was also found on the R:FR ratio. At clear sky conditions, the average R:FR ratio at 10° of solar elevation angle was 1.2, while it was close to 1.0 on cloudy days, and decreased strongly at solar angles <10° (Figure 2).

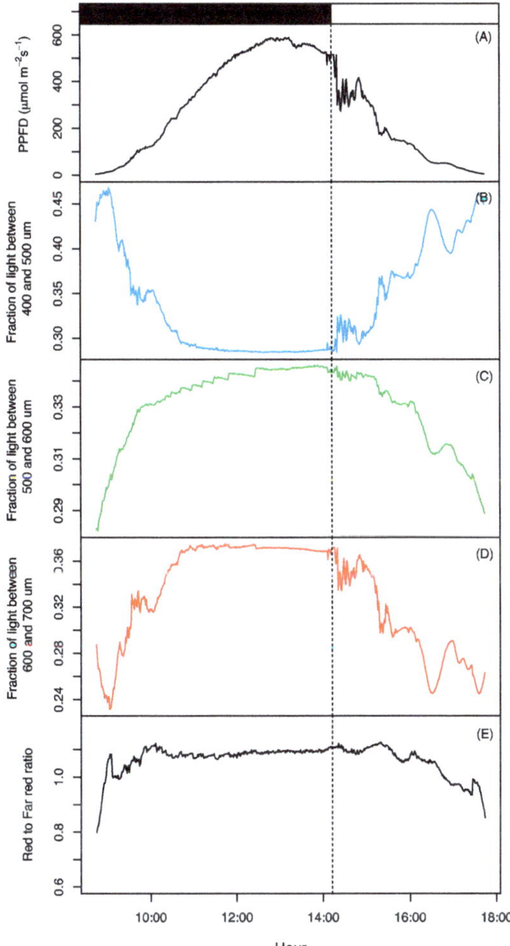

Figure 1. Changes in light quantity and quality as fraction of the photosynthetic photon flux density (PPFD) during a diurnal course. (**A**) Total PPFD; (**B**) blue light fraction (from 400 to 500 nm); (**C**) green light fraction (from 500 to 600 nm); (**D**) red light fraction (from 600 to 700 nm). (**E**) Red to far red (R:FR) ratio. The values are from a single, representative day with varying weather conditions with clear sky conditions until 14:15 (left hand side of the dotted vertical line) and partially overcast conditions during afternoon and evening (right hand side of the dotted vertical line). The data were recorded on 25 November 2018.

The inflection points calculated as the maximum value of the first derivate of each curve, for B, R light, and R:FR ratio under clear sky conditions was at a solar elevation angle of 13° for B and R and 14° for the R:FR ratio. Light quality quickly approached very stable values at solar elevation angles higher than 20°. As stated above, at solar elevation angles beyond 20°, only moderate effects of cloud cover on any wavelength fraction were found, with a small increase of the fraction of B light, on average, from 27% to 29% and a small reduction of the fraction of R from 38% to 36% at solar angles between 20 and 60° (Figure 2). The G light fraction, on the other hand, reached values close to 35% independently of the weather conditions, with its inflection point close to 10°. Finally, the R:FR ratio did not differ significantly between sunny and overcast days at solar elevations angles between 20° and 50° and stayed at a constant average value of 1.1, while at solar angles >50°, the R:FR ratio was even slightly higher on overcast days compared to days without cloud cover (Figure 2. More detailed values in supplementary Table S2).

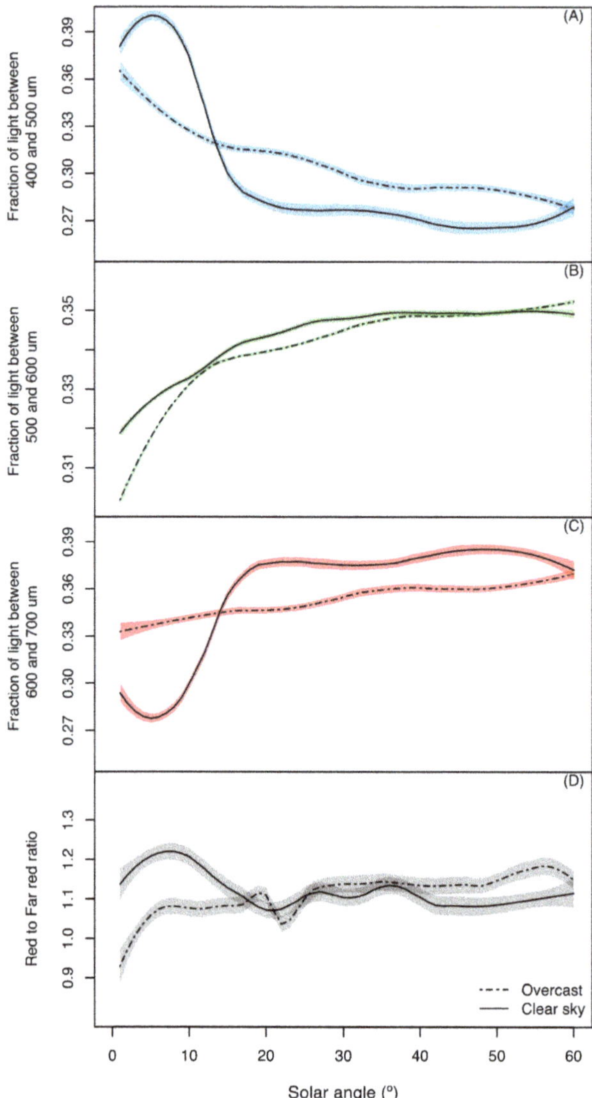

Figure 2. Changes in light quality as a fraction of the photosynthetic photon flux density (PPFD) depending of cloudiness (full line: Clear sky, dotted line: Overcast conditions) and the solar elevation angel. (**A**) Blue light fraction (from 400 to 500 nm), (**B**) green light fraction (from 500 to 600 nm), (**C**) red light fraction (from 600 to 700 nm). (**D**) Red to far red (R:FR) ratio. The lines represent the mean value of one day of each weather condition per month ($n = 12$; see methods for detail). Shaded areas correspond to the standard error of a locally estimated scatterplot smoothing (LOESS) fitted model.

3.3. Latitude Effects: Duration of Modified Light Quality and its Effect on Seedlings of Selected Tree Species

At higher latitudes, the period of daytime under modified sun light spectrum (i.e., solar angle below 20°) is significantly longer compared with lower latitudes, showing an exponential increase at higher latitudes. For example, at 30° N the maximum daily twilight period is reached as a single peak in mid-winter (Day 356) and does not exceed 5 h per day, while it shifts to the beginning of spring (Day 53) and the end of autumn (Day 292) at 60° N with a daily maximum duration of over 9 h (Figure 3).

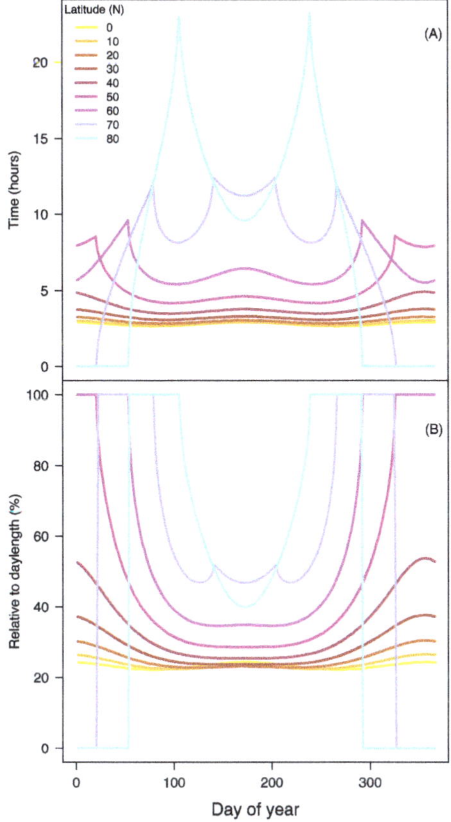

Figure 3. Estimated day length duration (**A**) and percentage relative to the total day length (**B**) of solar elevation angles between 0 and 20° for different latitudes.

For the seedlings of the selected tree species included in our analysis, both the light quality treatments and the latitudinal origin of the population significantly affected growth ($P_{\mathrm{value}} < 2.2 \times 10^{-16}$ and 0.02, respectively). However, no interaction between these two factors was found ($P_{\mathrm{value}} = 0.4$). The latitude effect is best described (fitted) as a quadratic effect (Figure 4A). Plants from higher latitudes had lower shoot elongation under the same light quality than southern ecotypes. In all studies higher R:FR ratios led to decreasing growth (Figure 4B). Additionally, the difference between the effects of the R and FR light treatments was more or less constant across trees from different latitudinal origin. For the light treatments, the best fit was a linear function after a logarithmic transformation, where plants treated with a larger fraction of FR light had larger elongation compared to trees treated with higher fractions of R light. Both factors were able to explain 82% of the variability (Figure 4. Available as 3D figure, Figure S1). A total of 54% of the variability was explained by the light treatments and the origin of the ecotypes could explain 38% when the different variables were tested independently.

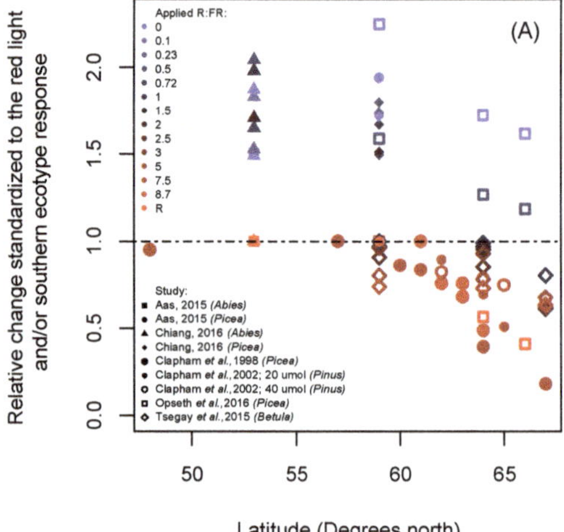

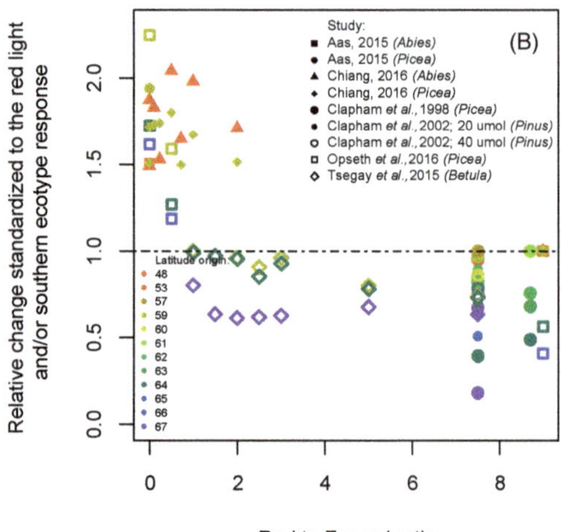

Figure 4. Effect of day light extension with different light qualities on seedling growth in selected temperate and boreal tree species: Relative changes of growth plotted (**A**) against the latitudinal origins under different red to far red ratios (R:FR ratios) and (**B**) against different R:FR ratios applied in trees from different latitudinal origins. The data were collected from work performed with seedlings of selected tree species (three evergreen conifers: *Picea abies*, *Pinus sylvestris*, *Abies lasiocarpa* [14–17,20–23] and one deciduous broadleaved species (*Betula pendula*) [19]), that were exposed to the different light quality treatments for 35–50 days. The additional legends give the first author, publication year, and tree genus. Data from Clapham et al. [23] were derived from two experiments with 20 and 40 µmol photons m^{-2} s^{-1} day extension light, respectively. The light treatments correspond to day extensions with different R:FR ratios and main day light periods of 9–12 h with similar light quantities in W m^{-2}. In the different experiments needle or plant height was measured as the growth response variable. Each study was standardized to the effect of red (R) light in the southern ecotype (when more than ecotype was included; see methods for details).

4. Discussion

Atmospheric constitution, e.g., the presence of clouds, can alter the composition of the light spectra. In our measurements, clouds increased the blue (B, 400–500 nm) light fraction by up to 10% in solar elevation angles above 20° through a reduction of direct light, which also led to a corresponding reduction of the red (R, 600–700 nm) light fraction in a similar magnitude. The green (G, 500–600 nm) light fraction was less affected by weather conditions, mainly due to a potentially 50% lower scattering compared to that of the B fraction [25]. R:FR ratios between weather conditions were not significantly different at high solar elevation angles but changed sharply at low solar angles.

It is well known that from solar elevation angles of –12° at the last two stages of twilight, i.e., the nautical and civil twilight (0° to –6° and –6° to –12°, respectively), the most substantial fraction of the spectra corresponds to the B light wavelength [1,12,26]. This is mainly due to a lack of direct radiation and a longer path length of the scattered sunlight through the atmosphere, which increases the probability of Rayleigh scattering of light by small atmospheric molecules and aerosols. With increasing solar elevation angles, there is an initial increase in the B light fraction together with a reduction of the fraction of R light (Figure 2). This initial increase in B light has been reported previously in a city environment at lower solar elevation angles than in our study [26]. This shift that was not present in a rural scenario, was explained by the presence of high-pressure sodium lamps as the city's main illumination source, were the timing for such a shift may accordingly depend on the city's illumination regime. The absence of this increase in rural scenarios and the low magnitude of this effect may indicate that this change should not play an important role as a biological signal in natural ecosystems. Once that the relative amount of direct light increases, a quick reduction of B light occurs, together with an increase of R light, mainly due to a shorter sunlight path length through the atmosphere that reduces the amount of B light refraction and therefore the B fraction in the sun spectra [2]. In contrast, G light tends to keep an asymptotic slower increase from lower to higher solar elevation angles with light quality reaching a steady state in solar angles higher than 10°.

The higher proportion of R light at twilight in cloud-covered conditions derives from the strong reflectance of R light from clouds into the lower atmosphere and the higher absorbance of B light by clouds. The intensity of this effect depends on the elevation of the clouds, its density, and its position on the horizon [27]. Many studies on radiation light quality with a more physical focus reported higher percentages of R light during sunrise and sunset than in the current study, mainly due to the direction of the used sensor and the aperture's angle. Zagury [27], for example, used a 25° aperture facing in the direction of the light source. This technical difference allows the sensor to detect exponentially more direct light and ignores the mostly diffuse light coming from other directions. The reduced amount of measured scattered light, which consists mostly of shorter wavelengths, therefore, increases the relative amount of R compared to FR light. Although the angle at which plants sense the light depends on the leaf angle, the measured values from the horizontal measurements and the inclusion of light coming from different angles as reported in our study here, are, on average, likely more similar to the light quality detected by plant leaves under natural conditions.

A very strong effect of the weather conditions at low solar elevation angles was found for the R:FR ratio, possibly partly explaining the larger differences between natural R:FR ratio measurements reported by previous authors (e.g., [1,28,29]). Although changes of the R:FR ratio in the presented magnitudes (Figure 2) have shown biological effects in short-term experiments with herbaceous plants [30], this effect has not been found in the few tree species investigated so far [15,16]. In contrast, many annual plants show high plasticity to lower fluctuation on the R:FR ratio conditions [31]. This may indicate that the previously investigated tree seedlings species may require several generations to adapt to the different R:FR ratios.

At higher latitudes, the period of daytime under modified sun light spectrum at twilight is exponentially longer compared with lower latitudes, especially during the spring and autumn. These prolonged periods of low solar elevation angle and the respective change of spectral light quality at higher latitudes might be used as pace-setting signals for plant biological processes in perennial plants, like bud break, growth, or bud set. In woody plant species that have broad latitudinal distributions, such as the trees used in our literature review, ecotypes from southern latitudes have been shown to require less radiation to keep growing than conspecific ecotypes from the northern distribution edge. For example, Mølmann et al. [14] tested different radiation intensities and showed that the effect of FR and R light treatments also depended on light intensity and plant origin, e.g., 1.7 W m^{-2} of FR light completely prevented bud set in more southern ecotypes (from 59 and 64° N latitude) of Norway spruce seedlings, while in a more northern ecotype (from 66.5° N latitude) only 43% bud set was reached. In all three ecotypes, lower light intensities, independent of the light quality, were not able to prevent bud set.

Several studies have suggested differential sensitivities of plant growth to light quantity and quality depending on their latitudinal origin [14–16,20–23] showing an interaction between the effect of light qualities and latitude [14,15]. The analysis of such data in the current study (see Figure 4A,B; Figure S1), also showed a distinct growth response to changed R:FR ratios depending on the latitudinal origin of the plants studied. The re-analysis of the combined data, clearly confirms the findings of the individual studies that an increased amount of FR light strongly promotes growth, compared to R light alone (Figure 4B). However, the previously observed interaction between the light treatments and the population origin was not found to be significant anymore (P_{value} = 0.4). Therefore, we propose that northern ecotypes tend to be as sensitive to light quality changes as southern ecotypes, but with higher light quantity requirements. The interaction between light quality and latitudinal origin on the growth of seedlings of temperate and boreal tree species proposed by previous authors, may actually be the result of two underlying, complementary responses: Firstly, an interaction between the light quantity and light quality (requiring higher amounts of FR light than R to reach similar results) and secondly, an interaction between the population origin and the light quantity (with northern ecotypes requiring more light than southern ecotypes). Remarkably, this is not contradictory to the results previously described by other authors, but a broader analysis may be needed to unequivocally reveal the relationship of the light quality and quantity requirements in tree seedlings from different latitudinal origins. Our analysis, together with our continuous light analysis suggests, that northern (or high latitude) ecotypes have adapted to longer photoperiods of modified light quality (mainly with respect to the R:FR ratio), which could have a more important role for biological processes like growth or bud set than in more southern (or low latitude) ecotypes of the same species. Although northern ecotypes may grow less under specific R:FR light conditions compared to southern ecotypes, at similar, non-saturating amounts of applied energy (in W m^{-2}), the amplitude of the effect of the light treatments between just R and FR light remain similar across the different latitudes (Figure 4A). This indicates that the accumulated amount of energy applied may play a more important role than the used light quality. Of course, effects of light quality and quantitiy on trees are occurring on top of other, fundamental environmental drivers, especially temperature were the effect of light quality have shown to be temperature dependent [13]. Thus, phenology and growth of trees species from boreal and temperate climates are regulated by temperature, light quantity, quality, and photoperiod, where the relative importance of each of these is likely dependent on the species origin latitude [32–35].

5. Conclusions

Here, we supply easily transferable continuous data of changes in light quality through the day and year in dependency of different weather conditions. These results are highly relevant from a plant biological perspective since the recorded wavelength areas are among the important determinants of plant growth and development. Such data is required to design LED systems simulating natural variation in light quality and quantity, which is becoming increasingly relevant today because an increasing number of plant growth facilities are using LED systems as the main source of radiation, and more natural light spectra are desirable.

Here, we also corroborate our hypothesis that the extended periods with modified light spectra at high latitudes correlates with the light requirements of seedlings of boreal and temperate tree species.

This suggest that in addition to other ecologically highly important factors such as temperature and photoperiod, changes in light quantity and quality play an important adaptive role on seedlings of woody plants at higher latitudes.

Supplementary Materials: The following material is available online at http://www.mdpi.com/1999-4907/10/8/610/s1, Figure S1: Effect of day extension with different red to far red ratios (R:FR ratios) on trees from different latitudinal origins, Table S1: Summary of the previous experiments used for the analysis, Table S2: Average light proportions of blue, green, and red and red to far red ratio under different solar elevation angles and two contrasting weather conditions.

Author Contributions: C.C., J.E.O., D.B. (Daniel Bånkestad), and G.H. designed the experiment; C.C. performed the experiment; C.C. and D.B. (David Basler) analyzed the data and prepared the figures; D.B. (David Basler) and J.E.O. validated the results; C.C. and G.H. wrote the initial draft of the paper. All authors provided inputs and suggestions and approved the manuscript for submission.

Funding: The present research was supported by PlantHUB - European Industrial Doctorate funded by the H2020 PROGRAMME Marie Curie Actions – People, Initial Training Networks (H2020-MSCA-ITN-2016). The program is managed by the Zurich-Basel Plant Science Center.

Conflicts of Interest: The authors declare no conflict of interest. The funders had no role in the design of the study; in the collection, analyses, or interpretation of data; in the writing of the manuscript, or in the decision to publish the results.

References

1. Smith, H. Light Quality, Photoperception, and Plant Strategy. *Annu. Rev. Plant. Physiol.* **1982**, *33*, 481–518. [CrossRef]

2. Garner, W.; Allard, H. Further studies in photoperiodism, the response of the plant to relative length of day and night. *J. Agric Res.* **1923**, *23*, 871–920.

3. Robertson, G.W. The Light Composition of Solar and Sky Spectra Available to Plants. *Ecology* **1966**, *47*, 640–643. [CrossRef]

4. Jenkins, G. The UV-B Photoreceptor UVR8: From structure to physiology. *Plant Cell* **2014**, *26*, 21–37. [CrossRef]

5. Hogewoning, S.; Trouwborst, G.; Maljaars, H.; Poorter, H.; Van Ieperen, W.; Harbinson, J. Blue light dose-responses of leaf photosynthesis, morphology, and chemical composition of Cucumis sativus growth under different combinations of red and blue light. *J. Exp. Bot.* **2010**, *61*, 3107–3117. [CrossRef]

6. Terashima, I.; Fujita, T.; Inoue, T.; Chow, W.S.; Oguchi, R. Green Light Drives Leaf Photosynthesis More Efficiently than Red Light in Strong White Light: Revisiting the Enigmatic Question of Why Leaves are Green. *Plant Cell Physiol.* **2009**, *50*, 684–697. [CrossRef]

7. Zhen, S.; Van Iersel, M.W. Far-red light is needed for efficient photochemistry and photosynthesis. *J. Plant Physiol.* **2017**, *209*, 115–122. [CrossRef]

8. Overdieck, D. CO_2-Gaswechsel und Transpiration von Sonnen-und Schattenblättern bei unterschiedlichen Strahlungsqualitäten. *Ber. Dtsch. Bot. Ges.* **1978**, *91*, 633–644.

9. Furuyama, S.; Ishigami, Y.; Hikosaka, S.; Goto, E. Effects of Blue/Red Ratio and Light Intensity on Photomorphogenesis and Photosynthesis of Red Leaf Lettuce. *Acta Hortic.* **2014**, *1037*, 317–322. [CrossRef]

10. Hernández, R.; Kubota, C. Physiological responses of cucumber seedlings under different blue and red photon flux ratios using LEDs. *Environ. Exp. Bot.* **2016**, *121*, 66–74. [CrossRef]

11. Bula, R.; Morrow, R.; Tibbitts, T.; Barta, D.; Ignatius, R.; Martin, T. Light-emitting Diodes as a Radiation Source for Plants. *Hortic. Sci.* **1991**, *26*, 203–205. [CrossRef]

12. Goldberg, B.; Klein, W. Variation in the spectral distribution of daylight at various geographical locations on the earth's surface. *SOL Energy* **1977**, *19*, 3–13. [CrossRef]

13. Clapham, D.H.; Dormling, I.; Ekberg, L.; Eriksson, G.; Qamaruddin, M.; Vince-Prue, D.; Vince-Prue, D. Latitudinal cline of requirement for far-red light for the photoperiodic control of budset and extension growth in Picea abies (Norway spruce). *Physiol. Plant* **1998**, *102*, 71–78. [CrossRef]

14. Mølmann, J.; Asante, D.; Jensen, J.; Krane, M.; Junttila, O.; Olsen, J. Low night temperature and inhibition of gibberellin biosynthesis override phytochrome action, and induce bud set and cold acclimation, but not dormancy in hybrid aspen. *Plant Cell Environ.* **2006**, *28*, 1579–1588. [CrossRef]

15. Opseth, L.; Holefors, A.; Rosnes, A.K.R.; Lee, Y.; Olsen, J.E. FTL2 expression preceding bud set corresponds with timing of bud set in Norway spruce under different light quality treatments. *Environ. Exp. Bot.* **2016**, *121*, 121–131. [CrossRef]

16. Chiang, C.; Aas, O.T.; Jetmundsen, M.R.; Lee, Y.; Torre, S.; Fløistad, I.S.; Olsen, J.E. Day Extension with Far-Red Light Enhances Growth of Subalpine Fir. (*Abies lasiocarpa* (Hooker) Nuttall) Seedlings. *Forests* **2018**, *9*, 175. [CrossRef]

17. Planck, M. On an improvement of Wien's Equation for the spectrum. *Ann. Phys.* **1900**, *1*, 730.

18. Sager, J.; Smith, W.; Edwards, J.; Cyr, K. Photosynthetic efficiency and phytochrome photoequlibria determination using spectral data. *Trans. ASAE* **1988**, *31*, 1882–1889. [CrossRef]

19. NOAA. NOAA Solar calculator. 2018. Available online: https://www.esrl.noaa.gov/gmd/grad/solcalc/ (accessed on 15 May 2019).

20. Aas, O. Effects of Light Quality and Temperature on Elongation Growth, Dormancy and Bud Burst in Norway Spruce (*Picea abies*) and Subalpine Fir (*Abies lasiocarpa*). Master's Thesis, Norges Miljø- og Biovitenskapelige Universitet, Ås, Norway, 2015.

21. Chiang, C. Interactive Effects of Light Quality and Temperature on Bud Set and Shoot Elongation in Norway spruce (*Picea abies*) and Subalpine fir (*Abies lasiocarpa*). Master's Thesis, Norges Miljø- og Biovitenskapelige Universitet, Ås, Norway, 2016.

22. Tsegay, B.A.; Lund, L.; Nilsen, J.; Olsen, J.E.; Molmann, J.M.; Ernsten, A.; Juntttila, O. Growth responses of Betula pendula ecotypes to red and far-red light. *Electron. J. Biotechnol.* **2005**, *8*. [CrossRef]

23. Clapham, D.H.; Ekberg, I.; Eriksson, G.; Norell, L.; Vince-Prue, D.; Vince-Prue, D. Requirement for far-red light to maintain secondary needle extension growth in northern but not southern populations of Pinus sylvestris (Scots pine). *Physiol. Plant* **2002**, *114*, 207–212. [CrossRef]

24. R Core Team. R: A Language and Environment for Statistical Computing. Available online: https://www.R-project.org/ (accessed on 23 July 2019).

25. Strutt, J.W. On the light from the sky, its polarization and color. *Philos. Mag.* **1871**, *41*, 107–120. [CrossRef]

26. Spitschan, M.; Aguirre, G.K.; Brainard, D.H.; Sweeney, A.M. Variation of outdoor illumination as a function of solar elevation and light pollution. *Sci. Rep.* **2016**, *6*, 26756. [CrossRef]

27. Zagury, F. The colors of the sky. *ACS* **2012**, *2*, 510–517. [CrossRef]

28. Ragni, M.; D'Alcala', M.R. Light as an information carrier underwater. *J. Plankton Res.* **2004**, *26*, 433–443. [CrossRef]

29. Yamada, A.; Tanigawa, T.; Suyama, T.; Matsuno, T.; Kunitake, T. Red: Far-red light ratio and far-red light integral promote or retard growth and flowering in *Eustoma grandiflorum* (Raf.) Shinn. *Sci. Hortic.* **2009**, *120*, 101–106. [CrossRef]

30. Hughes, J.E.; Morgan, D.C.; Lambton, P.A.; Black, C.R.; Smith, H. Photoperiodic time signals during twilight. *Plant Cell Environ.* **1984**, *7*, 269–277.

31. Morgan, D.C.; Smith, H. A systematic relationship between phytochrome-controlled development and species habitat, for plants grown in simulated natural radiation. *Planta* **1979**, *145*, 253–258. [CrossRef]

32. Vince-Prue, D.; Clapham, D.; Ekberg, I.; Norell, L. Circadian timekeeping for the photoperiodic control of budset in Picea abies (*Norway spruce*) Seedlings. *Biol. Rhytm. Res.* **2001**, *32*, 479–487. [CrossRef]

33. Tanino, K.K.; Kalcsits, L.; Silim, S.; Kendall, E.; Gray, G.R. Temperature-driven plasticity in growth cessation and dormancy development in deciduous woody plants: A working hypothesis suggesting how molecular and cellular function is affected by temperature during dormancy induction. *Plant Mol. Biol.* **2010**, *73*, 49–65. [CrossRef]

34. Olsen, J.E. Light and temperature sensing and signaling in induction of bud dormancy in woody plants. *Plant Mol. Biol.* **2010**, *73*, 37–47. [CrossRef]

35. Olsen, J.; Lee, Y. Trees and Boreal forests. In *Temperature Adaptation in a Changing Climate: Nature at Risk*; Storey, K., Tanino, K., Eds.; CABI: Wallingford, UK, 2012; Volume 3, pp. 160–178.

 *forests* 

Article

Photosynthetic Performance in *Pinus canariensis* at Semiarid Treeline: Phenotype Variability to Cope with Stressful Environment

Águeda María González-Rodríguez *, Patricia Brito, Jose Roberto Lorenzo and María Soledad Jiménez

Department of Botany, Ecology and Plant Physiology, University of La Laguna, Apdo. 456, 38200 Tenerife, Spain; pbrito@ull.es (P.B.); rloren@ull.es (J.R.L.); sjimenez@ull.es (M.S.J.)
* Correspondence: aglerod@ull.es

Received: 1 August 2019; Accepted: 25 September 2019; Published: 27 September 2019

Abstract: Low temperatures represent the most important environmental stress for plants at the treeline ecotone; however, drought periods at the semiarid treeline could modify photosynthetic performance patterns. Gas exchange, chlorophyll fluorescence, photosynthetic pigments, and α-tocopherol were measured in a *Pinus canariensis* forest located at a semiarid treeline forest at 2070 m altitude over a whole year. The level of summer drought, caused by an extended period without rain and very low previous rainfall, was remarkable during the study. Furthermore, the cold season showed extraordinarily low temperatures, which persisted for five months. All of these factors combined made the study period an extraordinary opportunity to improve our understanding of photosynthetic performance in a drought-affected treeline ecotone. A high dynamism in all the measured parameters was detected, showing robust changes over the year. Maximum photosynthesis and optimal values were concentrated over a short period in spring. Beyond that, fine regulation in stomatal closure, high WUEi with a great plasticity, and changes in pigments and antioxidative components prevented dehydration during drought. In winter, a strong chronic photoinhibition was detected, and α-tocopherol and β-carotene acquired a main role as protective molecules, accompanied by morphological variations as changes in specific leaf areas to avoid freezing. The recovery in the next spring, i.e., after these extreme environmental conditions returned to normal, showed a strategy based on the breakdown of pigments and lower photosynthetic functions during the winter, and rebuilding and regreening. So, a high level of plasticity, together with some structural and physiological adaptations, make *P. canariensis* able to cope with stresses at the treeline. Nevertheless, the carbon gain was more limited by drought than by low temperatures and more extended droughts predicted in future climate change scenarios may strongly affect this forest.

Keywords: drought; Mediterranean climate; photoinhibition; photosynthetic pigments; tocopherol

---

1. Introduction

At high-mountain habitats and treeline ecotones, many environmental factors may be described as stressful for plants. In general, from a world-wide perspective, the most important is freezing, referring to temperatures which are so low that plant tissues suffer irreversible damage due to ice formation [1]. Nevertheless, other factors are frequently detected.

One of the most challenging stresses that should be closely studied is photoinhibition. Photoinhibition is a result of the overexcitation of the photosynthetic apparatus when a large amount of light energy is trapped by chlorophylls and an excess cannot be safely dissipated [2]. This overexcitation carries with it the potential to damage the photosystem II (PSII) [3]. This circumstance can arise only through exposure to high levels of light, but it acquires a leading role in cases where additional stress factors limit photosynthesis, thereby, reducing the light saturation.

In treeline forests, this photoinhibition occurs mainly during low temperatures which are frequent in the cold seasons. Chilling temperatures cause the loss of metabolic activity and impairments in photosynthesis; therefore, plants in this ecotone are more susceptible to photoinhibition [4].

The decrease in photosynthetic rates, and therefore the enhancement in photoinhibition risk, are not only caused by low temperatures. Decreases in photosynthetic rates in response to environmental constraints are highly species- and environment-specific [5]. Furthermore, limitations could be enhanced by regional climatic features, such as drought prevalence. Low soil water availability is a serious factor for photosynthesis, and although drought has not been considered as a specific tree-line condition [1], it may severely affect tree behavior.

All of these environmental restrictions have been shown to greatly impact the photosynthetic performance of trees. Plants have acquired a complete set of several mechanisms to maintain a balance between the energy input and its utilization, such as light avoidance (for example, leaves and/or chloroplast movements), the screening of radiation (mainly by phenolic components), the dissipation of absorbed light energy, the cyclic electron flow around photosystem I (PSI), the photorespiration pathway, and the scavenging systems of reactive oxygen species (ROS). The latter includes multiple enzymes (i.e., superoxide dismutase, ascorbate peroxidase …) and antioxidants as α-tocopherol and carotenoids, such as zeaxanthin, neoxanthin, and lutein [6–10]. Furthermore, carotenoids are involved in thermal energy dissipation. In this case, the excess of absorbed energy is dissipated by the light-harvesting complexes as harmless, longer wavelengths with the conversion of violaxanthin (V) to zeaxanthin (Z) via antheraxanthin (A) (VAZ cycle) [3].

In these respects, photooxidative stress and photoinhibition play important roles in treeline environmental conditions. Trees growing at high altitudes have developed efficient systems with high concentrations of antioxidants and photoprotective carotenoids in order to obtain better protection [11]. As environmental conditions are incessantly changing and limiting the presence of optimal conditions for plants, all the protection mechanisms have to be continuously adapted to this changing environment. As a consequence, plants which are able to display continuous and reversible changes in their physiologies and morphologies in response to punctual or cyclic changes may have a clear selective advantage [12].

In the subtropical oceanic Canary Islands, the treeline shows a semiarid Mediterranean climate which receives less rainfall than the lowland areas, mainly due to the existence of a quasi-permanent inversion layer [13,14]. High temperatures occur in the summer, while freezing and chilling temperatures are commonly registered for several months; furthermore, despite the islands' oceanic localization, thermal oscillations are high. Due to all this, the Canary Islands treeline, with the endemic *Pinus canariensis* (Chr. Sm. ex DC in Buc) as the only tree species, may be considered a great natural laboratory to test the responses of trees to the environmental constraints which are representative of high altitudes, including under drought conditions.

Furthermore, treelines are considered very vulnerable to climate change. In future conditions, recurrent droughts may be present, and warmer temperatures are expected to be up to three times higher than the global average rate of warming [15,16]. So, the variability in photosynthetic and functional performance in the Canary Islands treeline could serve as an initial proxy to understand how an ecosystem will respond to the forecasted climate changes. A large amount of literature on treelines on a global scale exists; nevertheless, studies carried out on island treelines are scarce [17]. Several studies on water and CO_2 fluxes in this ecotone have been undertaken in recent years [18–21]; nevertheless, seasonal photosynthetic activity and physiological parameters have received little attention [22].

Although, the treeline position on large-scale is determined primarily by thermal limitations [23], the aim of this study was to test whether an additional constraint, i.e., drought, could modify photosynthetic performance patterns. Furthermore, the plasticity of the physiological features under these specific environmental conditions was also evaluated. Finally, all the results were viewed within the context of a changing global climate, in which warmer temperatures and recurrent drought conditions are expected.

2. Materials and Methods

2.1. Study Site and Meteorological Conditions

This study was conducted in a treeline forest located in Las Cañadas of Teide National Park, Tenerife (28°18′21.5″ N, 16°34′5.8″ W; Canary Islands, Spain), in an extreme distribution area of the pine forest. The study plot was at an elevation of 2070 m above sea level, where the dominant species was *P. canariensis*, with an almost non-existent understory. After reforestation in 1950–1960, the canopy reached 17 m in height in 2008, with a stand density of 291 trees ha^{-1}, a basal stem area of 53.8 m^2 ha^{-1}, and a leaf area index (LAI) of 3.4 m^2 m^{-2}. More details concerning the stand and climatic conditions are described in [20].

The main climatic features are characterized by a semiarid Mediterranean climate with a mean annual precipitation of 368 mm, where the drought period lasts from June to August, and precipitation is concentrated in late autumn and winter. The drought period coincides with the highest temperatures, while the minimum temperatures (subzero) are reached in the rainy period [20].

The climatic conditions over the course of the year were followed from an automatic meteorological station (MiniCube VV/VX16, EMS, Brno, CZ) located in a clearing at the edge of the experimental plot. Global radiation (EMS11, EMS, Brno, CZ), precipitation (MetOne370/376, EMS, Brno, CZ), air temperature (Tair), and relative air humidity (RH) (EMS33, EMS, Brno, CZ) at 1.5 m above ground were taken every minute and 30-min averages were recorded in a data logger (ModuLog 1029, EMS, Brno, CZ). The vapor pressure deficit (VPD) was calculated based on air temperature and relative air humidity. In addition, soil water potential (SWP) and soil water content (SWC) were also measured with three gypsum blocks and three soil moisture sensors (EC-10, Decagon Devices), respectively, to a depth of 30 cm at three locations; 30-min averages were recorded during the study period.

2.2. Measurements of Gas Exchange at Field Site

Leaf gas exchange rates of one-year-old needles were measured in sun-exposed branches in five representative trees at the field site. Measurements were done throughout the year, each 15 d or 20 d, on a total of 19 d, using a portable infrared gas analyzer (LCA-4, ADC, Hoddesdon, UK) equipped with a 6.25 cm^2 leaf chamber. In order to estimate differences in seasonal assimilation rates, measurements were always taken in the morning, where maximum rates have been recorded for this species [24]. The net CO$_2$ assimilation rate (An), transpiration (E), stomatal conductance (g$_s$) and intercellular CO$_2$ concentration (Ci) were calculated according to [25], and related to, the projected needle surface area. This area was estimated by measuring the length and diameter of the needle portion exposed in the leaf chamber, while ensuring that the projected leaf area of the fascicles was 300 mm^2 (nine to ten needles). During measurements, leaf temperature (TL) and photosynthetic photon flux density (PPFD) were also recorded. Moreover, the instantaneous water efficiency (WUEi) was calculated as the ratio between An and E (μmol CO$_2$ assimilated per mmol of water transpired) and intrinsic water use efficiency was calculated as An/g$_s$ (μmol mol^{-1}).

All measured data during the whole period were set together in a broad analysis in order to obtain optimal ranges of gas exchange variations with environmental factors. For this approach, the boundary line in a scatter diagram was estimated by selecting the optimal cause-and-effect relationship between two variables [24,26]. These relationships were calculated using the Origin 8 software (Origin Lab., USA). In the analysis of An versus PPFD, a light-response curve was fitted, using the quadratic equation of [27]:

$$An = \frac{\emptyset\,PPFD + An_{max} - \left(\sqrt{(\emptyset\,PPFD + An_{max})^2 - 4\emptyset\,PPFD\lambda\,An_{max}}\right)}{2k} - R_{day} \tag{1}$$

where An is net CO_2 assimilation ($\mu mol\ CO_2\ m^{-2}\ s^{-1}$), ø is the photosynthetic efficiency on a quantum basis as the initial slope of the light curve response, PPFD is photosynthetic photonic flux density ($\mu mol\ photons\ m^{-2}\ s^{-1}$), An_{max} is light-saturated rate of gross photosynthesis ($\mu mol\ CO_2\ m^{-2}\ s^{-1}$), k is convexity and Rday is dark respiration ($\mu mol\ CO_2\ m^{-2}\ s^{-1}$). All parameters were estimated using Photosyn Assistant software, version 1.1.2 (Dundee Scientific, Dundee, UK).

2.3. Chlorophyll Fluorescence Measurements

The chlorophyll fluorescence throughout the entire period was measured at midday using similar and very close needles to those used in the gas exchange measurements. These measurements were done with a portable fluorimeter (HandyPEA, Plant Efficiency Analyzer, Hansatech, UK) on dark, adapted needles (30 min) to determine the basal fluorescence (Fo). After a saturating red light pulse (650 nm, 3000 $\mu mol\ photons\ m^{-2}\ s^{-1}$) flashed by an array of six light-emitting diodes on a homogeneous irradiation area, the maximum fluorescence (Fm) was determined. F_V/F_M was calculated as the ratio (Fm − Fo)/Fm according to [28]. From this parameter, chronic photoinhibition (PIchr) was calculated as the percentage reduction in predawn F_V/F_M (F_V/F_M pd) relative to the annual maximum F_V/F_M measured in this species, and the dynamic photoinhibition (PIdyn) was calculated from midday F_V/F_M (F_V/F_M md) as the additional decrease [29]. Therefore:

$$PIchr = (F_V/F_M\ max - F_V/F_M\ pd)/(F_V/F_M\ max) \times 100 \tag{2}$$

$$PIdyn = (F_V/F_M\ pd - F_V/F_M\ md)/(F_V/F_M\ max) \times 100 \tag{3}$$

This approach is based on [30], that distinguishes between short- and long-term inhibitions. So, dynamic photoinhibition describes a fully reversible decrease in F_V/F_M, whereas chronic refers to a sustainable decrease in values at predawn. Total photoinhibition was calculated as the sum of chronic and dynamic photoinhibition.

Chlorophyll fluorescence transients induced by the saturating light pulse were recorded for 2 s, and the OJIP curve and derived parameters were calculated with PEAPlus V1.10 (Hansatech, UK). The fluorescence transient makes it possible to examine the single steps of the different movements in the initial phase of the electron transport chain. For example, energy fluxes through the PSII per active center of reaction (CR), denominated specific fluxes, or specific activities could be calculated where: ABS/CR is the effective antenna size of an active CR, TR_0/CR is the maximal trapping rate of PSII (maximal rate by which an exciton is trapped by the CR), DI_0/CR is the effective dissipation in an active CR, and ET_0/CR is the electron transport in an active CR. Furthermore, the performance index on the basis of absorption, PI_{ABS}, could be also calculated, considering three independent steps contributing to photosynthesis: relative number of CRs per unit of absorbing chlorophyll, the flux trapping ratio per unit of absorbed energy, and the electron transport flux ratio beyond QA^- per trapping [31,32]. In recent years, all these parameters have become a powerful tool with which to study different stresses [33].

2.4. Pigments and Tocopherol Contents

Photosynthetic pigments and tocopherol determinations were made on the same days as the gas exchange and fluorescence measurements. In the predawn, a central section of needles was collected, frozen in liquid N^2, and stored at −80 °C until analysis. Sections of the needles were pulverized in a mortar with liquid nitrogen and extracted in acetone with $CaCO_3$ in order to determine chloroplast pigments and tocopherol contents. Pigments and tocopherols were separated and quantified by reverse phase HPLC (Waters, Ireland). The chromatography was performed in a reverse phase Spherisorb ODS-1 C18 column (Waters, Ireland) and a Simetry C-18 guard column (Waters, Ireland). For the analysis of the pigments, the mobile phase was solvent A (acetonitrile: Water:methanol; 100:5:10 by vol.) and solvent B (acetone:etilacetate; 2:1 by vol.), where pigments were eluted using a lineal gradient 90% (A) to 10% (B) in 17 min, 17–28 min 80% (A) to 20% (B), and finally, returning within 5 min to the first condition described. All the pigments were assessed with a UV-VIS diode array detector (visible) at 440 nm (Waters 2996, Waters, Ireland). For the tocopherol analysis, a mobile phase with 1ml methanol was used and the detection was achieved with a fluorescence detector (Waters SFD 2475, Waters, Ireland), with excitation at 295 nm and emission at 325 nm [34].

2.5. SLA and Needles Water Potential Determinations

The specific leaf area (SLA; cm^2 projected area g^{-1} dry mass) was calculated in the same needles used for pigments and tocopherol analysis as the quotient of projected area and dry mass (72 h at 70 °C) in order to test the differences on morphological structure over the year.

In addition, needle water potential (Ψ) was measured from June to October (drought period) at predawn in order to obtain supplementary assessment of water status. For that, Ψ was measured in 15 samples (3 samples in 5 trees) using a pressure chamber (PMS Instruments Co., Corvallis, OR, USA).

2.6. Statistical Analysis

Differences in parameters were evaluated by season and/or days with analysis of variance (ANOVA). Homogeneity of variances was tested with Levene's test, and post hoc comparisons were computed according to Tukey test. Data were transformed in order to retain normality and homogeneity. A Tamhane test was used when homogeneity was not achieved. Differences at $P < 0.05$ were regarded as being statistically significant. Calculations were performed using the software package SPSS 21.0 (SPSS, Inc., Chicago, IL, USA).

3. Results

3.1. Meteorological Conditions

The main meteorological parameters followed the main features of the climatic conditions. In spring, mild temperatures were registered, with a daily mean T_{air} of 12.5 °C and with soil water availability at topsoil (SWP = −0.0524 MPa). In summer, the highest mean daily T_{air} were registered, with a maximum value of 31.3 °C (Figure 1), with associated high VPD values (mean daily VPD = 2.16 kPa). Furthermore, SWP displayed the lowest values, highlighting the top soil drought during the entire summer period (Table 1). After that, rains were registered during autumn and winter. During this rainy period, temperatures were low, with a mean daily T_{air} close to 10 °C and with remarkably cold temperatures from the end of October to March, where a mean daily T_{air} of 5.7 °C was registered, reaching temperatures even below 0 °C (minimum T_{air} = −5.66 °C) (Figure 1). During this period, SWP reached its maximum, with values very close to zero (Table 1).

Table 1. Mean vapor pressure deficit (VPD; kPa), mean global radiation (GR; W m^{-2}), mean air temperature (T_{air}; °C), and soil water potential at 30 cm depth (SWP; MPa) from 06:00 to 14:00 on the days in which the gas measurements were done at the study site. Values are mean ± standard deviation.

Season	VPD (kPa)	GR (W m^{-2})	T_{air} (°C)	SWP (MPa)
Spring	0.85 ± 0.54	645 ± 90	12.33 ± 3.45	−0.0662 ± 0.0742
Summer	2.16 ± 1.27	665 ± 111	21.63 ± 5.91	−1.1002 ± 0.0000
Cold season (Autumn-Winter)	0.82 ± 0.33	484 ± 77	10.56 ± 3.82	−0.0293 ± 0.0295
Next Spring	1.10 ± 0.34	713 ± 25	12.96 ± 2.29	−0.0202 ± 0.0012

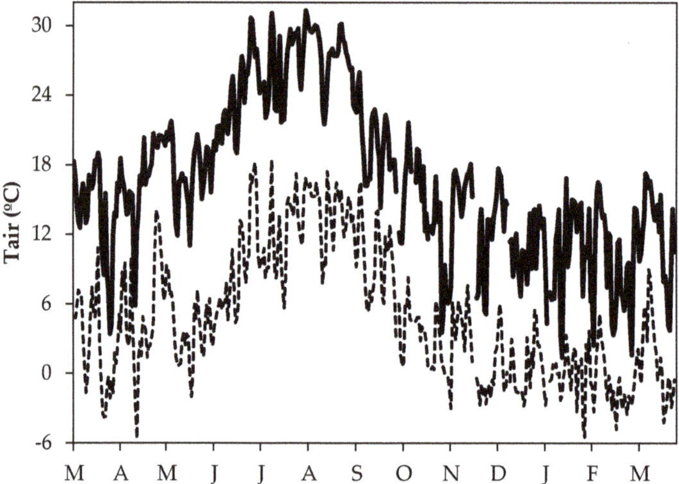

Figure 1. Seasonal course of daily maximum air temperature (solid line) and daily minimum air temperature (dotted line) during the study period.

Moreover, the high thermal oscillation over the entire year should be taken into account, with values reaching 16 °C on the days of the measurements, but were even higher throughout the entire year (Figure 1).

3.2. Broad Analysis of Gas Exchange Parameters: Detecting Optimal Values

Our measurements carried out on 5 trees and with a great variability of environmental factors displayed very variable data throughout the whole period. The plotting all data of each parameter versus meteorological variables and boundary lines of data clouds (lines at the upper surface of data), maximum rates, when other parameters were not limiting, were calculated. More than 2000 data points were used on the calculations, ensuring a suitable approximation to optimal values [24,35–37]. Therefore, the relationships between gas exchange parameters and meteorological parameters were tested by regressing the values at the edge of the body of data which enable a best fit equation of these variables pooled throughout the entire measurement period.

Maximum photosynthesis rates were detected at T_L = 26.4 °C, maintaining 50 % of the maximum An between 16.6 and 36.3 °C, showing a hump-shaped relationship between both parameters (Figure 2). E did not display a saturation tendency at high temperatures what was as clear as that of An, and increased exponentially, displaying the highest values at 30 °C (Figure 2). Stomatal conductance was maximal at 25.5 °C, maintaining a value of up to 80 % between 19.2 and 31.7 °C (Figure 2). Due to a reduction in g_s at high temperatures, An/g_s increased linearly up to 30 °C, but did not go beyond this threshold. In relation with VPD, An showed a maximum of 3 kPa, with 20% of the highest rates registered being between 1.5 and 4.1 kPa (Figure 3). E rates increased exponentially up to maximum values detected at 2.5 kPa, and after this, remained at a maximum of 5 kPa (Figure 3). The relationship between g_s and VPD did not present a clear pattern (Figure 3). Regarding g_s, An and E displayed maxima at around 200 mmol m^{-2} s^{-1}.

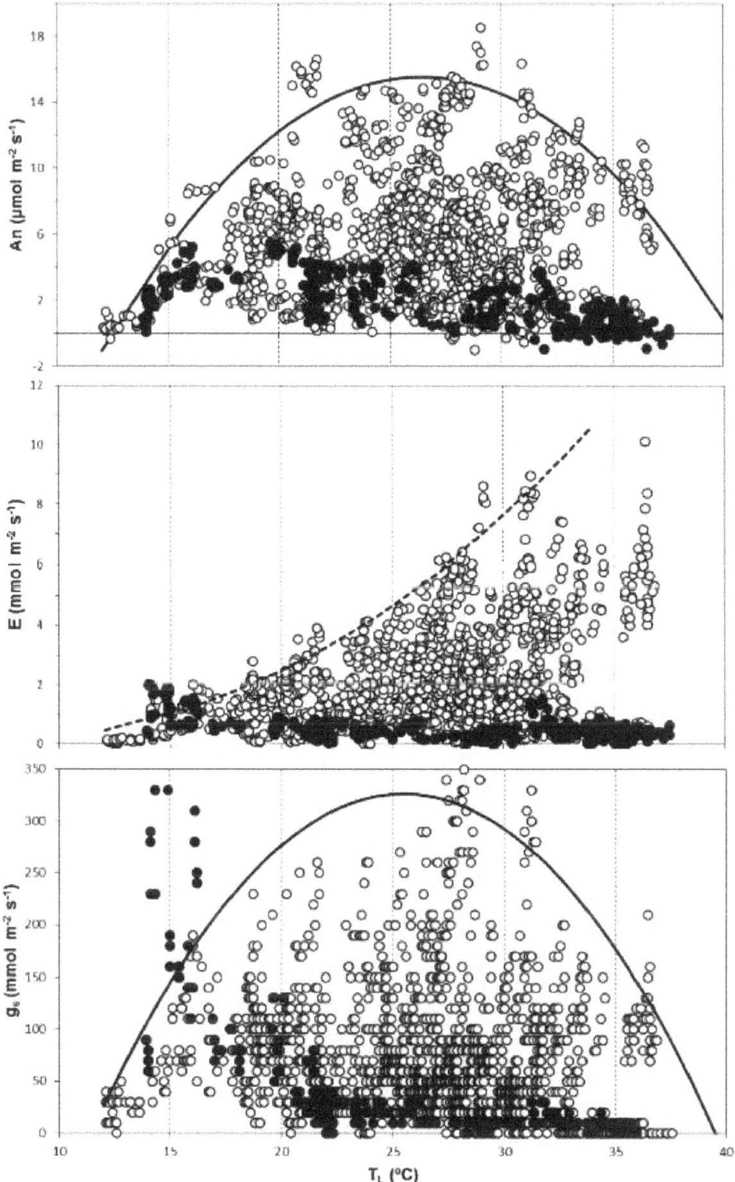

Figure 2. Regression analysis between leaf temperature (T_L) and net CO_2 assimilation rates (An: Top), transpiration rates (E: middle) and stomatal conductance (g_s: Bottom) using boundary lines including 99 % of data points. Individual measurements were collected from March to April. Data showed in black are from the summer period. Regression for entire data set: An = -0.080* T_L^2 + 4.224 * T_L − 40.268; E = −0.275 + 0.0012$T_L^{0.01}$; g_s = −1.65 T_L 2+ 84.26 T_L − 747.03, all P < 0.000.

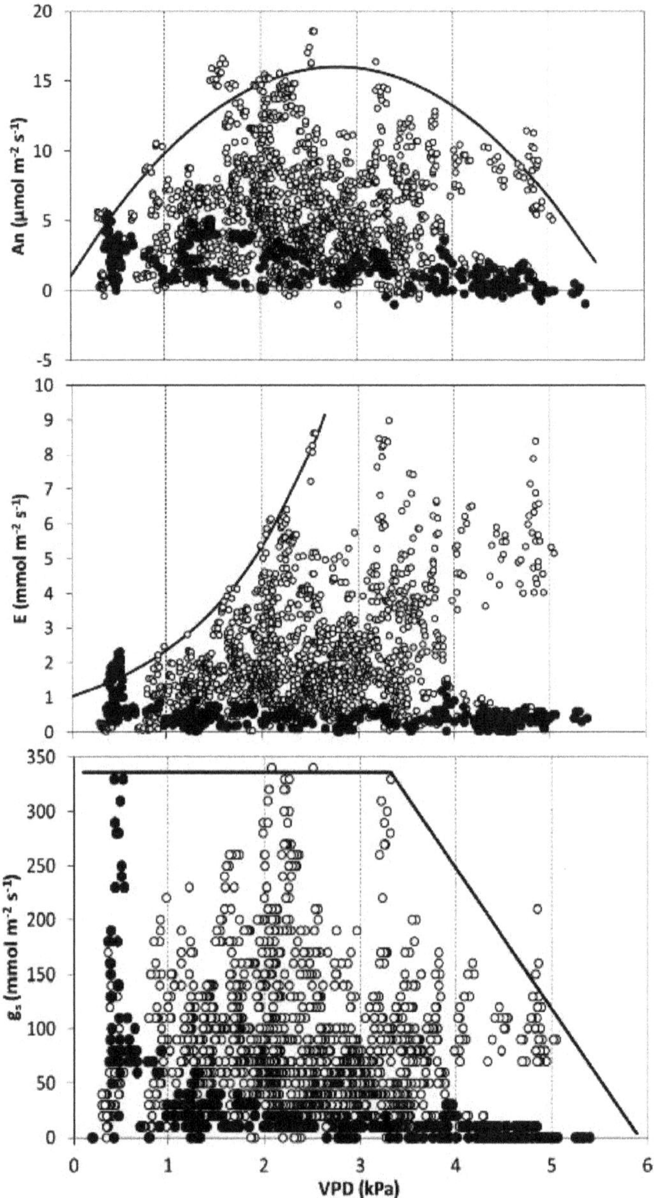

Figure 3. Regression analysis between vapor pressure deficit (VPD) and net CO_2 assimilation rates (An: Top), transpiration rates (E: middle), and stomatal conductance (g_s: bottom) using boundary lines including 99 % of data points. Individual measurements were collected from March to April. Data showed in black are from the summer period. Regression for entire data set: An = -1.91 VPD^2 + 10.7 VPD + 1.09 and E = 1.036 e$^{0.822\ \text{VPD}}$.

According to the Equation (1), An followed a typical light-response curve, where An$_{\text{max}}$ was 13.5 $\pm$ 0.92 μmol m^{-2} s^{-1} and Rday was 0.846 $\pm$ 0.49 μmol CO_2 m^{-2} s^{-1}. The light compensation point was very close to zero, and the light saturation point was 603 μmol CO_2 m^{-2} s^{-1}. The convexity (k) was 0.93 and ø of 0.021 ± 0.0033 μmol CO_2 $mmol^{-1}$ photons (data not shown).

3.3. Seasonal Pattern in Gas Exchange Parameters

Seasonal changes in mean An and E and their corresponding g_s and VPD at the time of measurement are shown in Figure 4. Maximum An and corresponding high E values were clearly reached in spring (9.75 ± 0.32 μmol m^{-2} s^{-1} and 5.06 ± 0.19 $mmol$ m^{-2} s^{-1}, respectively), where the maxima for g_s (166 ± 13.72 $mmol$ m^{-2} s^{-1}) (Figure 4) and c_i were also detected (224 ± 32 μmol mol^{-1}; data not shown). After that, minimum values on these parameters were detected by the highest leaf VPD (higher than 3 kPa); meanwhile, from late-spring to summer, An and E exhibited a marked depression and were very close to zero (Figure 4). During this drought period, stomatal closure was detected in the morning, and the maximum rates in g_s were 87 % lower than the spring maxima. Minimum seasonal mean values in c_i were also registered (132 ± 52 μmol mol^{-1}; data not shown).

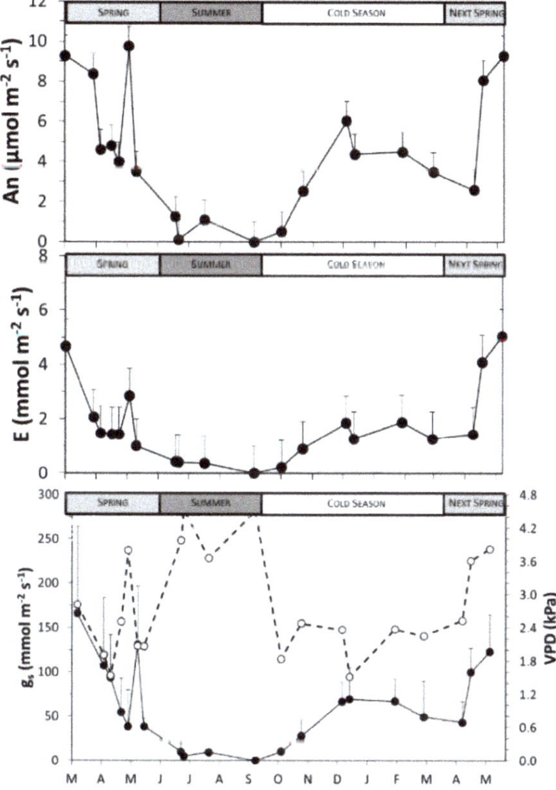

Figure 4. Seasonal course of daily net CO_2 assimilation rates (An: Top), transpiration (E: Middle) and stomatal conductance (g_s: Bottom, black circles) and vapor pressure deficit (VPD: Bottom, white circles). Data are means of fifteen needle fascicles ± standard deviation measured in the morning.

These reductions in gas exchange parameters during summer can be also observed in the broad analysis, where tendencies were undoubtedly different to the entire period (Black dots in Figures 2 and 3). Thus, An was much lower, and started a dramatic decline from 15 °C (−0.26 μmol m^{-2} s^{-1} per °C; Figure 2), while E was maintained close to zero, with values higher than 1 $mmol$ m^{-2} s^{-1} only being registered at around 15 °C (Figure 2). A similar response was observed with VPD in summer, where An dropped with the increase in VPD (-0.61 μmol m^{-2} s^{-1} per kPa; Figure 3), and E values remained very close to zero, with VPD higher than 1 kPa (Figure 3). Stomatal closure was detected, and higher values in g_s were only detected with the lowest T_L and VPD (Figures 2 and 3).

Additionally, when stomata were not closed in the early morning in the summer period, WUEi and An/g_s were higher than in the rest of the year with soil water availability (WUEi 6.15 vs. 3.44 μmol mol^{-1}; An/g_s 121.92 vs. 83 μmol mol^{-1}, respectively; $P < 0.01$). Thus, both parameters increased with high temperatures and VPD, reaching saturated values at 27 °C and 2.5 kPa (data not shown).

In the cold season, autumn and winter, rates in An and E were lower than would be expected based on TL recorded. In calculating the An$_{max}$ expected by regression using T_L measured (from regression showed in Figure 2), noticeable low values were detected on days where the T_{air} overnight was below zero. On these days, the reductions were between 40%–95% of An$_{max}$. The data obtained in the next spring should be highlighted, where the minimum T_{air} overnight was -0.02 °C, and low values in An were registered (2.58 μmol m^{-2} s^{-1}) despite high T_L (27.4 °C) (Figures 2 and 4).

3.4. SLA and Needle Water Potential

The values of specific leaf area (SLA) varied between 37.8 and 46.2 cm^2 g^{-1}. Moreover, they were significantly different, with a drastic reduction in autumn when temperatures started to decrease (Figure 5).

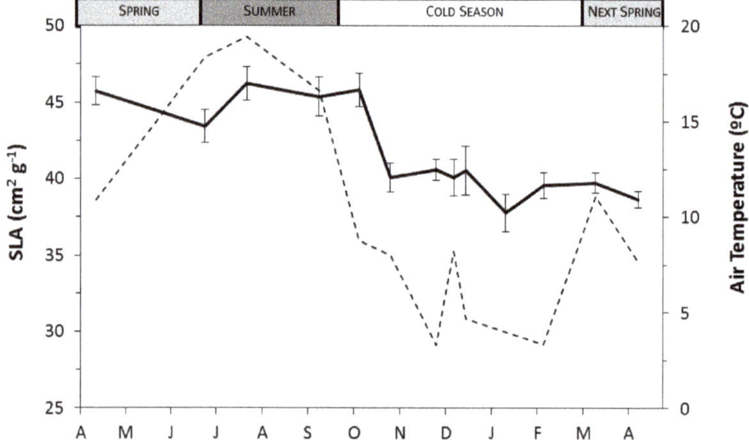

Figure 5. Specific leaf area (SLA, cm^2 projected area g^{-1} dry mass) in *P. canariensis* needles (solid line) and mean daily air temperature in previous 5-days of measurement day (dotted line). Linear regression SLA: y = -0.6645x + 46.467 (R^2 = 0.73, $P < 0.05$). Significant differences in SLA are shown with letters ($P < 0.05$).

Needle water potential ($\Psi_{needles}$) at predawn in the summer period showed lower values throughout the progress of the drought, with minimum values of -1.63 ± 0.33 MPa; nevertheless, $\Psi_{needles}$ did not present any significant differences (all $P > 0.05$; data not shown).

3.5. Seasonal Pattern in Chlorophyll Fluorescence

Maximum values in predawn F_V/F_M were measured during late spring and early summer, reaching the theoretical maximum (0.82 ± 0.02), while a greater reduction was detected in autumn and winter (0.650 ± 0.06). These maximum rates declined throughout the day, with sunlight exposition showing the typical midday depression (data not shown).

Selected representative days are shown in Figure 6, where total photoinhibition (chronic and dynamic), performance index (PI_{abs}), and flux to energy dissipation per active reaction center (DI_0/CR) are displayed. Chronic photoinhibition was non-existent at the beginning of the summer, with corresponding maximum PI_{abs} at predawn. Throughout the cold season, chronic photoinhibition and DI_0/CR increased progressively, with the maximum in late winter, where T_{air} overnight were below to zero. Correspondingly, PI_{abs} at predawn decreased over winter.

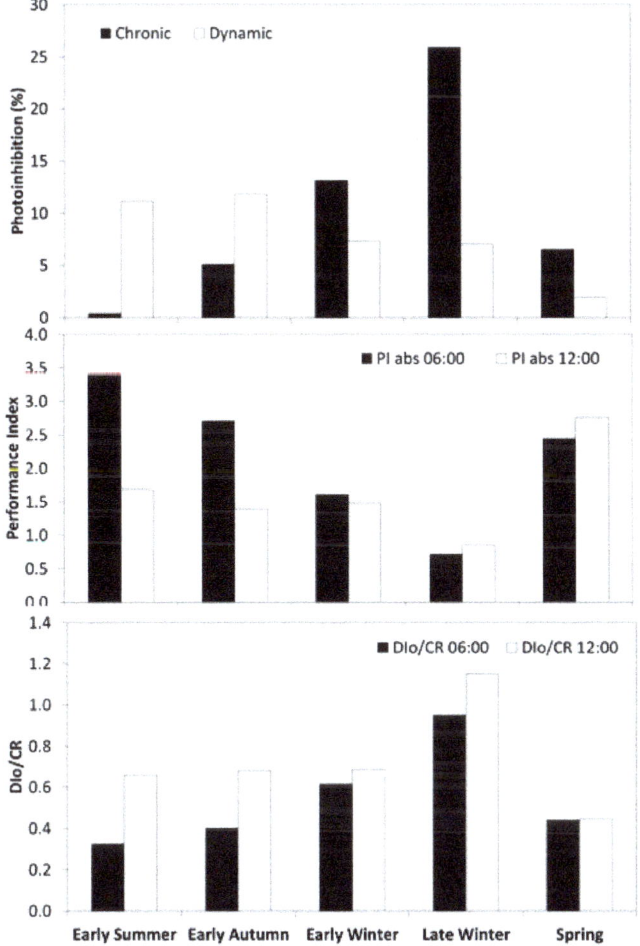

Figure 6. Top: Chronic (black columns) and dynamic photoinhibition (white columns) in percentage during representative days for the seasons. Medium: Performance index (PI abs) at predawn (06:00; black) and midday (12:00: white) on the same days represented on top figure. Bottom: Flux to energy dissipation per reaction center (DIo/CR) at predawn (06:00; black) and midday (12:00; white).

Throughout the year, dynamic photoinhibition was detected; this was emphasized under stressful conditions (drought and cold temperatures). The relevance of the midday depression was remarkable in late winter, with minimum values in PI_{abs}, matching an increase in the dissipation flux (DI_0/CR) (Figure 6) and the decrease in electron transport flux (ET_0/CR) (data not shown)

Despite having undergone double stress, the needles were able to recovery, and chronic photoinhibition was reduced in the next spring. Furthermore, PI_{abs} remained relatively high, even at midday, with similar values to those at predawn (Figure 6). DI_0/CR at predawn and midday were similarly low (Figure 6). Moreover, functional reaction centers (measured as CR/CS) showed the maximum value in spring (CR/CS = 240), corroborating the recovery, while in the previous summer, the value was 16 % lower, and in the previous winter, it showed an even greater reduction (25%; data not shown).

3.6. Seasonal Changes in Pigments and α-tocopherol

The maximum chlorophyll concentration was reached in the spring (578 ± 133 µmol m^{-2}) and decresed throughout the year, with minimum values in winter representing a decline of 15 % (Table 2). Chlorophyll values recovered slightly with the onset of the next spring, and these changes were significantly identified in Chl b but not in Chl a (Table 2). Hence, the Chl a/b ratio varied throughout the year, maintaining minimum values in spring and summer and increasing when cold temperatures appeared. In the next spring, the ratio was lower than in winter, but did not reach the initial values obtained in the previous spring (Table 2).

Minimum values in lutein, β-carotene and total VAZ were detected in the spring (Table 2). Lutein slightly increased in summer and was greater when cold temperatures appeared in autumn (late-October), maintaining these high values until the next spring (Table 2). The identical tendency was followed by total carotenoids, where minimum values were detected in the spring (Table 2). They increased throughout the seasons, including in the next spring, despite the recovery in total chlorophylls. In contrast, β-carotene reached maximum values in winter, higher than those in autumn, and no recovery was observed in following next spring (Table 2). In fact, β-carotene played the main role among the total carotenoids, showing an increase from 24% in spring and summer to 31 % in late winter (Figure 7). After minimum values in spring, VAZ increased approximately by 30 % in summer, and these values were maintained throughout the measurement period. In contrast to the other carotenoids, neoxanthin showed maximum values in spring and decreased in summer, autumn and winter, recuperating maximum values in the following spring (Table 2).

Regarding A+Z/VAZ, high values were observed throughout the year (0.44), reaching even higher values in winter (0.80), highlighting the great Z values which were observed at predawn (Table 2). This ratio was strongly inversely correlated with F_V/F_M at predawn (Pearson $R = 0.83$, $P < 0.01$).

α-Tocopherol was at its lowest in spring and doubled in summer. The maximum values were detected in autumn, increasing by a factor of 3.5 relative to spring; after that, they began to decrease, reaching significantly lower values in the following spring (Table 2).

Table 2. Pigment and tocopherol composition in *P. canariensis* needles (mean ± sd) where: chlorophyll *a* (Chl *a*; µmol m^{-2}); chlorophyll b (Chl *b*; µmol m^{-2}); total chlorophyll (Chl (*a* + *b*), µmol m^{-2}); chlorophyll *a* to chlorophyll *b* ratio (Chl *a/b*); total carotenoids per chlorophyll (Tot Carot/Chl); Lutein per chlorophyll (L/Chl); β-Carotene per chlorophyll (β-Car); Total VAZ (V: Violaxanthin, A: Anteraxanthin, Z: Zeaxanthin) per chlorophyll (VAZ/Chl); Neoxanthin per chlorophyll (N/Chl); Anteraxanthin and Zeaxanthin to total VAZ ratio (A+Z/VAZ); and α-Tocopherol per chlorophyll (α-Toc/Chl). All data per chlorophyll are shown in mmol mol^{-1} Chl *a* + *b*. Different letters denote statistically significant differences at *P* < 0.05 after Duncan test over the seasons.

Pigments and Tocopherol	Spring	Summer	Autumn	Winter	Next Spring	Annual Mean
Chl *a*	416 ± 94 a	381 ± 122 a	361 ± 110 a	361 ± 95 a	363 ± 107 a	368 ± 107
Chl *b*	161 ± 40 a	144 ± 48 ab	125 ± 40 bc	118 ± 34 c	130 ± 37 bc	130 ± 42
Chl (*a* + *b*)	578 ± 133 a	525 ± 170 ab	491 ± 144 ab	474 ± 132 b	493 ± 130 ab	498 ± 148
Chl *a/b*	2.60 ± 0.14 a	2.66 ± 0.21 a	2.89 ± 0.18 b	3.07 ± 0.21 c	2.82 ± 0.22 b	2.87 ± 0.26
Tot Carot/Chl	287 ± 32 a	337 ± 40 b	374 ± 49 c	397 ± 54 c	388 ± 55 c	370 ± 56
L/Chl	112 ± 13 a	127 ± 20 b	150 ± 16 c	157 ± 18 c	154 ± 19 c	145 ± 22
β-Car/Chl	69 ± 8 a	78 ± 11 b	97 ± 21 c	115 ± 25 d	111 ± 22 d	98 ± 25
VAZ/Chl	63 ± 16 a	89 ± 24 b	87 ± 22 b	88 ± 23 b	85 ± 19 b	87 ± 23
N/Chl	36.42 ± 1.47 a	34.65 ± 2.64 b	34.58 ± 3.08 b	33.98 ± 3.10 b	35.25 ± 2.50 ab	35.52 ± 2.92
A−Z/VAZ	0.51 ± 0.14 a	0.21 ± 0.14 b	0.51 ± 0.24 a	0.54 ± 0.19 a	0.49 ± 0.17 a	0.44 ± 0.23
α-Toc/Chl	98 ± 82 a	197 ± 104 b	341 ± 153 d	295 ± 117 cd	259 ± 117 c	274 ± 141

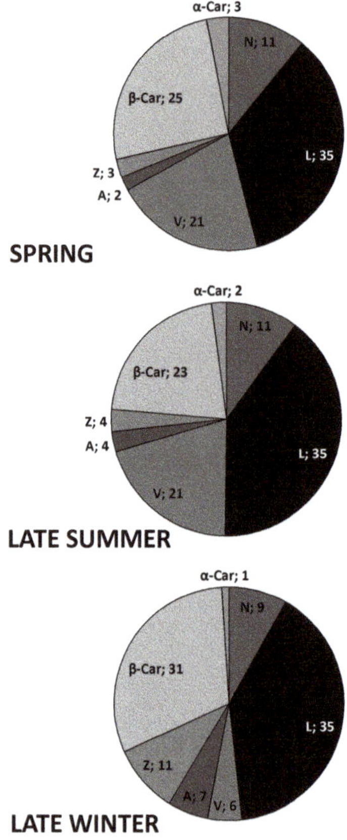

Figure 7. Composition of each carotenoids (N, Neoxanthin; L, Lutein; V, Violaxanthin; A, Anteraxanthin; Z, Zeaxanthin, β-Car, β-Carotene and α-Car, α-Carotene) in percentage in spring, late summer, and late winter.

4. Discussion

The main physiological parameters, including gas exchange, chlorophyll fluorescence, photosynthetic pigments and α-tocopherol, as well as SLA and needles water potential, were measured in a *P. canariensis* forest located at a semiarid treeline forest in Las Cañadas (Tenerife) at 2070 m altitude over the course of one full year. During the year, the summer drought was caused by an extensive rainless period from the end of May to the end of September, with very low previous rainfall (216 mm) [21]. Furthermore, the cold season showed extraordinarily low T_{air}, which persisted, with daily means close to 5 °C and minimum values close to −6 °C. All these conditions meant that the studied period as an extraordinary opportunity to improve our understanding of photosynthetic strategies in a treeline forest under the principal environmental constraints (temperature and drought).

Maximum photosynthetic rates were in the upper range measured in this species in the wind-trade area at a lower altitude (17 μmol m^{-2} s^{-1} [24], and in the range described for other Mediterranean conifers and *Pinus* species growing in similar latitudinal areas [5,38], suggesting no limitations in the net CO_2 uptake of *Pinus canariensis* at treeline compared to lower elevations [1,11]. Optimal T_L for photosynthesis was 26 °C, being very close to a previously identified value for this species (25 °C) [37], which includes, at the upper limit described, evergreen conifers and Mediterranean plants [39]. Nonetheless, the optimum range in which to maintain 50% A_{max} was highly reduced, showing a lower range (17–36 °C) than that previously described (11–40 °C) [24]. Maximum E displayed higher values than those described for other evergreen conifers, as previously documented for this species [24,37]. Similar to An, a maximum E was detected 5 °C lower than optimum temperature described for this species; above the threshold value of 30 °C, E did not increase. These narrowed optimums are frequently detected at field conditions where T_{air} is accompanied by additional environmental constraints [5], as it is in this treeline ecotone. These reductions in An and E were perfectly matched with a decrease in gs that remained at 50% between 16–35 °C. Maximum g_s values were high, reaching 350 mmol m^{-2} s^{-1}; nevertheless, higher rates have even been described for other *Pinus* species growing in humid habitats as *P. banksiana* and *P. ellioti* (500 mmol m^{-2} s^{-1}) [38].

The light saturation point was slightly higher at the treeline than in exposed needles for this species (600 vs. 550 µmol m^{-2} s^{-1}) [40], with similar values being observed in respiration rates. This saturation at higher PPFD is characteristic for plants with sun-type adaptations [39], in accordance with the high radiation and scarcity of clouds in this ecotone.

Gas exchange parameters and their meteorological optimal values displayed extraordinary seasonality. Maximum An and E corresponded with the main growing season (spring) [21], in which there was a high demand for carbohydrates. After that, from early summer, rates very close to zero caused by complete stomatal limitations with g_s close to zero at VPD>1kPa were detected. In this period, both high VPD values and soil drought occurred, and these, combined with previous low precipitation [20], resulted in this strong reduction not being detected in the wind-trade distribution area [37]. The deep root system of this species could not prevent this strong stomatal closure. However, a reduction in gas exchange was detected, and trees did not seem to be affected by severe dehydration, as Ψ_{needle} did not reach low values in September, thereby avoiding the tissue dehydration by earlier stomatal closure.

Despite this, early-morning stomatal opening was detected, permitting gas exchange which generated higher rates in WUEi during the drought period. This increase has been documented for other Mediterranean species under drought conditions [41]. Additionally, WUEi displayed high values during the entire period. WUEi tends to be high for pines due to efficiency in heat dissipation as a result of the species' needle anatomy being much more efficient than narrow leaves [38]. This, together with the special stomatal complex with a singular epistomatal chamber in *P. canariensis* needles [42] displayed a fine regulation in stomatal closure and high values in WUEi.

After summer, an increase in An and E was detected in the cold season; however, maximum rates were not reached. This was not only due to low evaporative demand; chilling temperatures were also a relatively major limitation, showing the vulnerability of this species to sustained winter photoinhibition. This depression at low temperatures has been documented in Mediterranean areas, and it seems to be a consequence of non-stomatal factors due to its assimilation being highly depressed by low temperatures. Nevertheless, at present, there is not sufficient data to assess the relative significance of the underlying mechanisms [5].

Linked to gas exchange variability, needles also showed morphological plasticity. SLA values measured were low for values described for species growing in hot and dry sites, but were within the upper limit described for semiarid species [43,44]. This lower surface-to volume ratio could be associated with higher photosynthetic WUEi [45]. Moreover, significant changes in SLA were detected; the daily mean T_{air} from September (20 °C) was reduced to 5 °C at the end October, and these cooler temperatures were closely related to a significant reduction in SLA. Low SLA at low temperatures could reduce the incidence and severity of cold temperature stress by slowing down the rate of freezing [46]. These modifications are commonly associated with changes in leaf density due to the thickening of cuticles and epidermises, with increased functional sclerenquima investments and smaller and more tightly-packed mesophyll cells with thicker cell walls [44]. This plasticity in a heterogeneous environment may play an adaptive role in a strong seasonal climate [26]; nevertheless, in Mediterranean ecosystems, it has been recognized as a means to cope with drought [17].

F_V/F_M values also showed variability. At midday, values were depressed, showing a pronounced dynamic photoinhibition; for that reason, *P. canariensis* was classified as a photoinhibition-tolerant species by [4]. Decreases in CO_2 assimilation rates during the drought did not severely affect the fluorescence parameters, so the decrease in the utilization of ATP and NADPH in the photosynthetic metabolism should be compensated for by other sinks, such as water-water cycling and photorespiration [48]. Freezing and chilling temperatures resulted in permanent declines in F_V/F_M and PIabs from early-autumn throughout early-spring. These reductions were maintained at predawn, showing a chronic photoinhibition that indicated a needle photochemistry that had been significantly affected. The chronic photoinhibition increased as the winter progressed, suggesting accumulative damage. Similar reductions during winter have been observed in many other conifer species [49–51]. This chronic photoinhibition coincided with the maintenance of xanthophyll deepoxidation at predawn, and high concentrations in VAZ pigments showing a high relevance in the thermal dissipation of excess energy, as also indicated by high DI_0/CR. Such deepoxidation retention overnight has been detected in other winter stressed Mediterranean plants [52,53]. This retaining at low temperatures might play a photoprotective role as a mechanism to avoid harm by heat dissipation, even in the early morning [54,55].

Nevertheless, this chronic sustained photoinhibition could not be explained by the high Z alone. The much higher ABS/CR detected in winter might indicate a decrease in the concentration of active reaction centers, rather than an increase of light absorbance [33]. For that, some degree of accumulative damage may have also existed, which was also indicated by a decrease in functional centers and reduced chlorophyll contents. However, non-persistent winter damage was verified during the consecutive growing season, where a general restoration of photosynthetic functions, fluorescence parameters, and pigments (chlorophyll and neoxanthin) were observed in the following (warmer) spring. This is in accordance with recovery studies in other conifer species [56,57]. Several studies have demonstrated strong inverse correlations when faced with environmental constraints, such as low temperatures and/or drought, and needle longevity, so as to optimize resource availability and carbon balance [58,59], which has also been recognized in *P. canariensis* [60]. In addition, photoprotective mechanisms related to changes in pigments and α-tocopherol may allow trees to overcome cold stressful periods and to recover on spring. Pigment contents and their dynamics were characteristic for plants from high radiation habitats, and were within the range expected by healthy plants [61]. The decrease in chlorophyll concentration detected in winter is a common strategy to reduce the light capture, and has been largely documented in the Mediterranean area, including pine species [62]. This decrease was accompanied by an increase in neoxanthin and Chl *a/b* ratio indicating a higher degradation on light harvesting complexes [63]. Nevertheless, a slight recovery on chlorophylls and neoxathin was detected in the next spring, coinciding with the retrieval of maximum An.

During the entire period carotenoids rates were within the maximum rates described for plants [61], showing a robust acclimation. The highest values were detected in the cold season, indicating an upregulation of carotenoids to cope with low T_{air} combined with high light. β-carotene played the main role in this increase. This pigment can directly quench singlet oxygen or can prevent the formation of chlorophyll triplets in the excited state; it has also been associated with chlorophyll situated at reaction centers in PSI and PSII, where achieves photoprotection [64]. Due to the fact that singlet oxygen generated by PSII is largely responsible for most of the non-enzymatic lipid peroxidation on the thylakoid membrane [65], this role is of relevance under high photoinhibition. In summer, the changes in pigments were similar to those in winter, but this drought effect was not as pronounced as the decrease in temperatures.

Previous studies have shown that α-tocopherol did not increase in needles of *P. canariensis* under more stressful environmental conditions [34]; this lipophilic molecule displayed a dynamic role at the treeline. The rate increased throughout the year; this is a common response in plants due to leaf ageing [66]. Nevertheless, our higher observed values in autumn than in spring indicate a distinctive role of its protection mechanism under low temperatures. Moreover, tocopherols have been described as being particularly important in response to severe photo-oxidative stress when other protection mechanisms are nonexistent or are failing [67]. This could be the case in the winter of the year of study, where higher photoinhibition was detected. Other authors have proposed that tocopherol plays a more crucial role in low-temperature adaptation than in photoprotection [68]; this could also explain why the highest values were detected in autumn, i.e., with the onset of low temperatures.

5. Conclusions

Great dynamism in the potential photosynthetic performance was detected in a *P. canariensis* treeline forest. Optimal values were concentrated over a short period in spring. Beyond that, fine regulation in stomatal closure, high WUEi with a great plasticity, and changes in pigments and antioxidative components prevented dehydration during the drought. In winter, *P. canariensis* had to face robust chronic photoinhibition, with α-tocopherol and β-carotene playing the main roles. Morphological variations in needles were also detected as a means to prevent freezing. Moreover, the breakdown of pigments during the winter reduced the risk of overexcitation and photodamage, making rebuilding and regreening possible in the next spring. Taking these factors into account, a set of wide-ranging, structural and physiological adaptations make this species very resistant to stress.

In this treeline, forest carbon gain takes place mainly over a short period in spring. On a long-term scale, it was more limited by drought than by low temperatures, as has been described in Mediterranean areas [69]. For this reason, this semiarid forest, situated on the edge of this ecological threshold, may experience dramatic consequences due to ongoing climatic changes, where increasing temperatures and extended drought periods are predicted [15,70].

Author Contributions: Conceptualization, A.M.G-R. and P.B.; methodology, M.S.J. and A.M.G-R.; software, P.B.; validation, M.S.J., P.B., J.R.L. and A.M.G-R.; formal analysis, P.B.; investigation, A.M.G-R. and P.B.; resources, J.R.L.; data curation, A.M.G-R.; writing—original draft preparation, A.M.G-R. and P.B.; writing—review and editing, A.M.G-R.; visualization, A.M.G-R.; supervision, M.S.J., P.B., J.R.L. and A.M.G-R.; project administration, M.S.J. and A.M.G-R.; funding acquisition, M.S.J.

Funding: This work was supported by the Spanish Government (CGL2006-10210/BOS, CGL2010-21366-C04-04 MCI) and University of La Laguna (ULL) both co-financed by FEDER. P.B. received a fellowship from Canarian Agency for Research, Innovation and Information Society (ACIISI) co-financed by FEDER.

Acknowledgments: The authors express their gratitude to the National Park's Network for permission to work in the Teide National Park.

Conflicts of Interest: The authors declare no conflict of interest.

References

1. Körner, C. *Alpine Treelines. Functional Ecology of the Global High Elevation Tree Limits*; Springer: Basel, Switzerland, 2012; p. 219.

2. Silva-Cancino, M.C.; Esteban, R.; Artetxe, U.; García-Plazaola, J.I. Patterns of spatio-temporal distribution of winter chronic photoinhibition in leaves of three evergreen Mediterranean species with contrasting acclimation responses. *Physiol. Plant.* **2012**, *144*, 289–301. [CrossRef] [PubMed]

3. Takahashi, S.; Badger, M.R. Photoprotection in plants: A new light on photosystem II damage. *Trends Plant Sci.* **2011**, *16*, 53–60. [CrossRef] [PubMed]

4. Martínez-Ferri, E.; Manrique, E.; Valladares, F.; Balaguer, L. Winter photoinhibition in the field involves different processes in four co-occurring Mediterranean tree species. *Tree Physiol.* **2004**, *24*, 981–990. [CrossRef] [PubMed]

5. Flexas, J.; Diaz-Espejo, A.; Gagoa, J.; Galléa, A.; Galmés, J.; Gulías, J.; Medrano, H. Photosynthetic limitations in Mediterranean plants: A review. *Env. Exp. Bot.* **2014**, *103*, 12–23. [CrossRef]

6. Asada, K. Production and scavenging of reactive oxygen species in chloroplasts and their functions. *Plant Physiol.* **2006**, *141*, 391–396. [CrossRef] [PubMed]

7. Dall'Osto, L.; Cazzaniga, S.; North, H.; Marion-Poll, A.; Bassi, R. The Arabidopsis aba4-1 mutant reveals a specific function for neoxanthin in protection against photooxidative stress. *Plant Cell* **2007**, *19*, 1048–1064. [CrossRef] [PubMed]

8.   Havaux, M.; Niyogi, K.K. The violaxanthin cycle protects plants from photooxidative damage by more than one mechanism. *Proc. Natl. Acad. Sci. USA* **1999**, *96*, 8762–8767. [CrossRef] [PubMed]

9.   Havaux, M.; Eymery, F.; Porfirova, S.; Rey, P.; Dörmann, P. Vitamin E protects against photoinhibition and photooxidative stress in *Arabidopsis thaliana*. *Plant Cell* **2005**, *17*, 3451–3469. [CrossRef]

10. Peng, C.L.; Lin, Z.-F.; Su, Y.-Z.; Lin, G.-Z.; Dou, H.-Y.; Zhao, C.-X. The antioxidative function of lutein: Electron spin resonance studies and chemical detection. *Funct. Plant Biol.* **2006**, *33*, 839–846. [CrossRef]

11. Wieser, G.; Tausz, M. *Trees at Their Upper Limit: Treelife Limitation at the Alpine Timberline*; Springer: Dordrecht, The Netherlands, 2007; p. 225.

12. Piersma, T.; Drent, J. Phenotypic flexibility and the evolution of organismal design. *Trends Ecol. Evol.* **2003**, *18*, 228–233. [CrossRef]

13. Fernández-Palacios, F.M.; de Nicolás, J.P. Altitudinal pattern of vegetation variation on Tenerife. *J. Veg. Sci.* **1995**, *6*, 183–190. [CrossRef]

14. Sperling, F.N.; Washington, R.; Whittaker, R.J. Future Climate Change of the Subtropical North Atlantic: Implications for the Cloud Forests of Tenerife. *Clim. Chang.* **2004**, *65*, 103–123. [CrossRef]

15. IPCC. *Intergovernmental Panel of Climate Change (IPCC) Synthesis Report, Contribution of Working Groups I, II and III to the Fifth Assessment Report of the Intergovernmental Panel on Climate Change 2014*; IPCC: Geneva, Switzerland, 2014. [CrossRef]

16. Expósito, F.J.; González, A.; Pérez, J.C.; Díaz, J.P.; Taima, D. High-resolution future projections of temperature and precipitation in the Canary Islands. *J. Clim.* **2015**, *28*, 7846–7856. [CrossRef]

17. Severin, D.H.I.; Anthelme, F.; Harter, D.E.V.; Jentsch, A.; Lotter, E.; Steinbauer, M.J.; Beierkuhnlein, C. Patterns of island treeline elevation—A global perspective. *Ecography* **2015**, *38*, 001–010.

18. Brito, P.; Morales, D.; Wieser, G.; Jiménez, M.S. Spatial and seasonal variations in stem CO_2 efflux of *Pinus canariensis* at their upper distribution limit. *Trees* **2010**, *24*, 523–531. [CrossRef]

19. Brito, P.; Jiménez, M.S.; Morales, D.; Wieser, G. Assessment of ecosystem CO_2 efflux and its components in a *Pinus canariensis* forest at the treeline. *Trees* **2013**, *27*, 999–1009. [CrossRef]

20. Brito, P.; Lorenzo, J.R.; González-Rodríguez, A.M.; Morales, D.; Wieser, G.; Jiménez, M.S. Canopy transpiration of a *Pinus canariensis* forest at the tree line: Implications for its distribution under predicted climate warming. *Eur. J. Res.* **2014**, *133*, 491–500. [CrossRef]

21. Brito, P.; Lorenzo, J.R.; González-Rodríguez, A.M.; Morales, D.; Wieser, G.; Jiménez, M.S. Canopy transpiration of a semi arid *Pinus canariensis* forest at a treeline ecotone in two hydrologically contrasting years. *Agric. For. Meteorol.* **2015**, *201*, 120–127. [CrossRef]

22. Gieger, T.; Leuschner, C. Altitudinal change in needle water relations of *Pinus canariensis* and possible evidence of a drought-induced alpine timberline on Mt. Teide, Tenerife. *Flora* **2004**, *199*, 100–109. [CrossRef]

23. Paulsen, J.; Körner, C. A climate-based model to predict potential treeline position around the globe. *Alp. Bot.* **2014**, *124*, 1–12. [CrossRef]

24. Peters, J.; Morales, D.; Jiménez, M.S. Gas exchange characteristics of *Pinus canariensis* needles in a forest stand on Tenerife, Canary Islands. *Trees* **2003**, *17*, 492–500. [CrossRef]

25. Von Caemmerer, S.; Farquhar, G.D. Some relationships between the biochemistry of photosynthesis and the gas exchange of leaves. *Planta* **1981**, *153*, 376–397. [CrossRef] [PubMed]

26. Gratani, L. Plant Phenotypic Plasticity in Response to Environmental Factors. *Adv. Bot.* **2014**, *2018*, 17. [CrossRef]

27. Prioul, J.L.; Chartier, P. Partitioning of transfer and carboxylation components of intracellular resistance to photosynthetic CO_2 fixation: A critical analysis of the methods used. *Ann. Bot.* **1977**, *41*, 789–800. [CrossRef]

28. Genty, B.; Briantais, J.M.; Baker, N.R. The relationship between the quantum yield of photosynthetic electron-transport and quenching of chlorophyll fluorescence. *Biochim. Biophys. Acta* **1989**, *990*, 87–92. [CrossRef]

29. Werner, C.; Correia, O.; Beyschlag, W. Characteristic patterns of chronic and dynamic photoinhibition of different functional groups in a Mediterranean ecosystem. *Funct. Plant Biol.* **2002**, *29*, 999–1011. [CrossRef]

30. Osmond, C.B. What is photoinhibition? Some insights from comparison of shade and sun plants. In *Photoinhibition of Photosynthesis: From Molecular Mechanisms to the Field*; Baker, N.R., Bowyer, J.R., Eds.; BIOS Scientific Publishers: Lancaster, PA, USA, 1994; pp. 1–24.

31. Strasser, R.J.; Tsimilli-Michael, M.; Srivastava, A. Analysis of the chlorophyll a fluorescence transient. In *Chlorophyll a Fluorescence: A Signature of Photosynthesis*; Papageorgiou, G.C., Govindjee, R., Eds.; Springer: Dordrecht, The Netherlands, 2004; pp. 321–362.

32. Strasser, R.J.; Tsimilli-Michael, M.; Qiang, S.; Goltsev, V. Simultaneous in vivo recording of prompt and delayed fluorescence and 820-nm reflection changes during drying and after rehydration of the resurrection plant *Haberlea rhodopensis. Biochim. Biophys. Acta* **2010**, *1797*, 1313–1326. [CrossRef] [PubMed]

33. Pflug, E.; Brüggemann, W. Frost-acclimation of photosynthesis in overwintering Mediterranean holm oak, grown in Central Europe. *Int. J. Plant Biol.* **2012**, 3. [CrossRef]

34. Tausz, M.; Wonisch, A.; Grill, D.; Morales, D.; Jiménez, M.S. Measuring antioxidants in tree species in the natural environment: From sampling to data evaluation. *J. Exp. Bot.* **2003**, *387*, 1505–1510. [CrossRef]

35. Morales, D.; Peters, J.; Jiménez, M.S.; Tausz, M.; Wonisch, A.; Grill, D. Gas exchange of irrigated and non-irrigated *Pinus canariensis* seedlings growing outdoors in La Laguna, Canary Islands, Spain. *Naturforsch* **1999**, *54*, 693–697. [CrossRef]

36. González-Rodríguez, A.M.; Morales, D.; Jimenez, M.S. Leaf gas exchange characteristics of a Canarian laurel forest tree species (*Persea indica* (L.) Spreng) under natural conditions. *J. Plant Phys.* **2002**, *159*, 695–704.

37. Peters, J.; González-Rodríguez, A.M.; Jiménez, M.S.; Morales, D.; Wieser, G. Influence of canopy position, needle age and season on the foliar gas exchange of *Pinus canariensis. Eur. J. Res.* **2008**, *127*, 293–299. [CrossRef]

38. Rundel, P.W.; Yoder, B.J. Ecophysiology of Pinus. In *Ecology and Biogeography of Pinus*; Richardson, D.M., Ed.; Cambridge University Press: Cambridge, UK, 2000; pp. 296–323.

39. Larcher, W. *Physiological Plant Ecology. Ecophysiology and Stress Physiology of Functional Groups*, 4th ed.; Springer: Basel, Switzerland, 2003; p. 488.

40. Peters, J. Ecofisiología del pino canario. Ph.D. Dissertation, University of La Laguna, Santa Cruz de Tenerife, Spain, 2000.

41. Gulías, J.; Flexas, J.; Mus, M.; Cifre, J.; Lefi, E.; Medrano, H. Relationship between maximum leaf photosynthesis, nitrogen content and specific leaf area in Balearic endemic and non endemic Mediterranean species. *Ann. Bot.* **2003**, *92*, 215–222. [CrossRef] [PubMed]

42. Zellnig, G.; Peters, J.; Jiménez, M.S.; Morales, D.; Grill, D.; Perktold, A. Three-Dimensional reconstruction of the stomatal complex in *Pinus canariensis* needles using serial sections. *Plant Biol.* **2002**, *4*, 70–76. [CrossRef]

43. Galmés, J.; Flexas, J.; Medrano, H.; Niinemets, Ü.; Valladares, F. Ecophysiology of photosynthesis in semi-arid cenvironments. In *Terrestrial Photosynthesis in a Changing Environment. A Molecular, Physiological and Ecological Approach*; Flexas, J., Loreto, F., Medrano, H., Eds.; Cambridge University Press: Cambridge, UK, 2012; pp. 448–464.

44. Niinemets, Ü. Climatic controls of leaf dry mass per area, density, and thickness in trees and shrubs at the global scale. *Ecology* **2001**, *82*, 453–469. [CrossRef]

45. Paula, S.; Pausas, J.G. Leaf traits and resprouting ability in the Mediterranean basin. *Funct. Ecol.* **2006**, *20*, 941–947. [CrossRef]

46. Poorter, H.; Niinemets, Ü.; Poorter, L.; Wright, I.J.; Villar, R. Causes and consequences of variation in leaf mass per area (LMA): A meta-analysis. *New Phytol.* **2009**, *182*, 565–588. [CrossRef] [PubMed]

47. Zunzunegui, M.; Barradas, M.C.D.; Ain-Lhout, F.; Alvarez-Cansino, L.; Esquivias, M.P.; Novo, F.G. Seasonal physiological plasticity and recovery capacity after summer stress in Mediterranean scrub communities. *Plant Ecol.* **2011**, *212*, 127–142. [CrossRef]

48. Baker, N.R.; Rosenqvist, E. Applications of chlorophyll fluorescence can improve crop production strategies: An examination of future possibilities. *J. Exp. Bot.* **2004**, *403*, 1607–1621. [CrossRef]

49. Nippert, J.B.; Duursma, R.A.; Marshall, J.D. Seasonal variation in photosynthetic capacity of montane conifers. *Funct. Ecol.* **2004**, *18*, 876–886. [CrossRef]

50. Porcar-Castell, A.; Juurola, E.; Ensminger, I.; Berninger, F.; Hari, P.; Nikinmaa, E. Seasonal acclimation of photosystem II in *Pinus sylvestris*. II. Using the rate constants of sustained thermal energy dissipation and photochemistry to study the effect of the light environment. *Tree Physiol.* **2008**, *28*, 1483–1491. [CrossRef] [PubMed]

51. Robakowski, P.; Wyka, T. Winter photoinhibition in needles of *Taxus baccata* seedlings acclimated to different light levels. *Photosynthetica* **2009**, *47*, 527–535. [CrossRef]

52. García-Plazaola, J.I.; Olano, J.M.; Hernández, A.; Becerril, J.M. Photoprotection in evergreen Mediterranean plants during sudden periods of intense cold weather. *Trees-Struct. Funct.* **2003**, *17*, 285–291.

53. Müller, M.; Hernández, I.; Alegre, L.; Munné-Bosch, S. Enhanced alpha-tocopherol quinone levels and xanthophyll cycle de-epoxidation in rosemary plants exposed to water deficit during a Mediterranean winter. *J. Plant Physiol.* **2006**, *163*, 601–606. [CrossRef] [PubMed]

54. Somersalo, S.; Krause, H.G. Photoinhibition at chilling temperatures and effects of freezing stress on cold acclimated spinach leaves in the field. A fluorescence study. *Physiol. Plant.* **1990**, *79*, 617–622. [CrossRef] [PubMed]

55. Robakowski, P. Susceptibility to low-temperature photoinhibition in three conifers differing in successional status. *Tree Physiol.* **2005**, *25*, 1151–1160. [CrossRef] [PubMed]

56. Ottander, C.; Öquist, G. Recovery of photosynthesis in winter-stressed Scots pine. *Plant Cell Env.* **1991**, *14*, 345–349. [CrossRef]

57. Westin, J.; Sundblad, L.G.; Hällgren, J.E. Seasonal variation in photochemical activity and hardiness in clones of Norway spruce (*Picea abies*). *Tree Physiol.* **1995**, *15*, 685–689. [CrossRef]

58. Oleksyn, J.; Reich, P.B.; Tjoelker, M.G.; Chalupka, W. Biogeographic differences in shoot elongation pattern among European Scots pine populations. *Ecol. Man.* **2001**, *148*, 207–220. [CrossRef]

59. Sheffield, M.C.P.; Gagnon, J.L.; Jack, S.B.; McConville, D.J. Phenological patterns of mature longleaf pine (*Pinus palustris* Miller) under two different soil moisture regimes. *For. Ecol. Manag.* **2003**, *179*, 157–167. [CrossRef]

60. Gasulla, F.; Gómez de Nova, O.; Barreno, E. Relaciones entre la variabilidad fenológica de *Pinus canariensis*, el clima y la concentración foliar de los nutrientes. *Cuad. Soc. Esp. Cienc.* **2004**, *20*, 141–146.

61. Esteban, R.; Barrutia, O.; Artetxe, U.; Fernández-Marín, B.; Hernández, A.; García-Plazaola, J.I. Internal and external factors affecting photosynthetic pigment composition in plants: A meta-analytical approach. *New Phytol.* **2015**, *206*, 268–280. [CrossRef] [PubMed]

62. Elvira, S.; Alonso, R.; Castillo, F.J.; Gimeno, B.S. On the response of pigments and antioxidants of *Pinus halepensis* seedlings to Mediterranean climatic factors and long-term ozone exposure. *New Phytol.* **1998**, *138*, 419–432. [CrossRef]

63. Lichtenthaler, H.K.; Babani, F. Light adaptation and senescence of the photosynthetic apparatus. Changes in pigment composition, chlorophyll fluorescence parameters and photosynthetic activity. *Adv. Photosynth. Respir.* **2004**, *19*, 713–736.

64. Telfer, A. Singlet Oxygen Production by PSII Under light stress: Mechanism, detection and the protective role of β-Carotene. *Plant Cell Physiol.* **2014**, *55*, 1216–1223. [CrossRef]

65. Triantaphylidès, C.; Krischke, M.; Hoeberichts, F.A.; Ksas, B.; Gresser, G.; Havaux, M.; Van Breusegem, F.; Martin, J.M. Singlet oxygen is the major reactive oxygen species involved in photooxidative damage to plants. *Plant Physiol.* **2008**, *148*, 960–968. [CrossRef] [PubMed]

66. Munné-Bosch, S. The role of a-tocopherol in plant stress tolerance. *J. Plant Phys.* **2005**, *162*, 743–748. [CrossRef]

67. Falk, J.; Munné-Bosch, S. Tocochromanol functions in plants: Antioxidation and beyond. *J. Exp. Bot.* **2010**, *61*, 1549–1566. [CrossRef]

68. Maeda, H.; Song, W.; Sage, S.L.; Della Pennaa, D. Tocopherols play a crucial role in low-temperature adaptation and phloem loading in *Arabidopsis*. *Plant Cell* **2006**, *18*, 2710–2732. [CrossRef]

69. Flexas, J.; Gulías, J.; Jonasson, S.; Medrano, H.; Mus, M. Seasonal patterns and control of gas exchange in local populations of the Mediterranean evergreen shrub *Pistacia lentiscus* L. *Acta Oecol.* **2001**, *22*, 33–43. [CrossRef]

70. Martín, J.L.; Bethencourt, J.; Cuevas-Agulló, E. Assessment of global warming on the island of Tenerife, Canary Islands (Spain). Trends in minimum, maximum and mean temperatures since 1944. *Clim. Chang.* **2012**, *114*, 343–355. [CrossRef]

Article

Photosynthetic Pigments in Siberian Pine and Fir under Climate Warming and Shift of the Timberline

Nina Pakharkova, Irina Borisova, Ruslan Sharafutdinov and Vladimir Gavrikov *

Chair Ecology and Nature Management, Siberian Federal University, Svobodny pr. 79, Krasnoyarsk 660041, Russia; nina.pakharkova@yandex.ru (N.P.); irina_borisova77@mail.ru (I.B.); ruslanate@mail.ru (R.S.)
* Correspondence: vgavrikov@sfu-kras.ru; Tel.: +7-913-042-4304

Received: 20 November 2019; Accepted: 2 January 2020; Published: 4 January 2020

Abstract: *Research Highlights:* For the first time, the *Pinus sibirica* Du Tour and *Abies sibirica* L. conifer forest at the West Sayan ridge timberline has been explored to reveal which species is likely to react to climate change and a shift of the timberline. Such a shift may modify the ecological functions of the forests. *Background and Objectives:* Long-term climate change has become obvious in the mountains of southern Siberia. Specifically, a half-century rise in annual mean temperatures has been observed, while precipitation remains unchanged. Trees growing at the timberline are likely to strongly react to climate alterations. The objective was to estimate which of the two species sharing the same habitat would benefit from climate alteration and shifting of the timberline. *Materials and Methods:* At several altitudes (from 1413 to 1724 m a.s.l.), samples of *P. sibirica* and *A. sibirica* needles have been collected and contents of chlorophyll *a* and *b* as well as carotenoids were measured in June 2019. The temperature of needles of the two species was measured in both cloudy and sunny weather conditions. *Results:* The studied species have been shown to have different patterns of pigment variations with the growth of altitude. The decline of chlorophylls and carotenoids was more pronounced in *P. sibirica* (ratio at timberline ca. 2.2) than in *A. sibirica* (ratio ca. 3.1). Accordingly, the electron transport rate decreased more strongly in *P. sibirica* at the timberline (ca. 37.2 µmol of electrons/m^{-2} s^{-1}) than in *A. sibirica* (56.9 µmol of electrons/m^{-2} s^{-1}). The temperatures of needles in both cloudy and sunny weather were higher in *A. sibirica* (10.5 and 43.3 °C, respectively) than in *P. sibirica* (3.8 and 24.2 °C, respectively). *Conclusions:* The considered physiological and ecological traits show that *P. sibirica* is better protected from higher-altitude hazards (excess insolation, rise of temperature etc.) than *A. sibirica*. *P. sibirica* may be therefore a more likely winner than *A. sibirica* in the movement of the mountain timberline under climate warming in the area.

Keywords: timberline; higher altitude; chlorophyll; carotenoids; climate change; *Pinus sibirica*; *Abies sibirica*

---

1. Introduction

In various regions of the globe, climate warming has become a matter of fact. Forest vegetation may react more weakly or strongly to the change, but those forests occupying the very extreme edges of distribution are expected to be more sensitive to the climatic trend [1]. The shift of the forested zone borderline is an apparent indicator of climate change and of important events shaping the outlook of landscapes. The ability of trees to move up the timberline will to a great extent arise from their ability to acclimate and grow.

Broadly seen, the system of tree photoreceptors ensures the start of acclimation to overwinter freezing temperatures [2]. Habitat warming may, however, distort the work of photoperiod receptors. As has been demonstrated in experiments with a warming of spruce bog communities [3], the trees extend their growth period but may fail to cue the photoperiod, which leads to premature loss of frost hardiness and tissue damage.

The growth potential of higher altitudes trees is often studied from the viewpoint of how their photosynthetic system functions under extremely harsh conditions. In subalpine *Abies faxoniana* Rehder & E.H. Wilson., the content of chlorophyll in mature trees needles was reported to be lower than that in juvenile trees [4]. The juvenile fir trees were seen as more vulnerable to climate change because altitude exerts a sufficient impact on their leaf traits. The seedlings of another fir species, Abies alba Mill., were subjected to experimental temperature increases [5]. The heating stress led to a decrease of chlorophyll and carotenoid pigments in the seedling leaves.

Poulos et al. [6] explored a stratification pattern in four pine species across the elevation gradient in the mountains of southern California. They found that both the lowest elevation species (*Pinus attenuata Lem.*) and highest elevation one (*Pinus contorta* Doug.) had high carotenoid and anthocyanin indices, which are considered to be a result of their acclimation to high ultraviolet radiation in upper habitats and also to drought stress in foothills. In another mountainous area, the Appalachian, the high-elevation *Abies fraseri* (Pursh) Poir. was compared to the lower-elevation *Picea rubens* Sarg. [7]. It was found that both green (chlorophyll) and yellow (carotenoids) pigments showed a higher content in *A. fraseri*.

Photosynthetic pigments may be of importance in assessing the general health of forest stands. Kopačková et al. [8] reported the carotenoid-to-chlorophyll ratio to be sensitive to soil geochemical conditions, and it could serve as a non-specific indicator of tree stress.

In the northern macro-slope of West Sajan Mountains, the timberline belt is dominated by two conifer species, Siberian pine (*Pinus sibirica* Du Tour) and Siberian fir (*Abies sibirica* L.). The species often grow in a mixture, but at the very elevated edge of the distribution the fir tends to form bush-like spatial clumps, while the pine presents lone-standing trees. Compared to Siberian pine the fir is more seldom an object of research, probably because of its much lower commercial value. However, it plays an important ecological role in covering large areas of mountainous slopes in wet northern Asian climates.

Sobchak and Zotikova [9] measured the content of pigments in differently aged needles of Siberian pine. The sampled 60–80 year-old trees grew at the altitude 2350 m a.s.l. in the vicinity of a glacier in the Altai Mountains. As the needles grow old, the content of both chlorophyll and carotenoid pigments first rises to achieve a maximum at the age of 4 years and then decreases significantly. The authors noted that the decrease of carotenoid content was weaker, which they attributed to the protection function of the pigments. Measurements of photochemical activity of chloroplasts showed that the two- and three-year-old needles were the most photosynthetically active.

Another study in the Altai Mountains [10] considered a 14-km-long transect from a plain site to 2110 m a.s.l. which corresponded to open woodland at the timberline. In two-year-old Siberian pine needles the largest pigment pool was recorded at lower altitudes, with chlorophyll *a* comprising the largest share in the pool. At the highest altitudes, however, green pigments were sufficiently smaller in content whereas carotenoids were at their maximum compared the whole transect. Also, the species exhibits definite variations of the pigment content during the seasons of the year. For Siberian pine that was an introduced species to mountainous sites of the Far East, it has been shown [11] that the green pigments achieve their maximum in August and then steadily decrease in autumn and the early winter months. Inversely, carotenoids were at a minimum in August and at a maximum in November (ca. 0.32 μkg/mkg, green weight). At the same time, the total pool of pigments stayed relatively stable (1.6–1.7 μkg/mkg, green weight) from May to November.

In the northern part of the Siberian pine area (61°–62° northern latitude), the two- and three-year-old needles of the trees exhibit a maximal photosynthetic activity [12]. In the spring–summer season, the share of green pigments decreased while that of yellow pigments increased closer to the dormant period in winter.

The pool of photosynthetic pigments in Siberian fir has been the focus of several kinds of research. Silkina [13] analyzed one- to five-year-old needles of the fir with respect to green pigments. Throughout the growing season, the content of both chlorophyll *a* and *b* was maximal in three–four year-old needles of older trees, ca. 2–4 μkg/mkg dry weight. Meanwhile, the pool of pigments was higher in five-year-old needles of younger fir undergrowth. Across the season, a maximal content of the pigments was observed in May, after which the content gradually decreased until next March. The author has also recommended using three–four year-old needles from the lower part of the crown because the needles of this location showed a maximal pigment content among other locations.

The seasonal dynamics of the pigments was followed in two-year-old fir needles [14]. The authors reported that the green pigments were at a minimum in March and May (ca. 2.4–2.5 μkg/mkg dry weight) and grew until December (3.82 μkg/mkg dry weight). The same was true for the yellow pigments, which were at a minimum in May (0.5 μkg/mkg dry weight) and at a maximum in December (0.83 μkg/mkg dry weight). Measurements of chlorophyll fluorescence have shown that maximal quantum yield of photochemical activity was maximal in July and minimal in March. Also, the fir needles showed higher levels of CO_2 exchange in summer months in comparison with some other conifers, spruce, and juniper [14].

It is widely believed that temperature is the limiting factor that inhibits or speeds up the growth in trees at higher altitudes. Presumably, the timberline shifts upward when the climate is warming. In the case of a monospecies tree community, only one species will occupy higher altitudes. In a multiple species case, however, climate change may not only shift the timberline but also change the species distribution. The aims of the study were, therefore, (1) to evaluate photosynthetic traits in Siberian pine and Siberian fir sharing the same habitats at the mountainous timberline in the West Sayan ridge, and on that basis (2) to make a prognosis of how the climate change would alter the forest communities at the timberline. Specifically, we explored which parts of the trees or the stages of their life cycles led to the formation of the upper timberline in the area.

2. Materials and Methods

2.1. Area of Research

The study area is located in the central part of the Western Sayan Mountains (Figure 1). After the Köppen classification, the climate of the area is continental (Dwb). Several continental air masses, Atlantic, Arctic, and central Asian, overlap over the area. Summertime lasts for ca. two months. Average July temperature varies from 17 °C to 6 °C while the annual mean over the area is –3.9 °C (Olen'ya Rechka weather station). Every 100 m of elevation up the slope brings a decrease of temperature by 0.6 °C. The annual sum of precipitation reaches 1480 mm at higher elevations and 800 mm in a low mountain belt. Half of the precipitation falls from June to August. The depth of snow cover is over 1.5 m and sometimes reaches up to 2 m. The surface relief is rather dissected, with altitudes varying from 400 to 2740 m. The higher mountain belt bears visible signs of Pleistocene glaciations, while dome-shaped peaks not touched by glaciers are common in the middle mountain belt. The river net is dense and branchy, with the rivers receiving water from both glaciers in spring and precipitation and permafrost in summer. The duration of the growing season is from 100 to 120 days.

The background soils of the area are Hyperskeletic Histic Leptosols Humic, Folic Leptic Skeletic Histosols Arenic Hyperorganic, Leptic Skeletic Cambisols Arenic Ferric Humic, and Leptic Skeletic Cambic Phaeozems Arenic Hyperhumic (World Reference Base [15]). Leptosols are formed on the west slope on eluvial facies (altitude 1630–1730 m). The structure and granulometrical composition were not determined due to the absolute dominance of the mineral material in the profile. The depth of the soil profile was no more than 35 cm.

Histosols are formed on the west slope on trans-accumulative facies (altitude 1560–1630 m). No gley materials are found but the profile is water-saturated. This soil has a light granulometrical composition. In terms of pH of the aqueous extract, the soil varies from highly-acid in organic horizons to slightly acid in mineral horizons. The soil profile has high rock fragment content. The depth of the profile is no more than 40 cm.

Cambisols are formed on the west slope on transeluvial facies (altitude 1475–1560 m). An illuvial-ferruginous diagnostic horizon with iron neoplasms and highly acidic reactions is characteristic of the soil type. The soil profile has also a high content rock fragments. The depth of the profile is no more than 35 cm.

Phaeozems are formed on the west slope on transeluvial facies, but at lower altitudes (1340–1475 m). A hyperhumic diagnostic horizon and highly acidic reactions are characteristic of the soil. The soil profile is strongly differentiated into organic and metamorphic (BM) horizons because of light granulometrical composition. The profile has also a high content rock fragments. The depth of the profile is no more than 45 cm.

Regarding humus abundance, the soils of the research area are described as having very high humus content (10%–17%), and according to the pH value of the aqueous extract, they are strongly acidic (pH 4.3–4.5). All the investigated soils are light in particle size distribution and belong to a cohesive sand group. It can be assumed therefore that all the soils have a low absorption capacity associated with a low content of finely dispersed fractions that constitute the soil-absorbing complex.

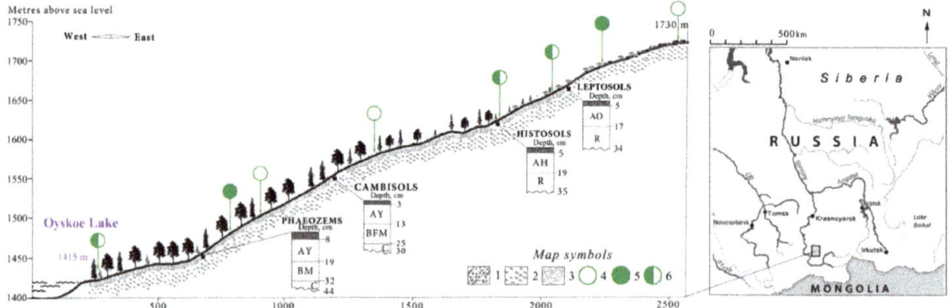

Figure 1. A graphic representation of area of research. The mountain top on the upper right has the coordinates N52.845872, E93.274658. Legend: 1—detritus; 2—metamorphic schist; 3—moraine loam; 4—*Pinus sibirica* sample points; 5—*Abies sibirica* sample points; 6—sample points for both *P. sibirica* and *A. sibirica*; AO = the coarse humic horizon; AY = the gray humic horizon; AH = the hyperhumic horizon; BFM = the ferruginous metamorphic horizon; BM = structural-metamorphic horizon; C = substratum, un-weathered geologic material; R = underlying bedrock.

In our study, a slope with a uniform climb and minimal microrelief was surveyed, which allowed us to consider the factor of increasing altitude to be a dominant one, as well the related changes in temperature, humidity, light, and other derivative parameters. The sampling points were taken at more or less equal distances from one another but taking into account the presence of trees of close age, ca. 40–50 years. The sampling of 50-cm-long shoots was done from a lower part of the crowns, with only vegetative shoots taken (no presence of female or male cones).

2.2. Climatic Data

The data on climate have been taken from station Olen'ya Rechka that operates under index 29974 as a part of the Regional Synoptic Net (Region II) of the World Meteorological Organization. For the goals of the study, mean annual temperatures and total annual precipitation data were taken to check if a climate change was significant for the area of research. The range of the data was from 1967 to 2017.

2.3. Measurements of Pigment Content

To find out the pigment composition in needles, the second-year growth needles of a peripheral shoot at the bottom of the crown was sampled. Because Siberian pine and fir grow often disjunctively at a higher elevation, the sampling followed sometimes an ad hoc approach—where both pine and fir might have been sampled at the same location/elevation. In a few instances, however, pine and fir trees were sampled at different but still close locations (see Figure 1).

The cut shoots were delivered to the laboratory in dense paper bags within a day, and then the shoots were placed in vessels with water and were kept in the same light and temperature conditions.

The content of chlorophylls and carotenoids was quantified in an alcohol extract on a SPEKOL 1300 spectrophotometer (Analytik Jena AG, Jena, Germany) using wavelengths of 440.5, 649, 654, and 665 nm. The calculations were carried out according to the formulas of Smith and Benitez [16].

Having established the concentration of the pigment in the extract, its content in the test needles was found through the formula:

$$F = (V \times C)/P$$

where F is the pigment content in the plant material, µkg/mkg; V is the volume of the extract, l; C is the pigment concentration, µkg/l; and P is the weight of sample material, mkg. All the data were recalculated to dry weight.

2.4. Measurement of Leaf Fluorescence

The analysis of chlorophyll fluorescence (ChlF) provides a large amount of information on the physiology of conifers, especially concerning the response of plants to various environmental factors. This approach is based on the fact that the light energy absorbed by chlorophyll *a* molecules in the photosystem II (PSII) antenna has three alternative paths: it may be used by the PSII reaction center (RC) in photosynthesis, dissipate as heat, or emit as light with a shifted wavelength (ChlF). The alternatives correlate with the condition of the plants as well as with other ecophysiological factors affecting the plants [17,18]. Chlorophyll fluorescence measurements are very sensitive; various tools and analytical methods have been developed that can be used both on individual leaf and as remote sensing [18–21].

In this study, the induced chlorophyll fluorescence was measured using a JUNIOR-PAM handheld fluorimeter (Heinz Walz GmbH, Effeltrich, Germany). The electron transport rate (ETR) coefficient was taken as the main indicator, in µmol of electrons/m^{-2} s^{-1}, calculated through the WinControl-3 software (Heinz Walz GmbH, Effeltrich, Germany).

Because chlorophyll fluorescence is a measure of re-emitted light (in the red wavebands) from PSII, any ambient light can interfere with the measurement of fluorescence and thus many early systems had to be used in darkness and/or highly controlled light environments. This issue was overcome by the development of modulating systems where the light used to induce fluorescence (the measuring beam) is applied at a known frequency (modulated) and the detector is set to measure at the same frequency [22]. In this way, the detector will only measure fluorescence that results from excitation by the measuring beam and will not permit interference from ambient light. The clear advantage of this is that measurements can be made without darkening the leaf.

The fluorescence parameters of two-year-old needles were measured directly on the sampling day after dark adaptation of shoots under room conditions.

Additionally, the surface temperature of the needles and soil was measured in the field using a Flir E5 thermal imager (Flir Systems, Tallinn, Estonia). The geographical coordinates and the altitudes were found with the help of satellite receivers RTK-GPS system Trimble R8S CSM (Trimble Navigation, Sunnyvale, California, USA). The RTK-GPS system has the accuracy of ca. 50 cm at measurements of altitudes.

3. Results

As follows from Figure 2, there are definite climatic changes in the higher altitude areas of West Sayan ridge. In the period of the last 50 years, mean annual temperatures rose by ca. 2 °C, while large year-to-year variations still took place. In the same period, the total annual precipitation varied greatly but no significant trend was observed.

The highest amount of chlorophyll *a* in the needles of *P. sibirica* was found in the needles of trees on sampling point 1 in the lower part of the slope; in its middle part at heights of 1505–1637 m a.s.l. (sampling points 2 to 4) the amount of chlorophyll *a* in the needles remains constant; then, from 1637 m upward the chlorophyll content is reduced (Table 1).

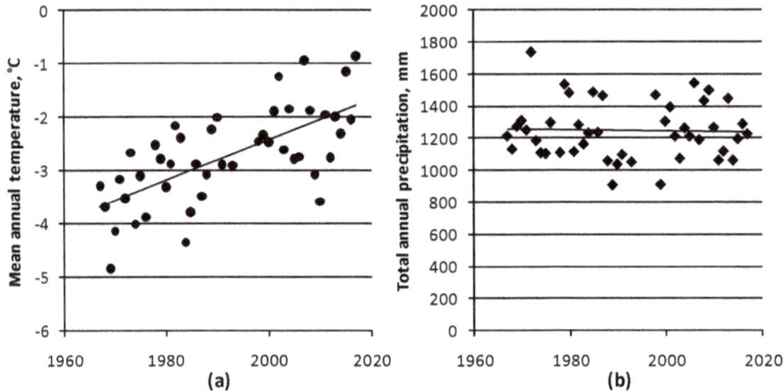

Figure 2. Long-term climatic trends over the area of research: (**a**) Annual mean temperatures; (**b**) Total annual precipitation. Straight lines give an idea of linear components of the trends. The slope of the temperature trend (~0.038) is significant at $p < 0.05$, $R^2 = 0.45$. The slope of precipitation trend (~−0.343) is not significant.

Table 1. A summary of the measured values of the pool of photosynthetic pigments at various altitudes in the vicinity of the timberline of West Sayan ridge. Sp = sample point, Chl = chlorophyll.

	Sp [1], Altitude a.s.l.,	Sp1	Sp2	Sp3	Sp4	Sp5	Sp6
P. sibirica	m	1418	1505	1567	1637	1660	1724
	Chl *a* (μkg/mkg)	1941 ± 0.069 [2]	1741 ± 0.063	1786 ± 0.065	1773 ± 0.065	1489 ± 0.066	1329 ± 0.047
	Chl *b* (μkg/mkg)	0.790 ± 0.026	0.744 ± 0.031	0.772 ± 0.028	0.782 ± 0.028	0.798 ± 0.019	0.799 ± 0.018
	Carotenoids (μkg/mkg)	0.681 ± 0.024	0.624 ± 0.022	0.575 ± 0.020	0.638 ± 0.023	0.695 ± 0.025	0.952 ± 0.034
	Sp, Altitude a.s.l.,	Sp1	Sp2	Sp3	Sp4	Sp5	
A. sibirica	m	1413	1475	1640	1660	1704	
	Chl *a* (μkg/mkg)	0.894 ± 0.031	1605 ± 0.056	2022 ± 0.128	2150 ± 0.052	1508 ± 0.047	
	Chl *b* (μkg/mkg)	0.348 ± 0.012	0.643 ± 0.020	0.897 ± 0.032	0.954 ± 0.032	0.635 ± 0.023	
	Carotenoids (μkg/mkg)	0.375 ± 0.013	0.555 ± 0.020	0.508 ± 0.018	0.811 ± 0.029	0.685 ± 0.024	

[1] Sample point index. [2] Standard deviation.

For other species of the *Pinus* genus, in particular, for *P. canariensis,* a decrease in the chlorophyll content in the treeline zone in winter was also found, which reduced the risk of overexcitation photoinhibition and photodamage, making rebuilding of photosynthetic activity possible in spring [23].

The amount of chlorophyll *b* remains constant at all points of sampling. The content of carotenoids in the needles of *P. sibirica* begins to increase from sampling point 3 (1567 m a.s.l.) upward and reaches a maximum at the point 6 (1724 m a.s.l.), where the trees transform into a shrub form.

In the needles of *A. sibirica,* an increase in the entire pigment pool from sampling point 1 (1413 m a.s.l.) to point 4 (1660 m a.s.l.) is observed; by sampling point 5 (1704 m a.s.l.) the pigment content is decreasing. Above this elevation, fir is not found.

The data obtained show that at the beginning of the growing season the higher the altitude the smaller the amount of chlorophylls in the needles of *P. sibirica,* while the amounts of carotenoids increase (Figure 3). In *A. sibirica* however, the ratio has a pronounced peak in middle elevations.

The chlorophyll *b* content in fir remains almost constant. Accordingly, the ratio of chlorophyll *a* to chlorophyll *b* in the needles decreases with altitude (Figure 4). In *P. sibirica*, the decrease is obviously more expressed.

Up the slope, the electron transport rate (ETR) varies in both *P. sibirica* and *A. sibirica* (Figure 5). However, after peaking at ca. 1567 m a.s.l. the *P. sibirica* ETR decreases up to the timberline. The ETR value in *A. sibirica* peaks higher, at ca. 1660 m a.s.l., and remains bigger at the timberline (ca. 1704 m a.s.l.) than that of *P. sibirica*. These data are related to chlorophyll *a* content (Table 1) and give evidence that *A. sibirica* needles have higher photosynthetic activity at altitudes of 1640 and 1660 m a.s.l.

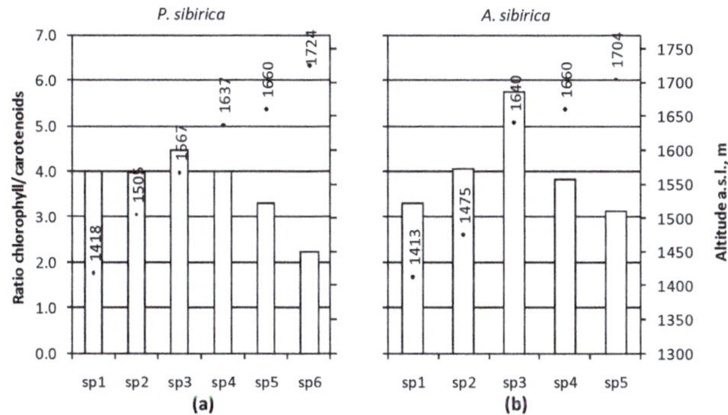

Figure 3. Change of ratio chlorophyll/carotenoids (white bars) along the study slope: (**a**) *P. sibirica*; (**b**) *A. sibirica*. sp1–sp6 denote the numbers of sample points. Vertically oriented numbers are altitudes of the sample points.

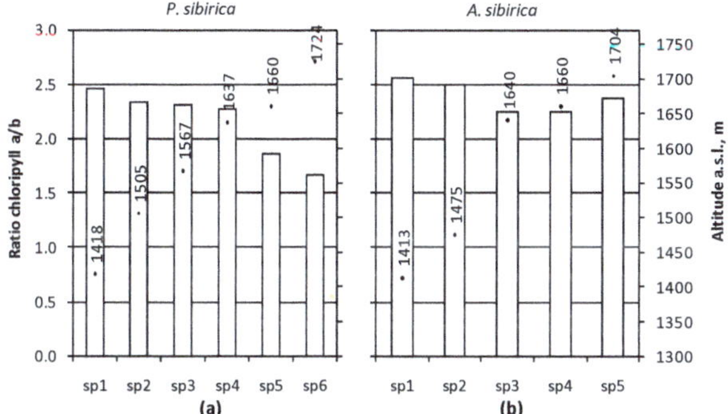

Figure 4. Change of ratio chlorophyll *a*/chlorophyll *b* (white bars) along the study slope: (**a**) *P. sibirica*; (**b**) *A. sibirica*. sp1–sp6 denote the numbers of sample points. Vertically oriented numbers are altitudes of the sample points.

Thus, right after the disappearance of the snow cover, there is an increase in altitude and a decrease in photosynthetic activity and chlorophyll content in the needles of *P. sibirica*, as well as an increase in the content of carotenoids. This feature contributes to the preservation of the photosynthetic apparatus at low temperatures and high insolation conditions. Adaptations of the kind allow *P. sibirica* to climb up the slopes under conditions of climate change, increasing the elevation of the timberline. For *A. sibirica*, such a clear adaptive picture is not observed throughout the entire slope, but when approaching the timberline, similar tendencies can be seen in the needles of fir trees.

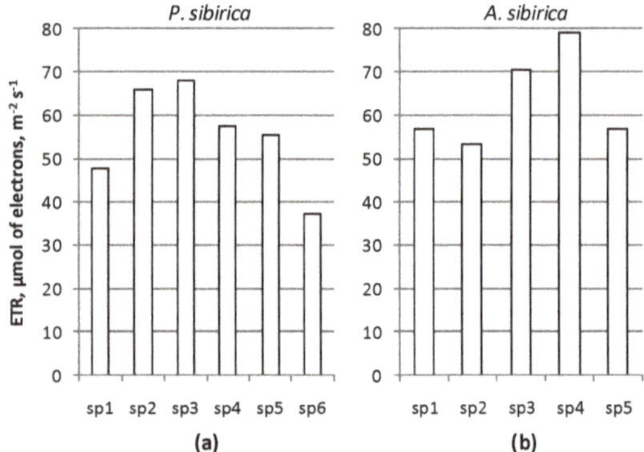

Figure 5. Change of electron transport rate (ETR) (white bars) along the study slope: (**a**) *P. sibirica*; (**b**) *A. sibirica*. sp1–sp6 denote the numbers of sample points.

At the time of measurements, the temperature of the litter was slightly positive, while the temperature of the soil under the litter had a temperature of about −5 °C (Figure 6).

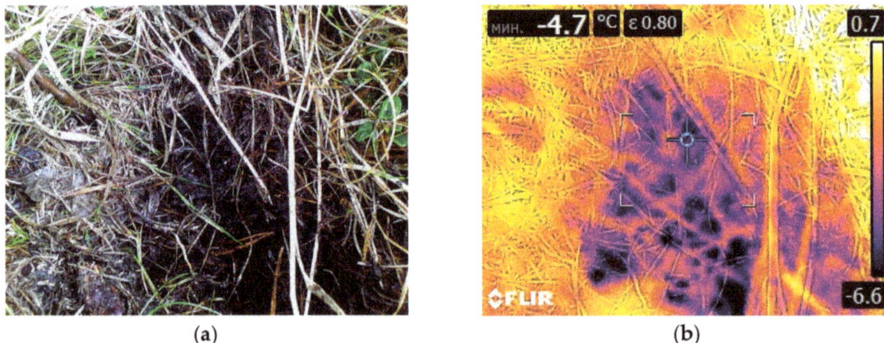

Figure 6. Photograph (**a**) and thermal image (**b**) of the soil under litter. The both images were received as single-step with the help of thermal imager Flir E5 under mean daily ambient temperature of +9.4 °C (data from the Olen'ya Rechka weather station).

At daytime, mean air temperatures rose to about +12.5 °C (according to data of Olen'ya Rechka weather station located at 1404 m a.s.l.). As per the thermal image data (Figure 7), under identical weather conditions of a cloudy morning the temperature of *A. sibirica* needles is much higher than that of *P. sibirica* (+10.5 °C and +3.8 °C, correspondingly).

On sunny days, when maximum air temperatures rose to about +18.9 °C, these differences become even more pronounced; temperatures of +43.3 °C for *A. sibirica* needles and +24.2 °C for *P. sibirica* needles were recorded. The lower needle temperatures in *P. sibirica* can be explained by the higher transpiration rate.

Figure 7. Photographs (**a**) and thermal images (**b**) of the young needles in *P. sibirica* and *A. sibirica* under cloudy (upper four images) and sunny (lower four images) weather.

4. Discussion

Altogether, the regional climate is becoming warmer and probably slightly drier because of rising evapotranspiration. These climatic changes are likely to influence the trees' functioning, especially those occupying the very edge of their distribution in the mountains, known as the timberline.

Since timberlines not always be straight but are often fragmented [24,25], the question is whether this fragmentation is the result of the climatic features of a given area or is related to the steepness of the slope, the type of soil, and other heterogeneities of a particular landscape. Harsch and Bader [26] distinguish four primary treeline forms: diffuse, abrupt, island, and krummholz. Growth limitation is dominant only at the diffuse treeline, which is the form that has most frequently responded as expected to growing-season warming, whereas the other forms are controlled by dieback and seedling mortality and are relatively unresponsive.

In our study, a slope with a uniform climb and minimal microrelief was surveyed, which allowed us to consider the factor of increasing altitude to be a dominant one, as well the related changes in temperature, humidity, light, and other derivative parameters.

Mountain plants generally acclimatize to high insolation [27–29] and extreme temperatures [30,31]. Temperature-related phenomena are highly likely to be determinative factors of the position of the forest boundary on a global scale [32,33].

As a rule, even extremely low winter temperatures are not a survival problem for taxa adapted to temperate zones [34–37]. Some authors reported that the treeline advance was positively associated with winter warming and not significantly associated with summer warming [38–40]. During winter dormancy, the vital activity of trees decreases sharply. Therefore, physiological processes under a gradual temperature effect are practically not affected during this period. However, in spring during the transition to the seasonal growth, plants become most vulnerable to temperature fluctuations. Negative effects of winter stress and damage to recruitment, tree survival, and growth appear to contradict the dominance of summer growth control [41–43].

In the spring on the upper edge of the forest, there are two types of threats that are not directly related to the low negative temperatures typical of winter: (1) photoinhibition caused by a combined effect of increasing solar radiation and low temperature, and (2) the so-called "winter drying" (Figure 8).

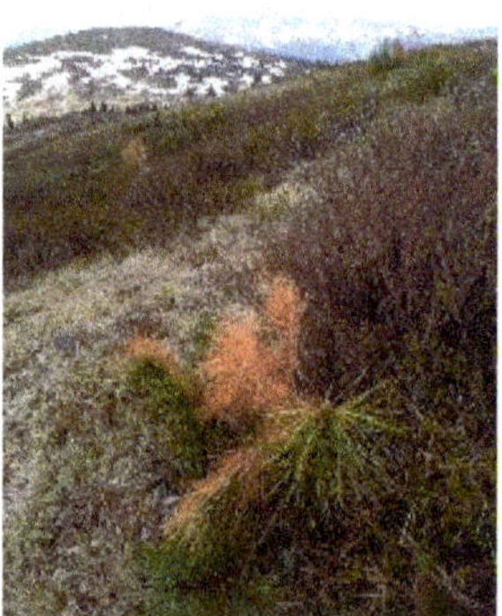

Figure 8. Altitude 1728 m a.s.l., *P. sibirica* sapling growing above the timberline. Brown dead needles are likely a result of winter desiccation.

Night frosts followed by a combination of strong solar radiation at negative or slightly positive temperatures in the daytime are typical situations in which breaking or even destruction of the photosystem by excess energy can occur. The UV radiation is not a tree-specific stress factor, given the well-designed protective measures developed by plants, but its effect can be seen by a decreased ratio of chlorophylls to carotenoids [44]. Carotenoids perform many important functions in the process of photosynthesis: they provide additional pigments at the absorbing of solar energy and they protect the photosystem. The protective function includes a quenching of triplet chlorophyll and singlet oxygen, protection of the reaction center of the photosystem from powerful energy flows at high light intensities, and protection of the lipid phase of thylakoid membranes from overoxidation.

In spring, an increase in the carotenoid pool with a lower chlorophyll content may allow *P. sibirica* saplings from higher altitudes to avoid photoinhibition. Also, by reducing the speed of electron transport (Figure 5), the protection of the needles from water deficiency and physiological desiccation is ensured. The desiccation is known to result from the inaccessibility of soil moisture during the period of low temperatures. In the needles of middle altitude *A. sibirica*, a peak of the concentration of pigments is found. In the upward direction, their content decreases and the electron transport speed decreases as well (Figure 5), which indicates a slowing down in general photosynthetic activity. Moreover, the share of *A. sibirica* carotenoids in the total pool of its pigments grows with altitude, similarly to *P. sibirica*, but the growth is weaker (Figure 3).

Some authors believe that acclimation to high insolation at low temperatures reduces the ability for primary photosynthetic reactions, that is, the ability to assimilate carbon and synthesize sugar. Through this acclimation, the plant can use high-intensity light more efficiently while limiting photosynthesis at low light flux without changing the quantum yield [29,31]. Thus, the decrease in the electron transport rate (ETR) in the needles with altitude (Figure 5) is, apparently, a transitional reaction of acclimation at the beginning of the growing season in both *P. sibirica* and *A. sibirica*.

Trees, like any plants, have a variable tissue temperature compared to the air temperature. Under direct insolation, alpine plants can warm up 10–20 °C above air temperature and sometimes even more [45].

For temperate zone plants, the temperature optimum for photosynthesis processes is known to be in the range of 25–30 °C, while higher temperatures lead to a decrease in the rate of photosynthesis. Therefore, heating of needles up to 40 °C and above is unfavorable and is associated most likely with a lack of water for cooling by transpiration. The root system in *P. sibirica* is located superficially, while in *A. sibirica* it lies in deeper layers of the soil, where there is still no accessible (in the liquid phase) water at this time.

The exceptionally low snow cover is a norm in the upper part of the slope and on the peak during cold weather. This leads to deep freezing of the soils which are also as usual shallow ones. The prerequisites are thus created under which the drying out of needles at the end of winter becomes very likely. This is the result of low positive temperatures at daytime (<10 °C) and a light frost at night (>−5 °C), which are often considered to be critical for plant activity [46,47].

In the middle and lower parts of the slope, the soils can be separated from the atmosphere by a layer of snow that also causes trees to experience the warm temperatures of the canopy, while the soils are still cool or even frozen [48]. Furthermore, if thick layers of moss isolate the ground during the summer, the soil does not thaw deep enough or thaw too slowly [49,50].

On the slope studied, trees and undergrowth of *P. sibirica* were located more or less singly, while the undergrowth of *A. sibirica* was distributed as clumps confined to more gentle parts of the microrelief. Apparently, this distribution of *A. sibirica* is due to the species tending to show a loci where more moisture accumulates during the growing season because the *A. sibirica* is less drought-resistant than *P. sibirica*.

However, some authors note that colder soil temperatures under woody vegetation, compared e.g., with grass or shrubs, are a common phenomenon for any forest type [51,52]. This observation would indicate a positive effect of more sparsely located trees on the growth near the timberline, while clumping of forest in densely standing trees would be unfavorable for root temperatures. After the classification by Harsch and Bader [26] *P. sibirica* trees form a diffuse treeline form, whereas an abrupt treeline form or island treeline form are rather characteristic of *A. sibirica*.

5. Conclusions

To conclude, under conditions of possible climate warming in the mountains of Southern Siberia, *P. sibirica* will have an advantage in colonizing the zone above the timberline. Some features of the species, both physiological and biochemical, will ensure its wider distribution; these features include a higher content of carotenoids, characteristic of light-coniferous photophilous plants, and a lower rate of photosynthesis at low temperatures. Morphologically, the surface root system is located in the upper, more warmed up soil layers, which allows the species to have more water available during this period and, despite the high level of insolation, to maintain the optimal temperature of the needles to avoid desiccation. As a result, *P. sibirica* may be a more likely winner than *A. sibirica* in driving up the mountain timberline under climate warming in the area.

Author Contributions: Conceptualization, N.P. and V.G.; methodology, N.P.; investigation, N.P. and I.B.; data curation, N.P., I.B., and R.S.; writing—original draft preparation, N.P.; writing—review and editing, N.P., R.S., and V.G.; visualization, R.S. and V.G.; project administration, V.G.; funding acquisition, V.G. All authors have read and agreed to the published version of the manuscript.

Funding: This research was funded by Russian Foundation for Basic Research, Government of Krasnoyarsk Territory, Krasnoyarsk Regional Fund of Science, to the research project: "Prognosis of region-specific responses of Siberian mountain forests to global environmental changes and of the landscape development trajectories for mitigation of environmental risks and an effective long-term planning in various economic sectors", grant number 18-45-240001, and by the Russian Foundation for Basic Research to the research project: "Lateholocene dynamics of Asia boreal forests at the background of changing geochemistry and climatic conditions", grant number 19-05-00091.

Acknowledgments: We thank two anonymous reviewers for helpful comments that helped to improve the manuscript.

Conflicts of Interest: The authors declare no conflict of interest.

References

1. Takahashi, K. Virtual issue: Alpine and subalpine plant communities: Importance of plant growth, reproduction and community assemblage processes for changing environments. *J. Plant Res.* **2018**, *131*, 891–894. [CrossRef]

2. Welling, A.; Palva, E.T. Molecular control of cold acclimation in trees. *Physiol. Plant* **2006**, *127*, 167–181. [CrossRef]

3. Richardson, A.D.; Hufkens, K.; Milliman, T.; Aubrecht, D.M.; Furze, M.E.; Seyednasrollah, B.; Krassovski, M.B.; Latimer, J.M.; Nettles, W.R.; Heiderman, R.R.; et al. Ecosystem warming extends vegetation activity but heightens vulnerability to cold temperatures. *Nature* **2018**, *560*, 368. [CrossRef]

4. Peng, G.; Wu, C.; Xu, X.; Yang, D. The age-related changes of leaf structure and biochemistry in juvenile and mature subalpine fir trees (*Abies faxoniana* Rehder & EH Wilson.) along an altitudinal gradient. *Pol. J. Ecol.* **2012**, *60*, 311–321.

5. Miron, M.S.; Cristea, V.; Sumalan, R.L. Physiological responses of European silver sir (*Abies alba* Mill.) seedlings to drought and overheating induced stress conditions. *J. Hortic.* **2018**, *22*, 115–120.

6. Poulos, H.M.; Berlyn, G.P.; Mills, S.A. Differential stress tolerance of four pines (*Pinaceae*) across the elevation gradient of the San Bernardino Mountains, Southern California, USA. *J. Torrey Bot. Soc.* **2012**, *139*, 96–109. [CrossRef]

7. Jordan, R.K. Comparative Ecophysiology of Two High-Elevation Southern Appalachian Conifers: The Importance of the Winter Season. Ph.D. Thesis, Appalachian State University, Boone, NC, USA, 2018.

8. Kopačková, V.; Lhotáková, Z.; Oulehle, F.; Albrechtová, J. Assessing forest health via linking the geochemical properties of a soil profile with the biochemical parameters of vegetation. *Int. J. Environ.* **2015**, *12*, 1987–2002. [CrossRef]

9. Sobchak, R.O.; Zotikova, A.P. Impact of high altitude conditions on anatomical and physiological characteristics of Siberian pine needles. *Vestnik Tomskogo Gosudarstvennogo Universiteta* **2009**, *326*, 200–202. (In Russian)

10. Bender, O.G.; Zotikova, A.P.; Velisevich, S.N. Water relation features and pigment complex state of Pinus sibirica Du Tour needles in the north-eastern Altai Mountains. *Vestnik Tomskogo Gosudarstvennogo Universiteta Biologia* **2009**, *3*, 63–72. (In Russian)

11. Titova, M.S. Seasonal dynamics of the pigments availability in needles of Siberian pine (Pinus Sibirica) and Korean pine (Pinus Koraiensis). *Vestnik Krasnoyarskogo Gosudarstvennogo Agrarnogo Universiteta* **2010**, *8*, 77–81. (In Russian)

12. Varlam, I.I.; Rusak, S.N.; Kazartseva, K.V. Seasonal changes of the Pinus sibirica pigment structure in the conditions of urboecosystems of northern territories (on the example of Surgut). *Ekologia Urbanisirovannyh Territorij* **2019**, *1*, 82–86. (In Russian)

13. Silkina, O.V. Complex Evaluation of Ecological and Physiological Parameters of Needles in *Abies sibirica* and *Picea abies* in the Process of Vegetation and Its Phytoproduction Activity. Ph.D. Thesis, University of Kazan, Kazan, Russia, 2006. (In Russian).

14. Golovko, T.K.; Yatsko, Y.N.; Dymova, O.V. Seasonal changes in the state of the photosynthetic apparatus in three boreal conifers in middle taiga of European North-East. *Khvoinye Borealnoy Zony* **2013**, *30*, 73–78. (In Russian)

15. IUSS Working Group WRB. *World Reference Base for Soil Resources 2014, Update 2015 International Soil Classification System for Naming Soils and Creating Legends for Soil Maps*; World Soil Resources Reports No. 106; FAO: Rome, Italy, 2015; Available online: http://www.fao.org/3/i3794en/I3794en.pdf (accessed on 12 November 2019).

16. Wintermans, J.E.G.; De Mots, A. Spectrophotometric Characteristics of chlorophyll a and b and their phaeophytins in ethanol. *Biochimica et Biophysica Acta* **1965**, *109*, 448–453. [CrossRef]

17. Maxwell, K.; Johnson, G.N. Chlorophyll fluorescence—A practical guide. *J. Exp. Bot.* **2000**, *51*, 659–668. [CrossRef]

18. Kalaji, H.M.; Schansker, G.; Ladle, R.J.; Goltsev, V.; Bosa, K.; Allakhverdiev, S.I.; Brestic, M.; Bussotti, F.; Calatayud, A.; Dąbrowski, P.; et al. Frequently asked questions about in vivo chlorophyll fluorescence: Practical issues. *Photosynth. Res.* **2014**, *122*, 121–158. [CrossRef]

19. Damm, A.; Guanter, L.; Paul-Limoges, E.; van der Tol, C.; Hueni, A.; Buchmann, N.; Eugster, W.; Ammann, C.; Schaepman, M.E. Far-red sun-induced chlorophyll fluorescence shows ecosystem-specific relationships to gross primary production: An assessment based on observational and modeling approaches. *Remote. Sens. Environ.* **2015**, *166*, 91–105. [CrossRef]

20. Meroni, M.; Rossini, M.; Guanter, L.; Alonso, L.; Rascher, U.; Colombo, R.; Moreno, J. Remote sensing of solar-induced chlorophyll fluorescence: Review of methods and applications. *Remote Sens. Environ.* **2009**, *113*, 2037–2051. [CrossRef]

21. Qiu, N.; Zhou, F.; Gu, Z.; Jia, S.; Wang, X. Photosynthetic functions and chlorophyll fast fluorescence characteristics of five Pinus species. *Chin. J. Appl. Ecol.* **2012**, *23*, 1181–1187.

22. Schreiber, U.; Schliwa, U.; Bilger, W. Continuous recording of photochemical and nonphotochemical chlorophyll fluorescence quenching with a new type of modulation fluorometer. *Photosynth. Res.* **1986**, *10*, 51–62. [CrossRef]

23. González-Rodríguez, A.; Brito, P.; Lorenzo, J.; Jiménez, M. Photosynthetic Performance in Pinus canariensis at Semiarid Treeline: Phenotype Variability to Cope with Stressful Environment. *Forests* **2019**, *10*, 845.

24. Holtmeier, F.-K. Die Höhengrenze der Gebirgswälder. In *Arbeiten Aus Dem Institut für Landschaftsökologie 8*; Westfälische Wilhelms-Universität: Münster, Germany, 2000.

25. Holtmeier, F.-K. *Mountain Timberlines: Ecology, Patchiness, and Dynamics (Vol. 36)*; Springer Science & Business Media: New York, NY, USA, 2009; p. 436.

26. Harsch, M.; Bader, M. Treeline form—A potential key to understanding treeline dynamics. *Glob. Ecol. Biogeogr.* **2011**, *20*, 582–596. [CrossRef]

27. Anderson, J.M.; Osmond, C.B. Shade-sun responses: Compromises between acclimation and photoinhibition. In *Photoinhibition*; Kyle, D.J., Osmond, C.B., Arntzen, C.J., Eds.; Elsevier: New York, NY, USA, 1987; pp. 1–37.

28. Bailey, S.; Walters, R.G.; Jansson, S.; Horton, P. Acclimation of Arabidopsis thaliana to the light environment: The existence of separate low light and high light responses. *Planta* **2001**, *213*, 794–801. [CrossRef]

29. Walters, R.G. Towards an understanding of photosynthetic acclimation. *J. Exp. Bot.* **2005**, *56*, 435–447. [CrossRef]

30. Huner, N.P.A.; Öquist, G.; Hurry, V.M.; Krol, M.; Falk, S.; Griffith, M. Photosynthesis, photoinhibition and low temperature acclimation in cold tolerant plants. *Photosynth. Res.* **1993**, *37*, 19–39. [CrossRef]

31. Huner, N.P.A.; Öquist, G.; Sarhan, F. Energy balance and acclimation to light and cold. *Trends Plant Sci.* **1998**, *3*, 224–230. [CrossRef]

32. Rossi, S.; Deslauriers, A.; Anfodillo, T.; Carraro, V. Evidence of threshold temperatures for xylogenesis in conifers at high altitudes. *Oecologia* **2007**, *152*, 1–12. [CrossRef]

33. Slatyer, R.O. Altitudinal variation in the photosynthetic characteristics of snow gum, Eucalyptus pauciflora Sieb. ex Spreng. VII. Relationship between gradients of field temperature and photosynthetic temperature optima in the Snowy Mountains area. *Aust. J. Bot.* **1978**, *26*, 111–121. [CrossRef]

34. Larcher, W. Frostresistenz. In *Handbuch der Pflanzenkrankheiten*; Rademacher, B., Ed.; Parey-Verlag: Berlin/Heidelberg, Germany, 1985; Volume 1, pp. 177–259.

35. Larcher, W. Winter stress in high mountains. In *Establishment and Tending of Subalpine Forest: Research and Management*; Turner, H., Tranquillini, W., Eds.; Swiss Federal Institute of Forestry Research: Birmensdorf, Switzerland, 1985; Volume 270, pp. 11–19.

36. Larcher, W. *Physiological Plant Ecology*, 4th ed.; Springer: Berlin/Heidelberg, Germany; New York, NY, USA, 2003; p. 513.

37. Sakai, A.; Larcher, W. *Frost Survival of Plants: Responses and Adaptation. Ecological Studies 62*; Springer: Berlin/Heidelberg, Germany; New York, NY, USA, 1987; p. 321.

38. Kullman, L. Tree line population monitoring of Pinus sylvestris in the Swedish Scandes, 1973–2005: Implications for treelinetheory and climatechangeecology. *J. Ecol.* **2007**, *95*, 41–52. [CrossRef]

39. Rössler, O.; Bräuning, A.; Löffler, J. Dynamics and driving forces of treeline fluctuation and regeneration in central Norway during the past decades. *Erdkunde* **2008**, *62*, 117–128. [CrossRef]

40. Harsch, M.A.; Hulme, P.E.; McGlone, M.S.; Duncan, R.P. Are treelines advancing? A global meta-analysis of treeline response to climate warming. *Ecol. Lett.* **2009**, *12*, 1040–1049. [CrossRef]

41. Tranquillini, W. *Physiological Ecology of the Alpine Timberline*; Springer: New York, NY, USA, 1979; p. 137.

42. Pereg, D.; Payette, S. Development of black spruce growth forms at treeline. *Plant Ecol.* **1998**, *138*, 137–147. [CrossRef]

43. Rickebusch, S.; Lischke, H.; Bugmann, H.; Guisan, A.; Zimmermann, N.E. Understanding the low-temperature limitations to forest growth through calibration of a forest dynamics model with tree-ring data. *Forest Ecol. Manag.* **2007**, *246*, 251–263. [CrossRef]

44. Germino, M.J.; Smith, W.K. Sky exposure, crown architecture, and low-temperature photoinhibition in conifer seedlings at alpine treeline. *Plant Cell Environ.* **1999**, *22*, 407–415. [CrossRef]

45. Körner, C. *Alpine Plant Life: Functional Plant Ecology of High Mountain Ecosystems; with 47 Tables*; Springer Science & Business Media: Berlin/Heidelberg, Germany; New York, NY, USA, 2003; p. 359.

46. Körner, C. Winter crop growth at low temperature may hold the answer for alpine treeline formation. *Plant Ecol. Divers* **2008**, *1*, 3–11. [CrossRef]

47. Sutinen, M.L.; Ritari, A.; Holappa, T.; Kujala, K. Seasonal changes in soil temperature and in the frost hardiness of Scots pine (*Pinus sylvestris*) roots under subarctic conditions. *Can. J. For. Res.* **1998**, *28*, 946–950. [CrossRef]

48. Körner, C. Treelines will be understood once the functional difference between a tree and a shrub is. *Ambio* **2012**, *41*, 197–206. [CrossRef]

49. Ballard, T.M. Subalpine soil temperature regimes in southwestern Brithish Columbia. *Arct. Alp. Res.* **1972**, *4*, 139–146. [CrossRef]

50. Scott, P.A.; Bentley, C.V.; Fayle, D.C.F.; Hansell, R.I.C. Crown forms and shoot elongation of white spruce at the treeline, Churchill, Manitoba, Canada. *Arct. Alp. Res.* **1987**, *19*, 175–186. [CrossRef]

51. Körner, C.; Paulsen, J.; Pelaez-Riedl, S. A bioclimatic characterisation of Europe's alpine areas. In *Alpine Biodiversity in Europe*; Nagy, L., Grabherr, G., Körner, C., Thompson, D.B.A., Eds.; Springer: Berlin/Heidelberg, Germany; New York, NY, USA, 2003; Volume 167, pp. 13–28.

52. Körner, C.; Paulsen, J. A world-wide study of high altitude treeline temperatures. *J. Biogeogr.* **2004**, *31*, 713–732. [CrossRef]

Article

Arctic Greening Caused by Warming Contributes to Compositional Changes of Mycobiota at the Polar Urals

Anton G. Shiryaev [1,*], Pavel A. Moiseev [2], Ursula Peintner [3], Nadezhda M. Devi [2], Vladimir V. Kukarskih [2] and Vladimir V. Elsakov [4]

[1] Department of Vegetation and Mycobiota Biodiversity, Institute of Plant and Animal Ecology, Urals Branch of the Russian Academy of Sciences, 8 march str., 202, 620144 Ekaterinburg, Russia
[2] Department of Dendrochronology, Institute of Plant and Animal Ecology, Urals Branch of the Russian Academy of Sciences, 8 march str., 202, 620144 Ekaterinburg, Russia; moiseev@ipae.uran.ru (P.A.M.); nadya@ipae.uran.ru (N.M.D.); voloduke@mail.ru (V.V.K.)
[3] Institute of Microbiology, Innsbruck University, Technikerstr. 25, 6020 Innsbruck, Austria; ursula.peintner@uibk.ac.at
[4] Department of Northern Flora and Vegetation, Institute of Biology, Komi Scientific Centre of Urals Branch of the Russian Academy of Sciences, Kommunisticheskaya str., 28, 167982 Syktyvkar, Russia; elsakov@ib.komisc.ru
* Correspondence: anton.g.shiryaev@gmail.com

Received: 19 October 2019; Accepted: 4 December 2019; Published: 6 December 2019

Abstract: The long-term influence of climate change on spatio-temporal dynamics of the Polar mycobiota was analyzed on the eastern macro slope of the Polar Urals (Sob River valley and Mountain Slantsevaya) over a period of 60 years. The anthropogenic impact is minimal in the study area. Effects of environmental warming were addressed as changes in treeline and forest communities (greening of the vegetation). With warming, permafrost is beginning to thaw, and as it thaws, it decomposes. Therefore, we also included depth of soil thawing and litter decomposition in our study. Particular attention was paid to the reaction of aphyllophoroid fungal communities concerning these factors. Our results provide evidence for drastic changes in the mycobiota due to global warming. Fungal community composition followed changes of the vegetation, which was transforming from forest-tundra to northern boreal type forests during the last 60 years. Key fungal groups of the ongoing borealization and important indicator species are discussed. Increased economic activity in the area may lead to deforestation, destruction of swamps, and meadows. However, this special environment provides important services such as carbon sequestration, soil formation, protecting against flood risks, and filtering of air. In this regard, we propose to include the studied territory in the Polarnouralsky Natural Park.

Keywords: treeline; climate change; fungal ecology; diversity; monitoring; NDVI; permafrost; remote sensing data

---

1. Introduction

Arctic greening is one of the world's most significant large-scale environmental responses to global climate change [1]. Numerous studies have reported rising average temperatures over the past century, with the most noticeable and rapid changes occurring at high latitudes [2]. Here, temperatures have risen by 0.6 °C per decade during the past 30 years, which is twice as fast as in average in the world [1,3,4]. Earth-derived satellite imagery indicates a global increase of the normalized differential vegetation index (NDVI, average value), over the past decades [5]. This value reflects the degree of 'greening' in high-latitudes areas, due to an increase of above ground phytomass production during the growing season [6]. Satellite imagery data from high latitudes of the Canadian Arctic, northern Alaska, and western Greenland, showed an increase in NDVI by up to 15% between 1982 and 2008 [6,7]. A temperature increase in the Arctic may cause shifts of vegetation boundaries by several kilometres northward in lowlands, and by hundreds of meters difference in altitude in the mountains, thus leading to significant changes in species composition and biomass production processes [2,8,9].

Changes in vegetation cover productivity and treelines are also obvious in the Russian Arctic [10–12]. The increase of NDVI associated with climate warming was 10–15% during last 15 years [12,13]. At the same time, the actual monitoring of the effects of climate change on the Arctic biota and on Arctic ecosystems cannot be limited to remote methods only: they give only an idea of large-scale vegetation processes (in spatial terms). After satellite verification, more detailed analyses of dynamic processes in Arctic ecosystems are required. Changes in vegetation are accounting for phytomass reserves and products. In order to obtain a more detailed picture, measurements of the current soil respiration and soil moisture as well as carbon transpiration data of the vegetation, differentiated by plant groups, are required. They should be complemented by an assessment of projective cover, measurements of the height of phytocenotic horizons (for calculating the specific density of phytomass), and other important data. The latest synthesis of data on the productivity of Arctic ecosystems, based on field measurements in the Russian Arctic, was carried out 20–30 years ago [14]. In that period, it was already clear that plant productivity of the Arctic vegetation started to increase. This was due to the advancement of the forest to the north, rising of tundra shrubs from the valley bottom to the upland, an increase in the proportion of sedges and cereals in the cover, a decrease in the coverage of lichens, etc. [15]. At the end of the last century and the beginning of the present, an increasingly deep thawing of Arctic soils was also observed, leading to an increase in the activity of soil and litter microbiota [7,16].

The reaction of different groups of organisms of the Arctic Flora and Fauna to the ongoing greening and treeline dynamics caused by climate change have been studied for a long time [12,15], but the ecologically highly significant fungi have rarely been included in such studies [17,18]. The dynamics of fungal diversity could be very indicative for climate transformation (e.g., indirectly through trophic or symbiontic interactions, etc.) [19–22]. The Fungi are a crucial part of the heterotrophic block of ecosystems. Also, in the Arctic, the functioning of fungi (especially of mycelia activity) strongly depends on the length of the vegetation period, the plant activity and the amount of phytomass.

The relationship between warming in the Arctic and fungal dynamics has been addressed for various groups of macromycetes [22–25]. However, the relationship between fungal diversity and substrate availability has been included in a few works only [26–30], thus not excluding the possibility that fungi might rather react to changes in the structure of the Arctic vegetation, than to the actual increase in temperature [31–33].

In the vast expanses of the Russian Arctic, the study of biodiversity and ecology of macromycetes has been conducted for almost a century [34]. However, the majority of such studies were short-term, lasting only one to two years. It is quite difficult to find long term mycological research areas, where also other parameters such as vegetation dynamics, soil properties, and climate data, were studied. This means that weather stations should be located in the same area, and the dynamics of other elements of the biota should be studied. In Russia, such a territory is the eastern macro slope of the Polar Urals. The Arctic Research Station of the Institute of Plant and Animal Ecology UrB RAS was established in the town Labytnangi (Yamal-Nenets Administrative District) in 1954. On the basis of this research station, ecological studies concerning various components of the high-latitude biota have been conducted for 60 years. Over the past 20 years, it has become extremely important to compare the results obtained in 1950–1960 with the current state of the Arctic biota. In particular, the treeline dynamics was studied in the region of Mount Rai-Iz, showing that since 1900 the timberline has risen by 60 m (relative to sea level). Moreover, the crown density and the height of the stand increased significantly [35]. Earlier, larch stands prevailed in the forest, but now spruce and other boreal plants are increasing [36]. Changes in the grass-shrub layer have also been studied in detail [37], and a connection of these changes with climate warming has been established [38]. Litter decomposition rate, soil thaw depth, and changes in floristic richness are also being investigated in the region [39–41]. Over the past 60 years, the structure of the plant cover has transformed from forest-tundra to north boreal forests in the area of town Labytnangi and Mountains Slantsevaya and Rai-Iz, and this could be associated to climate warming and increased rainfall, especially in winter [36]. Thus, various elements of the biota and bioclimatic parameters have changed very much in the area, posing the question whether or not the mycobiota reacts to such changes.

Thus, the aim of this study was to evaluate the response of the mycobiota to ongoing changes in climate and vegetation cover in the Polar Urals. We addressed the following questions: (1) How big is the impact of climate change occurring during the last 60 years on treeline position, forest areas, crown density, aboveground phytomass productivity, and permafrost thaw depth? (2) Was the associated mycobiota changing as well? If yes, which species and ecological or biogeographical groups of macromycetes were most affected by warming? Can selected groups of fungal species act as indicators of warming in arctic areas?

2. Materials and Methods

2.1. Study Area and Environmental Conditions

The study area is located where the Sob River cuts through the eastern slope of the Polar Urals at the territory of the Yamal-Nenets Autonomous District. The river flows from northwest to south-east, with the valley framed from west and south-west of Mountain Rai-Iz, and from north and north-east of Mountain Slantsevaya. The area is located 30 km north of the Arctic Circle (N 66°54′; E 65°44′) and the border between Europe and Asia, 60 km east of Vorkuta town (Komi Republic), 50 km west of Salekhard town (capital of the Yamal-Nenets Autonomous District) and 30 km west of Labytnangi town (Figure 1).

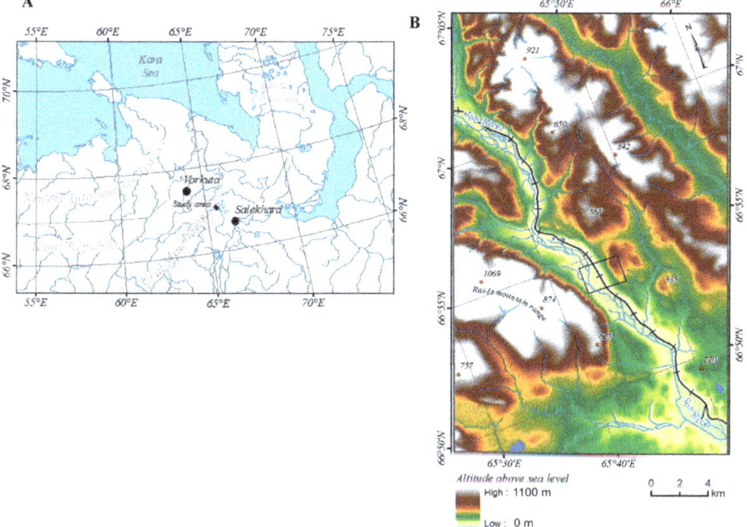

Figure 1. Map of the area under investigation in the Polar Urals (**A**). Detailed map of the study site, Sob River valley with Mountain Slantsevaya and Mountain Rai-Iz. The region addressed for NDVI studies marked as black rectangle (**B**). The black line along the Sob River represents the Vorkuta-Labytnangi railway.

The digital relief model of the study area was taken from ArcticDEM (http://www.pgc.umn.edu/data/arcticdem/). This model is used for topographic correction of figures of changes in crown density, NDVI, and types of vegetation [42].

The nearest weather station with a continuous series of meteorological observations is Salekhard, located 50 km south-east from the area under investigation. The average annual temperature is negative (−6.4 °C), as indicative for the harsh thermal regime in the region (Figure 2). Mean annual precipitation is 415 mm, solid precipitation reaches 45% of the total amount. The region is characterized by a complex wind regime with a predominance of the westerly winds. The average wind speed is 8.5–8.8 m/s in winter and 6.5–7.0 m/s in summer.

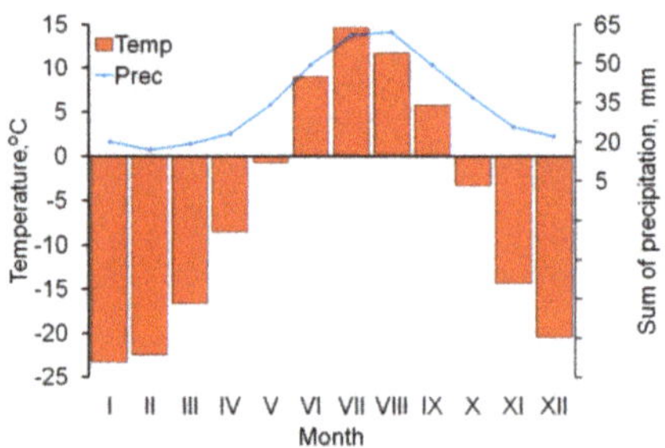

Figure 2. Climagram for Salekhard weather station (1892–2015). The blue line represents the long-term average sum of precipitation, red columns average monthly temperatures.

Anomalies of the 10-day course of surface air temperature and of the amount of precipitation were calculated according to the generalized data of VNIIGMI-WCD (meteo.ru) for the 65-year observation period (1960–2015) at Salekhard weather station.

Currently, the anthropogenic impact on nature in the study area is minimal. As a first important step for exploitation of the area, the Vorkuta–Labytnangi railway was built in the 1940s. However, there are no permanent settlements in the area of investigation, with the exception of a single house near the railway. Only two people permanently live in this valley of 6 km^2 at the moment (population density: 300 ha/individual). The territory is occasionally used during winter in reindeer husbandry. The pasture load on phytocenoses is low (64.1 ha/individual) compared to the average load for the Yamal-Nenets Autonomous District (59.0 ha/individual), or to adjacent farms of the Vorkuta region in the Komi Republic (46.4 ha/individual).

2.2. Litter Decomposition and Soil Defrost Estimation

The litter decomposition rate was estimated by weight loss of standard samples of pure cellulose (laboratory filter paper photo soft extra, Borregaard A/S Sarpsborg, Norway) during a fixed period of time. Despite the simplicity of the method, it is quite informative, since it allows direct measurements of velocity at specific points in space and is currently widely used in soil biology [43], with different modifications (for example, bait lamina test, litter bag test, cotton strip assay). Cellulose samples were placed in nylon mesh bags of 5×10 cm with a mesh size of 0.5 mm. They were deposited between the litter layer and the humus-accumulating layer (soil). Two series of experiments (1978–1980 and 2013–2015), each lasting three years, were carried. For both series, nylon mesh bags were deposited at the same places in the middle of August. Every year in August, mesh bags were replaced. Mesh bags were placed every 50 cm along a line (at the same height above the sea level) for 25 m, i.e., 50 packets in each line. Each year, three lines were laid along an altitudinal gradient: line one was situated on the treeline of 1978 (230 m a.s.l.); line two was situated in the forest on the slope (180 m a.s.l.); line three was established in the closed forest of the valley bottom (110 m a.s.l.). Thus, during one time series, the data of nine lines were analyzed, amounting for 450 mesh bags. These results in a total of 900 mesh bags for two series.

A study of the soil thaw depth was carried out at the same place (altitudes, transects, and during the same years: 1978–1980 and 2013–2015) as the study of litter decomposition, on the slope of the south-western exposure of Mountain Slantsevaya. Briefly, depth of the soil thaw was studied at 230 m, 180 m, and 110 m a.s.l. The measurements were carried out using a pole of 4 m length. Measurements were carried out on each altitudinal line in one-meter increments for a total of 25 m length. Additional excavations were carried out with a shovel if necessary. As in the case of mesh bags, the depth of soil thaw was studied in mid-August, when the soil melting was most pronounced during the summer period.

2.3. Studies of NDVI and Vegetation Cover

Changes in the vegetation cover of the region were estimated based on the normalized differencevegetation index (NDVI) extrapolated for summer from 1988–2017. To calculate the values, the albedo values spectral reflectance (SRefl) were used. To analyze changes, we selected summer images with similar shooting dates: Landsat TM4 (07.15.1988), Landsat TM5 (07.17.1994), Landsat 8 OLI (07.21.2013 and 07.07.2017).

The total change in crown density of tree species and partial shrub layer (the thickness of the snow cover reaches 0.5–1.0 m) was determined from winter images of Landsat TM5 (04.25.1987) and Landsat 8 OLI (04.23.2018).

The reliability to detect the correct index of crown density increases when using winter images, because manifold forms of the Earth's surface microrelief are masked by snow cover, foliage is lacking and lichen, moss, grass, shrub layers are covered, leading to a greater contrast of the studied components [44]. This effect was previously shown based on an analysis of changes in the density of stands of slope forests of Subpolar Urals [45]. Due to the low participation of tree species in the formation of the total reflection spectra, the open forest (with a canopy of less than 15%) was not considered. Of greatest interest are the changes in the class of woodlands (the canopy density of the stands is 15–30%). To characterize the dominant complexes of the vegetation cover, a supervised classification has been performed [46]. In the study area, mountain-tundra, forest-tundra, and north boreal communities are represented (Figure 3). Larch forests occupy the largest areas on the slopes of Mountains Slantsevaya and Rai-Iz while spruce forests occupy the Sob River valley. Above the forest boundary, bush communities are widespread with the predominance of *Betula nana* and *Salix spp.*

To assess the treeline dynamics (from 1956 to 2016), the following methodology is proposed. Earlier research on Mountain Konzhakovsky Kamen (Northern Urals, Russia) revealed that in 2005 the treeline of open forests (crown density is about 20%) was on average 18 m higher compared to the forest border displayed on the topographic map (M 1:25000) created in 1957 [47]. We also compared the treeline contours from a number of modern topographic maps (M: 1:25000) with the maps of woody vegetation that we created for individual peaks of the Southern and Northern Urals based on modern high-resolution satellite images (less than 1 m^2/pixel). We found that the treeline contours of forest communities displayed on topographic maps coincide well with the boundaries of stands with a crown density of 35–40% [48]. Based on this, we considered the treeline contours displayed in historical large-scale topographic maps of 1950 1960 and 1980–1990 as reliable for reconstruction of the upper border of stands with a crown density of 35–40%.

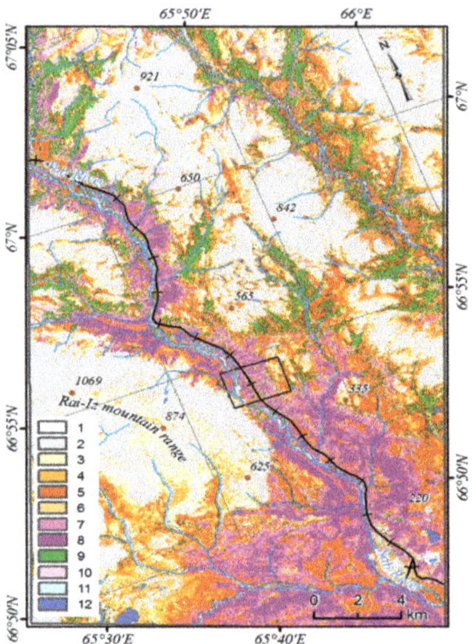

Figure 3. Map showing different types of vegetation of the Sob River valley area (July 2019).Legend:
1—sandy beaches; 2—stones, settlement lands; 3—shrub-lichen tundra; 4—shrub tundra; 5—shrubs
(different in height, but they can be combined to complexes with 3); 6—larch forests and woodlands;
7—spruce forests (dark coniferous forests) and mixed forests; 8—closed valley spruce forests; 9—willows
and alders; 10—meadows; 11—swamp complexes and boggy tundra; 12—water surfaces.

A quantitative assessment of changes of forest sites on the slopes of mountain ranges was carried
out using the SAS. Planet 190707 program (http://www.sasgis.org/). Using the "Add Polygon" tool
from the tab of the "Tags" main menu, the contours of all sections were selected based on images in
layers with modern (~2016) submeter resolution satellite imagery and topographic maps (~1956) of the
State GisCenter (M: 1: 25000), where the density of stands was higher than 35–40% (closed forests).
Information about the area and the length of the borders of the allocated plots in 1956 and 2016 was
taken from the "Label Information" tab in the "Label Management" block. The information obtained
was used to compare changes in the total area of closed forest plots in the research area from the end of
the 1950s to the present.

2.4. Data Sampling and Processing for Dendrochronological Studies

All trial plots were installed in August 2015 and 2019. In order to study the current state and to
reconstruct the phytomass reserves for different periods, a full recount of trees (n = 263) was performed
within six model plots of 20 × 20 m, located at 230 m a.s.l. in the forest. The trees were Siberian larch
(*Larix sibirica* Ledeb.) in association with Siberian spruce (*Picea obovata* Ledeb.). The size of the total
area of investigation was 1595 m^2. The dendrometric parameters (height, base diameter and crown
projection) of each tree were measured with roulette tape (accuracy 0.5 cm), telescopic ruler (accuracy 1
cm) and digital rangefinder (accuracy 0.3 m) (Table 1).

Table 1. Mean and standard errors of the main dendrometric parameters within the trial plots

	Dendrometric Parameters (n = 263)		
	Stem Base Diameter (cm)	**Stem Height (m)**	**Age (Years)**
Closed forest	19.92 ± 0.87	10.09 ± 0.36	125 ± 4

Each tree was cored to determine its age and tree ring width. Trees with a diameter > 2–3 cm were cored, using 400/5.15-mm increment borer (Haglöf, Sweden) at a height of up to 10 cm from the ground surface. Individuals with a diameter < 2 cm were cut at ground level. In total, 263 individuals were considered. Standing dead trees (< 3% of all trees) were also included to evaluate past stand dynamics. Further processing of the wood cores and discs was carried out according to standard dendrochronological techniques [49,50]. Tree-ring width was measured using a binocular microscope and a mechanic measure system with an accuracy of 0.01 mm (LINTAB, F. Rinn SA, Heidelberg, Germany). All samples were first cross-dated visually in TSAP programv4 (http://www.rinntech.com/) and then checked using the software package COFECHA [51]. Using tree-ring series and proportional method of geometric diameter determination [52], the historical stem diameters of all examined trees were reconstructed for each 20 years of the 20th century.

The tree biomass estimates were conducted at the border of the trial-plots. So-called model trees (n=33) were felled and sectioned. The stems' phytomass was determined with the hand scales with accuracy 50 grams. The weights of the leafless branches, needles, and generative organs were determined separately with a digital weight (accuracy 0.01 g). All subsamples were then oven-dried at 106 °C to stable weight. Drying time ranged from several hours to days. Then, the masses of all the fractions were summarized and allometric functions between the absolutely dry tree biomasses and the tree diameters at base height were conducted. Using the obtained dependence, the annual biomass accumulation of each tree was calculated. The calculations were made according to the following formula

$$D_n = (R_n/R_{final}) \times D_{final}$$

where D_n is the computed tree diameter in a certain year; R_n is distance from the tree core to a certain annual ring; R_{final} is the current tree radius; and D_{final} is the current tree diameter.

Taking the fractional structure of the model trees into account, the reserves of individual phytomass fractions were reconstructed as well. Accounting the size of the trial plots, these values were converted into stand biomass values (t/ha). The change in phytomass of the two main forest trees was evaluated: *Larix sibirica* and *Picea obovata*, for 120 years, starting from 1900, dividing this period into six time intervals of 20 years, each (1900–1919, 1920–1939, 1940–1959, 1960–1979, 1980–1999, 2000–2019).

2.5. Fungal Sampling

Aphyllophoroid fungi are used as model group of macromycetes. Data from 60 years of mycological monitoring in the Sob river valley and on the slopes of Mountains Slantsevaya and Rai-Iz are included in the study [53–56]. Mycologists have been working here since 1961. The history of work can be divided into three periods, each lasting 20 years:

(1) 1960–1979: mycologists from the Institute of Plant and Animal Ecology of the Ural Branch of the Russian Academy of Sciences (Sverdlovsk), V.L. Komarov Botanical Institute (Leningrad), as well as from the Institute of Zoology and Botany, Estonia (Tartu);

(2) 1980–1999: numerous studies were conducted by employees of various scientific organizations in Russia and Europe. In 1996, the fifth international symposium on Arctic and Alpine mycology was held [54], where mycologists from Finland, Denmark, Norway, Poland, Austria, and Italy participated. By the end of the 20th century, the eastern slope of the Polar Urals was the most studied region of the Russian Arctic in terms of the identified species richness of aphyllophoroid fungi [34].

(3) 2000–2019: work focussing on aphyllophoroid fungi is carried out by employees of the Institute of Plant and Animal Ecology UrB RAS (Ekaterinburg), Finnish Environment Institute (Helsinki) and the Institute of Biology of the Komi Scientific Centre UrB RAS (Syktyvkar).

These three stages coincide with the periods for studying the dynamics of woody phytomass. A list of aphyllophoroids fungi analyzed in this work is given in [55] and in (Table S1 and Figure S1). Undoubtedly, one must not underestimate the insufficient exploration of the territory at different historical stages. The more specialists work, the more high-quality information is accumulated. The list of fungi includes information each species concerning:

(1) Fruitbody types of aphyllophoroid fungi include three species-rich groups (corticioids, poroids, and clavarioids), and some smaller groups, e.g., cantharelloids, thelephoroids, and stipitate hericioids. These groups are good indicators of bioclimatic conditions: in the tundra without native woody substrates, clavarioids are dominant, in shrub tundra with many different small twigs, corticioids on *Betula nana* bushes as well as a lot of clavarioids on different grasses and herbs are dominating. In the boreal forests with many big coniferous and deciduous tree trunks, poroids have a big share of the aphyllophoroid species richness [56,57].

(2) Ecological strategies of aphyllophoroids include three basic types. Parasites (on alive trees, shrubs, grasses, and mosses), symbionts (mycorrhiza-forming and basidio-lichens), and saprobs (on deadwood, humus, and litter). Fungi growing on grass, leaves, needles, mixed decaying litter, twigs less than 2 cm in diameter, as well as on wood in its final decay stage 5 (representing individual fibers), or on a separate bark (for example, inside the birch bark) were defined as litter composers. A few fungi are mixotroph (e.g., genus *Tomentella*—mycorrhizal and litter saprobes, *Osteina obducta* (Berk.) Donk—wood and soil saprobs, *Typhula micans* (Pers.) Berthier—parasite and litter saprob, etc.). Therefore, the number of species in each of the time periods does not coincide with the total number of species identified in the corresponding time periods. To get a general picture, the shares were normalized to a common denominator (100%). In general, the ratio of these three groups varies significantly depending on the latitudinal-zonal position of the studied region.

(3) Height groups of fruitbodies are estimated only for clavarioids, cantharelloides, thelephoroids, and stipitate hydnoids, which have clearly distinguishable negative geotropic fruit bodies. They are divided into three groups: group I — fruit bodies up to 3 cm high, group II — up to 8 cm, and group III — more than 8 cm high. Small-fruiting species (I group) prevail in the arctic tundra, while in rich hemiboreal and nemoral forests with tall trunk trees, the proportion of large-fruiting species (III group) is high [58].

(4) Chorological groups were established based on a study of the distribution of aphyllophoroid fungi along the Urals, including the tundra vegetation zone, the Arctic, mountains, nemoral forests, steppe, and temperate deserts [55]. Four main groups are here proposed based on the confluence of fungal species in zonal units (not in the centre of the ecological optimum): Eurybionts — collected in most vegetation zones at the Urals, in forest areas, as well as in treeless areas like tundra, steppes or deserts. Arcto-alpine — growing in treeless alpine areas, as well as in treeless boreal habitats, like bogs and meadows. Boreal—found in the boreal regions, or boreal 'islands' in the nemoral zone. Forests — collected in most of the forest vegetation zones of the Urals. Aphyllophoroid fungal species are also analyzed by groups: northern — Arcto-alpine and eurybiontic (mainly represented in the tundra zone) and boreal-forest and boreal (most common in the taiga forests of the Urals).

Particular emphasis is placed on the discussion of newly appearing large-fruiting species that occurred in the third period of research (which could not have been missed earlier), for example, *Osteina obducta*, *Trametes suaveolens* (L.) Fr., *Cantharellus cibarius* Fr., etc. The collected fruit bodies were deposited in the following mycological collections: Estonian University of Life Sciences, Tartu (TAA); University of Helsinki (H); University of Copenhagen (K); Institute of Plant and Animal Ecology UrB RAS, Ekaterinburg (SVER); V.L. Komarov Botanical Institute, St. Petersburg RAS (LE); Siberian Institute of Plant Physiology SB RAS, Irkutsk (IRK); Institute of Biology Komi SC UrB RAS, Syktyvkar (SYKO); and Institute of North Development Problems SB RAS, Tyumen. Some specimens are deposited in the private collections of A. Shiryaev and H. Kotiranta.

3. Results

3.1. Climate Warming

Surface air temperatures increased by an average of 2–4 °C (Salekhard weather station), resulting in changes in plant growth phases and average soil temperatures. Steady intensive rise in temperatures (up to 6–11 °C from the average for a decade) was pronounced starting during the winter season of the first half of the1980s (Figure 4). The increase in summer temperatures (excess of 10-day anomalies of 5–9 °C from the average) shifted towards the end of the 1980s: the most intense changes occurred in late May, the first half of summer. After a short-term decline in 1995–2000, also the snow-free season exhibited a smooth expansion of periods with increased temperature. After 2010, temperature indicators characteristically increase for all seasons of the year. The average annual and average seasonal (except for the spring period) air temperatures increased by 0.44 °C over 10 years (for spring 0.64 °C for 10 years). Over the past 60 years, the average summer air temperature (June–August) increased by 2.0 °C (to +12.2 °C), and average winter temperatures (November–March) by 1.8 °C (to −18.6 °C) in the area of investigation.

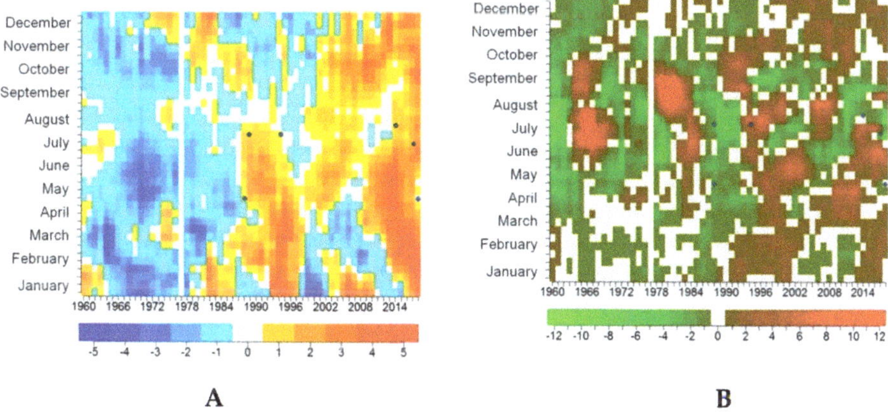

Figure 4. Two-dimensional diagrams of the anomalies of the 10-day course of surface air temperature, °C (**A**) and the amount of precipitation, mm (**B**) for Salekhard weather station for the time period 1960–2018. Dots indicate the sampling dates.

Along with warming, there was an increase in rainfall. For precipitation, the total annual precipitation increased by 10.8 mm over 10 years, but significant redistribution of indicators between seasons was observed: for winter, increase over 10 years was 2.2 mm, for spring 4.7 mm, for summer 2.1 mm, for autumn 1.2 mm. The amount of precipitation during summer and early autumn has short-term phases of increase during 1964–1968 and 1978–1980. Since the early 1990s, a precipitation increased steadily during all seasons, but with a shift to earlier periods of the year (Figure 4). Since 1960, the average annual rainfall in summer and winter increased by 30 mm and 49 mm, respectively. The last decade was the warmest and wettest for the entire period of research.

Due to these climatic changes, the duration of the growing season increased by 6–7 days in the study area. Similar indicators were also found for the slopes of Mountain Chernaya, located at the southern slope of Mountain Rai-Iz [35].

3.2. Changes in Litter Activity and Soil Thaw Depth

Due to climate warming and an increase in rainfall in the study area, the activity of soil and litter microbiota has significantly increased over the past 35 years (from 1978–1980 to 2013–2015). For the first (upper) level (230 m a.s.l.), where the upper border of the forest was located in 1900, the rate of cellulose degradation increased from 4.1% to 19.8%, i.e., almost five times (Table 2). For the middle and lower levels, this parameter nearly duplicated. A similar study was carried in the Polar Urals, in the valley of the Sob River in the late 1960s [39,40]. They found that the decomposition rate of pure cellulose varied in average values of 10–19%, which is in accordance to our results (18.5%). Climate is playing a significant role on litter decomposition. Initial litter decomposition is determined by temperature in cold biomes, and for every 10 °C increase in temperature, a doubling of microbial decomposition is anticipated [59]. Comparing our data with global estimates, it becomes evident that changes in decay rates are extraordinarily high in the Arctic.

Table 2. Parameters of the frequency distribution of cellulose degradation (% per year) in three altitudinal belts on Mountain Slantsevaya in 1978–1980 and 2013–2015

Parameter	Altitude and Years					
	230 m a.s.l.		180 m a.s.l.		110 m a.s.l.	
	1978–1980	2013–2015	1978–1980	2013–2015	1978–1980	2013–2015
Number of samples	150	150	150	150	150	150
Max, %	20.5	57.6	57.9	82.7	67.0	83.4
Mean value, %	4.1	19.8	13.2	24.5	18.5	29.1
Min, %	1.4	5.9	3.5	6.4	4.8	7.7
CV, %	115.6	78.7	74.3	63.0	60.3	51.4

Note: Max—maximal decay rate, Min—minimal rate, CV—coefficient of variation.

The average depth of soil thaw at the first (upper) level was 1.5 m during 1978–1980, and in 2013–2015 it was 3.4 m (Table 3). That means that over 35 years the soil was thawed 2.3 times deeper than at the beginning, reaching an average of 1.9 m at an altitude of 230 m a.s.l. At the second and third levels, the penetration depth became a meter deeper (1.4 times in both cases). Consequently, over the entire altitude profile, the soil began to thaw 1–2 meters deeper. For the late 1970s, a large difference in decay rates of cellulose was observed between three altitudinal belts, which was especially obvious between upper level 'near the upper border of the forest'(soil thaw: 1.3–1.6 m) and 'forest'(soil thaw: 2.3–2.8 m). At present, the soil thaw parameters for all three zones are close (varying in a narrow range of 3.2–4.0 m). It is worth noting that in the 2010s a similar increase in the thaw depth was reported for different types of soils in the vicinity of Labytnangi town (30 km east of Mountain Slantsevaya) [60]. Moreover, soil thaw depth also increased in the tundra of the East European plain, located 50 km west of the study [61]. Permafrost thawing results in increased emissions of all three important greenhouse gases—carbon dioxide, methane, and nitrous oxide [62]. The release of these gases is often linked to increased microbiological activity: enzyme activity—e.g., soil glucosidase increases—dramatically in warming environmental conditions such as thawing. With Arctic permafrost transforming into active soil layers with climate warming, activity of these exoenzymes substantially increase rates of carbon breakdown [63].

Table 3. Mean soil thaw (in meters) in three altitude belts on Mountain Slantsevaya during 1978–1980 and 2013–2015

Parameter	Altitude and Years					
	230 m a.s.l.		180 m a.s.l.		110 m a.s.l.	
	1978–1980	2013–2015	1978–1980	2013–2015	1978–1980	2013–2015
Number of measurements	75	75	75	75	75	75
Max, m	1.9	3.8	2.9	3.9	3.2	>4.0
Mean value, m	1.5	3.4	2.4	3.5	2.7	3.7
Min, m	1.0	2.6	1.9	3.0	2.3	3.4
CV, %	103.2	70.8	71.1	62.4	58.7	44.6

Note: Max—maximal soil thaw depth, Min—minimal depth.

3.3. Vegetation Cover Dynamics

The treeline has risen in response to warming and an increased rainfall, deeper thaw of the soil, and increased activity of the soil-litter microbiota. A comparison of historical data (where the density of stands was higher than 35–40% according to topographic maps of 1956) with our satellite images takenin 2016 revealed that the total area of closed forests in the Sob River valley increased by 30%, i.e., by 15.6 km^2, starting from 52.1 km^2 to 67.7 km^2 (Figure 5). This results in an increase in closed forest area of 0.095 km^2 per 1 km of the upper boundary since 1956.

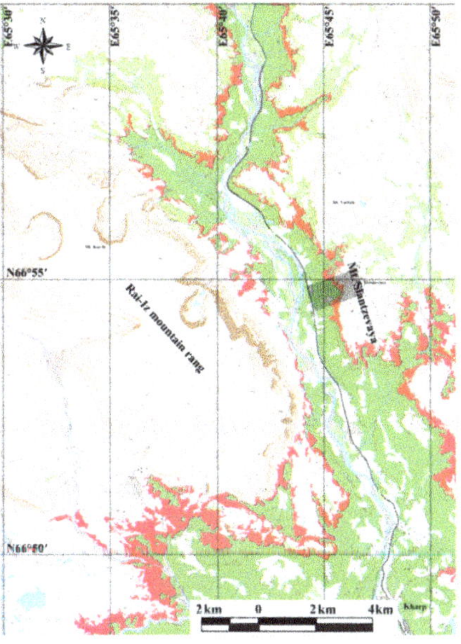

Figure 5. Treeline dynamics from 1956 to 2016 (green color—forest in 1956;red—forest rise to 2016; grey rectangle—study area).

Since 1956, the vertical shift of the treeline of forests with crown density of about 20% was 41 m (from 231 to 257 m a.s.l.) for closed forests 35 m (from 195 to 230 m a.s.l.). The horizontal shift for open forests was 290 m, for closed forests 520 m. These forest dynamics were confirmed comparison of photographs vegetation taken at different times in this area [36,47]. Almost all recent photographs show an increase in the productivity of pre-existing stands, which is expressed in an increase in the size of trees in height and in diameter, density of stands crowns [36]. The young generation of larch, formed during a favourable climatic period, differs markedly in shape and growth rate from older generations, especially the overmature ones, which endured and survived severe climatic conditions for most of their life. Currently, the basis of the wood layer in the study area is made up of young and middle-aged generations of trees. Almost two-thirds of the trees did not exceed the age of 120 years (Table S2).

Along with the increase in forest areas the growth of trees in height and in diameter has significantly increased during the last century [35]. The average annual radial growth was 0.83 mm during the past 70–90 years, about 4 times more than during the previous 60–80 years (0.21 mm).

An increase in density indicators was noted for sites with woody vegetation during the period 1987–2018, with lowland dark coniferous forests exhibiting the most pronounced increase in closure (more than 15–20%) (Figure 6). For larch, the most intensive renewal of larch growing on the south-eastern slope of the Mountain Rai-Iz massif occurred during the period 2004–2014, and was faster than detected for the previous period 1991–1999 [35].

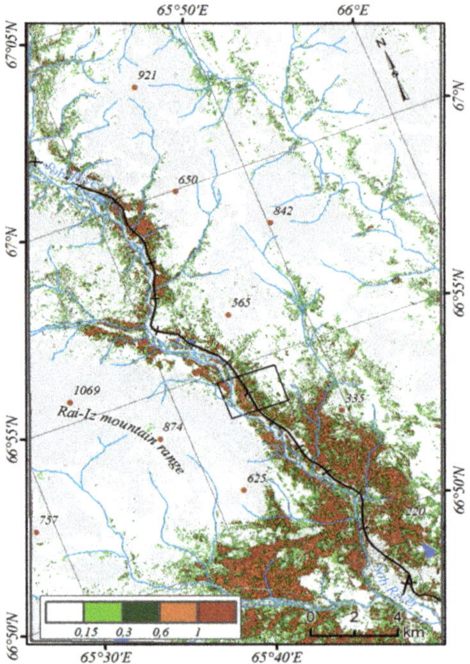

Figure 6. Total changes in the density of shrubs and trees (percent per year) based on winter satellite images taken on 25.04.1987 and 23.04.2018.

Repeated descriptions of trees on a constant altitudinal profile in the treeline ecotone were established in the early 1960s on the southern slope of Mountain Rai-Iz (Mountain Chernaya region). Changes in phytomass, number of trees, and stock, and stand density were monitored over the past 40 years [64]. The obtained data showed that there was a 2–5-fold increase in phytomass stand density in nearly all sites. Exceptions were areas which were almost treeless in the 1960s, where phytomass and stand density increased even more. Siberian spruce became established under the canopy of larch stands, of the lower part of the treeline ecotone of mountain slopes in the Polar Urals, and advanced of its upper distribution border into higher areas. Young spruce trees usually have a high increase in diameter and, especially, in height. Here, spruce successfully displaced larch, first resulting in a two-tier spruce-larch forest, then into a larch-spruce forest of the north-boreal type.

The NDVI generally increased in most of the studied territory. However, the total increase of NDVI values for all plant communities and the entire period 1988–2017 did not exceed 0.001 per year (Figure 4). A similar study carried out in boreal forests of the area also reported a similar increase in NDVI values [45,46], and reported a considerable variability between different biome types, maximum values for larch covered areas being significantly correlated to July temperature [63]. We found a distinct increase in NDVI for most plant communities addressed: e.g., the maximum increase in NDVI (up to 0.005 per year) was observed in spruce forests in the period 1988–1994, corresponding to the first phase of summer temperature increase (Figure 4). However, the intensity of changes varied significantly in subsequent years, thus causing the low average value of total NDVI increase. Since 1988, the maximum total increase of NDVI (14–25%) occurred in spruce forests, as well as in willow thickets with northern green alder growing in the valley and on the slopes of the mountains bordering the Sob River valley (Figure 7). The average of NDVI increased per model site was 15% over 26 years.

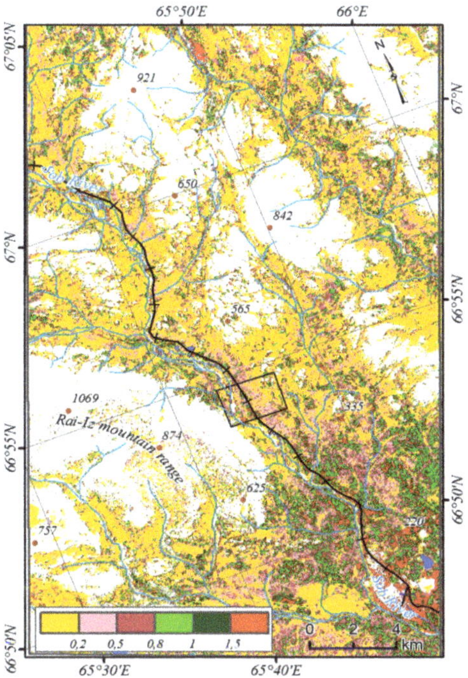

Figure 7. The total trend of NDVI changes during 1988–2017 (units/yr).

This proves a real greening effect of the region. On the slopes, the trees rise higher into the mountains, and low areas trees grow older, had higher crown densities and wider annual rings. This leads to an increase in phytomass volume of the main tree species (Figure 8). Phytomass of *Larix* and *Picea* increased very much in the area, where the upper border of the forest was located in 1900 during the past 60 years: *Picea* phytomass increased by a factor 21, from 0.22 to 4.64 t/ha, and *Larix* phytomass increased by a factor 3.5, from 28.34 to 97.60 t/ha.

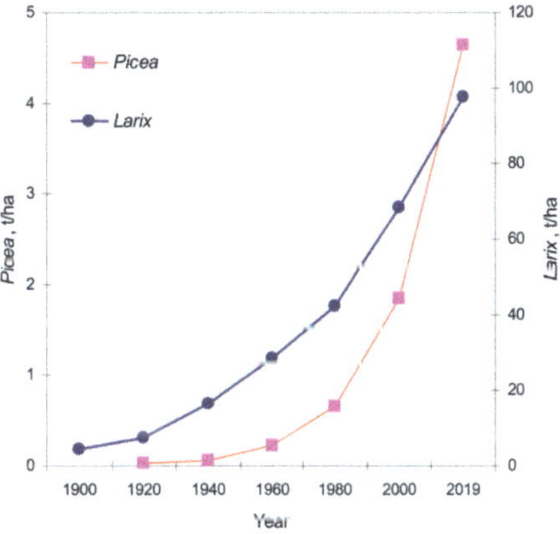

Figure 8. Average stock of aboveground phytomass for *Larix* and *Picea* (t/ha) from 1900 to 2019.

Phytomass dynamics was also studied for individual fractions of larch and spruce (trunks, branches, and needles). The trunk biomass was growing especially fast for spruce, accounting for 2.2 times the sum of the branches and needles (Table 4). Larch trunks had 1.8 times more biomass than its branches and needles. In general, for 2019, the aboveground larch phytomass was 21 times higher than that of spruce. According to our data, the aboveground phytomass of all arboreal species (including green alder, willow, rowan, and birch) is currently 102.25 t/ha, whereas in 1960 it was 28.56 t/ha, and in 1900 only 4.77 t/ha. In closed forest stands, which started to form during the second part of 19th century, annual production of phytomass was 0.16 t/ha per year in 1900, but currently it is 10 times higher, with an average rate of 1.6 t/ha per year. Since 1900, the contribution of spruce to the total phytomass increased by only 4.5%, and spruce phytomass increased drastically (395 times). This is because in the 1900s like in 2019, more than 95% of the aboveground woody phytomass was formed by larch in this region (Table S3).

Table 4. Stock of individual fractions in aboveground biomass of *Larix* and *Picea* (t/ha) from 1900 to 2019.

Year	*Larix*			*Picea*		
	Trunks	Branches	Needles	Trunks	Branches	Needles
1900	3.09	1.28	0.38	0.01	0.003	0.0006
1920	5.20	2.15	0.64	0.02	0.01	0.001
1940	10.78	4.47	1.32	0.04	0.01	0.003
1960	18.42	7.65	2.27	0.15	0.06	0.01
1980	27.42	11.44	3.38	0.44	0.17	0.04
2000	44.39	18.44	5.46	1.26	0.47	0.11
2019	63.44	26.35	7.81	3.18	1.19	0.27

The following tree range expansions have been observed in the floodplain of Ob River and the Ural Mountains: aspen (*Populus tremula* L.), gray alder (*Alnus incana* (L.) Moench), bird cherry (*Prunus padus* L., or *Padus racemosa* (Lam.) Gilib.), Siberian pine (*Pinus sibirica* Du Tour), and European pine (*Pinus sylvestris* L.). These trees are accompanied by coenotically related herbaceous plants. According to our observations, now the northern distributional border of some of them reaches just 120 km south of the towns Labytnangi, Salekhard, and the Arctic Circle, although 30 years ago it was located 300–350 km south. The vegetation of the study area was characterized as forest-tundra in the 1960s, and at the moment it has transformed into taiga [36], and with the advent of above-mentioned trees it may become middle boreal. The Arctic Research Station of Labytnangi town has a Botanical Garden. In the 1970s, exotic tree species like aspens, Siberian pines, alders, and caraganas were planted there. Over the past 20 years, there has been a sharp activation of the growth of these trees. Currently, aspen and Siberian pines have spread throughout the town and are actively settling in the surrounding tundra and forest-tundra landscapes. This further confirms the fact that the climatic conditions of the region are becoming more comfortable for the 'southern' trees. Saprobial fungi have already adapted to the availability of these new substrates: e.g., in 2018, the clavarioid fungus *Typhula erythropus* (Pers.) Fr. was detected on dead leaf of aspen in the tundra.

Along with warming in the region, precipitation, length of the growing season, soil thaw depth, and activity of soil microbiota increased. This resulted in rising of the timberline and increase in forest area and crown density, along with a maximal NDVI (Table S4).

3.4. Mycobiota Dynamics

The number of species of aphyllophoroid fungi detected doubled: starting from 127 species detected in the first period of the study (1960–1979), it increased to 206 in the second (1980–1999), and 257 in the third period (2000–2019) (Figure 9). A total of 263 species of fungi have been identified in the region over the entire period (Table S1), which makes up 26.3% of the number of aphyllophoroid fungi known in the Urals [53]. Each period includes aphyllophoroid species found only during this period.

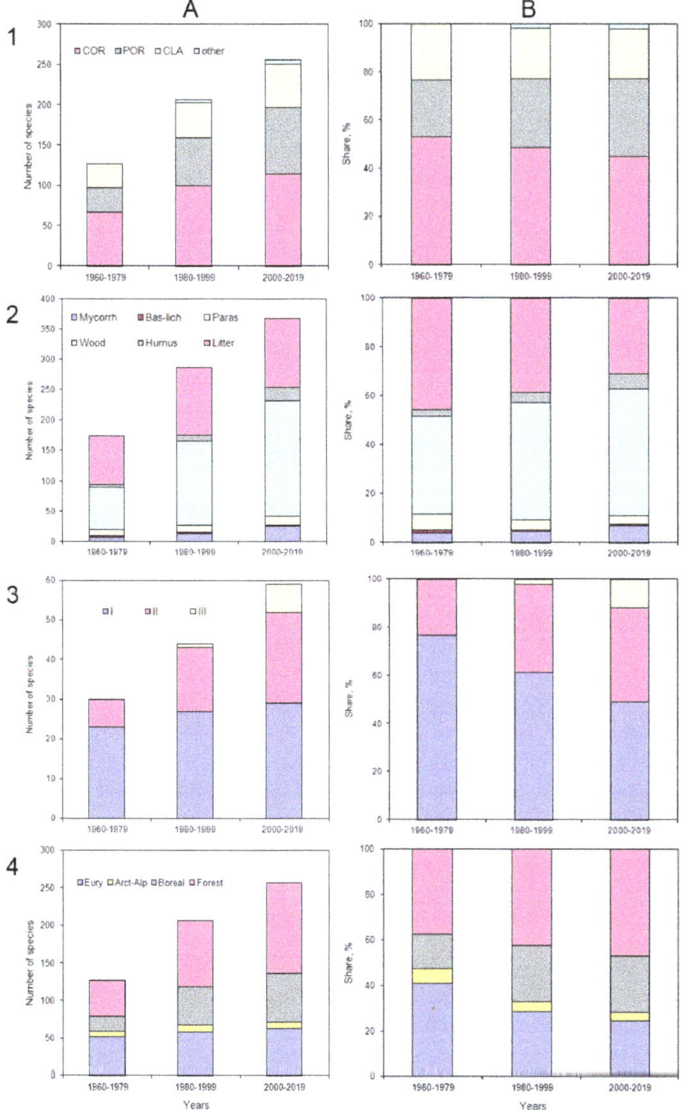

Figure 9. Dynamics of aphyllophoroid fungi during the last three periods of 20 years, each represented as (**A**) species richness and (**B**) their share in percent. (**1**) Fruit body types; (**2**) Ecological strategy; (**3**) Fruit body height; (**4**) Chorological groups of in the area of investigation. Abbreviations: Fruit body types: COR—corticioids, POR—poroids, CLA—clavarioids; Ecological strategy (trophic): Mycorrh—mycorrhiza-forming; Bas-Lich—basidial lichens; Paras—parasites; Wood—wood saprobs; Humus—humus saprobs; Litter—litter saprobs; Height groups: I—first (fruit bodies up to 3 cm high), II—second (3–8 cm), III—third (more than 8 cm); Chorological groups: Eury—eurybionts; Arct-Alp—arcto-alpine; Boreal—boreal; Forest—forest.

The ratio of the main fruitbody types changed significantly. The Corticioids as most divers group decreasedfrom 52.8% to 44.7%, and also Clavarioids decreased (from 23.6% to 20.9%). In contrast thereto, Poroids increased from 23.6% to 32.1% (Figure 9).

The number of aphyllophoroid fungal species forming fruit bodies on large deadwoodincreased from 69 to 190 (increasing from 39.9% to 51.9% of the total diversity).

During the last 20 years, many typical boreal polypores such *Osteina obducta*, *Laurilia sulcata* (Burt) Pouzar, *Phellinus alni* (Bondartsev) Parmasto, *P. ferrugineofuscus* (P. Karst.) Bourdot & Galzin, *Polyporus arcularius* (Batsch) Fr., *Ramaria apiculata* (Fr.) Donk, *Steccherinum luteoalbum* (P. Karst.) Vesterh., and *Trametes suaveolens* were first discovered in the area (Figure S1 and Table S5).

Litter inhabiting species increased from 79 to 113 species. Over the past 20 years, the following species were first identified: *Clavaria fumosa* Pers., *Lentaria epichnoa* (Fr.) Corner, *Macrotyphula tremula* Berthier, *Ramariopsis biformis* (G.F. Atk.) R.H. Petersen, *Ramaria subdecurrens* (Coker) Corner (Figure 9). An interesting discovery was the clavaroid fungus – *Clavariadelphus mucronatus* V.L. Wells & Kempton: its main distributional range corresponds to North America. It was found in Eurasia only once, on the Kamchatka Peninsula before this record [56]. Over the past 10 years, it was also reported fromparks and gardens of the Republics of Tatarstan and Udmurtia [65,66].

Only a limited number of ubiquitous aphyllophoroid fungi, which are widely distributed throughout the forest zones of the Urals, developed on deadwood, or on grass litter. Despite an increase in species richness, ubiquitous fungi decreased by almost 15% (from 45.7 to 30.9%).

Over 60 years, the proportion of mycorrhizal species has increased almost 2 times (from 4 to 7%), and humus saprobs 2.5 times (from 2.8 to 6.2%). Mycorrhizal fungi typical for boreal regions appeared for the first time, e.g., *Cantharellus cibarius* Fr., *Ramaria botrytis* (Pers.) Bourdot, *R. flavobrunnescens* (G.F. Atk.) Corner, *Thelephora terrestris* Ehrh. The most noticeable increase occurred during the third time period, when the soil was most thawed, and temperature, precipitation increased and woody phytomass increased most. Interestingly, the violet webcap, *Cortinarius violaceus* (L.) Gray, a rare agaricoid mycorrhizal fungus, which is included in various regional Red Books of Russia [67], occurred regularly on the slopes of Mountain Slantsevaya in 2019.

Unlike the mycorrhiza forming ones, the number and proportion of humus saprobs increased evenly throughout the study period. Fruit bodies of typically boreal fungi began to develop on the soil: *Albatrellus ovinus* (Schaeff.) Kotl. & Pouzar, *Coltricia perennis* (L.) Murrill, *Cantharellus cibarius*, *Hydnum umbilicatum* Peck, *Clavariadelphus pistillaris* (L.) Donk, etc. These species do not occur in the tundra and forest-tundra of Western Siberia and Eastern Europe, nor on the treeline or in the alpine areas of the Polar Urals.

With an increased diversity of herbaceous and woody plants species, the diversity of associated parasites increased (from 11 to 14 species). However, their proportion decreased almost halved, dropping from 6.4% to 3.5%. Diversity of moss parasites decreased (*Clavaria sphagnicola* Boud., *Ramariopsis subarctica* Pilát), tree parasites increased (on conifers *Laricifomes officinalis* (Vill.) Kotl. & Pouzar, *Phellinus laricis* (Jacz. ex Pilát) Pilát, *Laetiporus sulphureus* (Bull.) Murrill s.l.; on deciduous trees (*Chondrostereum purpureum* (Pers.) Pouzar, *Inonotus obliquus* (Fr.) Pilát), and grass parasites (*Typhula incarnata* Lasch) were detected for the first time. On willows, *Phellinus igniarius* (L.) Quél. expanded his territory from one single tree in the valley to 20% of all trees.

During the entire study period, only two basidio-lichens (*Multiclavula vernalis* (Schwein.) R.H. Petersen, *M. corynoides* (Peck) R.H. Petersen) were detected in the study area, and their proportion decreased from 1.2% to 0.5%.

For fungi with negatively geotropic basidiomata (Clavarioids, Chanterelles and stipitate Hydnoids), the number of species increased from 30 to 59 species. In the first period, more than 60% of all species belonged to size class 1 (less than 3 cm in height), while only 23.3% belonged to the second. By the third period, the proportion of small-body species decreased by almost 30%, and now makes up less than half of all species. On the other hand, for species of the second size class (from 3 to 8 cm in height) the share increased by 15% to 39%, and species of the third class newly appeared (11.9%).

Chorological groups changed during the last 60 years. The proportion of Arcto-alpine species was nearly halved (decreasing from 6.3% to 3.5%). *Ramariopsis subarctica* was less and less found, and in dry and hot 2017, it was not collected at all throughout the valley, although it was regularly seen before [56,67]. Also, *Multiclavula corynoides* and *M. vernalis* were less and less found on open soil, and the Arcto-alpine species *Datronia scutellata* (Schwein.) Gilb. & Ryvarden, *Peniophora aurantiaca* (Bres.) Höhn. & Litsch., *Plicatura nivea* (Fr.) P. Karst. decreased in the alder thickets. The only exception was *Clavaria sphagnicola*, another Arcto-alpine species growing in marshes, which did not reduce the number of fruiting bodies, even in hot and dry years.

Ubiquitous aphyllophoroids were the second largest group in the tundra. They diversity and proportions decreased by 16% (from 40.9 to 24.5%). Typical forest species increased in diversity from 37.8 to 47.1%, and their proportion increased from 15.0 to 24.9%. In general, 60 years ago the northern species accounted for almost half (47.2%), now only about four (28%) of aphyllophoroid species diversity. In contrast thereto, the proportion of taiga species increased by 25%.

4. Discussion

Fungi represent an important heterotrophic block of the ecosystem. The development of fungi sometimes requires specific sources of nutrients, and this needs to be considered when focusing on the effect of climate change on macromycetes. Based on our results, the diversity of aphyllophoroid fungi doubled during the last 60 years in the area. One first interpretation might be, that the first period (1960–1979) was understudied. A re-examination of samples collected during that period confirms that mainly widespread common species were identified. Upon careful examination, it turns out that none of these species grow on large-sized deadwood, on rich soil, or on mesophilic boreal herbs and grasses. The largest diversity of species was then detected on small deadwood, larch branches, shrubby willow, rowan and northern green alder. About 50–60 years ago, the study area was open larch forest tundra with a few spruce trees over poor permafrost soil. More than 80% of larches in area were less than 70 years old, and single spruce trees were 20–40 years old. This undoubtedly confirms that the fungi detected during the first period were adequately sampled, because due to shortage of large-sized deadwood, the habitat for species able to form large fruiting bodies was not present. The ubiquity of permafrost, with a thaw depth of only 1–1.5 m in summer, indicates that litter saprobial and mycorrhizal fungi were restricted under such conditions. Therefore, mycologists who worked in the 1960s in the Sob river valley and on the slopes of Mountain Slantsevaya were studying aphyllophoroid fungi at the arctic border of the forest. In this regard, the complex of fungi reported for the first period (1960–1979) is correct and reflects the bioclimatic conditions during that period.

More importantly, the ratio of the main fruitbody types changed significantly, and this can therefore be considered as an important indicator of bioclimatic conditions [27]. Corticioids decreased significantly and poroids increased. The ratio of corticoids to poroids appears to be a parameter typical for forest types. This was also confirmed by a comparison with earlier results for the Ural latitudinal-zonal transect [55,68]. An increase in the volume of large wood contributes to an increase in the diversity and proportion of poroids. Many poroid species need large deadwood for their growth [27,68]. Changes in bioclimatic conditions, especially between periods 2 and 3 go along withsignificant changes in the structure of the vegetation cover. Larch and spruce phytomass increased, and trees became older. Therefore, also the number of dead trunks increased and they were generally thicker, which significantly affected the species composition of wood-destroying fungi. Suitable habitats for the growthof typical forest polypores are now present in the area. During the last 20-year period, many boreal fungi such as *Osteina obducta* and *Laurilia sulcata* (Tables S2 and S5) were first discovered on large dead larch trunks. *Phellinus ferrugeneofuscus*, *Ramaria apiculata*, *Steccherinum luteoalbum*, and *Mucronella culou* (Alb. & Schwein.) Fr. appeared on the dead spruce wood. Birch (*Betula pendula*) grows now higher than 12 m, thus enabling the fructification of *Polyporus arcularius*. Phytomass produced in shrub belts also increased in the study area. The northern green alder (*Duschekia fruticosa* (Rupr.) Pouzar, Syn. *Alnus fruticosa* Rupr.), which is mainly responsible for such shrub thickets, can now reach up to 6–7 m in height [69]. Now, *Phellinus alni* can be regularly collected in these areas. Isolated findings were reported from 1980-1999, but now it occurs on a third of all *Duschekia* bushes. The maximum NDVI also increased in floodplain willows, which start to be older, taller and fatter. Willows (*Salix caprea*) up to 8 m high appeared along the Sob River, and in 2019, the poroid fungus *Trametes suaveolens* was first collected on them. Thirty years ago, the northern border for the distribution of this species was 200 km south of the towns Labytnangi and Salekhard, in typical northern boreal forests. The occurrence of wood-decomposing fungi undoubtedly reflects changes in the structure of vegetation cover, and characterizes the recent mycobiota as typical for taiga, and not for forest-tundra anymore [55,56].

The phytomass of needles increased, leading to an increase in the diversity and richness of litter (Table 3). The diversity of grassy litter has also increased. 30-50 years ago, many boreal herbaceous plants (*Aconitum excelsum* Reichb., *Athyrium filix-femina* (L.) Roth ex Mert., *Veratrum lobelianum* Bernh., *Anthriscus sylvestris* (L.) Hoffm.) were found only at the valley bottom. Now, due to warming, these species have expanded their territory to upland habitats and to the treeline. European blueberry (*Vaccinium myrtillus* L.) acts as an indicator of borealization of bioclimatic conditions in the region. In the 1960s, it was not present at all on Mountain Slantsevaya (E. Parmasto, pers. comm.), and in the 1980s it was found only at the valley bottom, but not the slopes of Mountain Slantsevaya (S. Obolevski, pers. comm.). Compositional and quantitative changes in the vegetation permitted for a significant increase in diversity of litter inhabiting fungi (Figure 9). Moreover, an increase in the soil thaw depth and concomitant increase in the activity of the soil microbiota also contributed to the appearance of Cantharelles and stipitate Hydnoids in such high-latitude permafrost regions [55]. The detected proportion of litter species is characteristic for complexes of aphyllophoroid fungi growing in the taiga.

Eurybiont aphyllophoroid fungi decreased by almost 15% during the last 60 years. A lower proportion of ubiquitous species is typical for boreal forests, as confirmed by similar carried out in the Urals [55,68].

An increase in phytomass and diversity of woody plants, together with deeper soil thaw and increase in litter activity, contributed to the increase of typical forest fungi forming mycorrhiza. Also, the proportion of mycorrhized roots increased in the study area due to warming [41]. The proportion of mycorrhizal fungi almost doubled during the last 60 years, and the species composition reflects atypical boreal mycobiota [55,58].

With an increased diversity of herbaceous and woody plants species, also the diversity of associated parasites increased, but their proportion decreased. Only parasites on trees and mosses were reported in previous periods, but now grass parasites appeared for the first time (*Typhula incarnata*). Grass parasites on are generally expanding their territory along the floodplain of the Ob River, and *T. ishikariensis* has already been reported from Salekhard in 2018. An increase in the number of tree and grass parasitic species was also reported from in other regions of the Arctic [70].

Basidio-lichenshave theirpeak of diversity in the tundra, especially in Arctic deserts [71]. Only two species were detected, and their proportion was <1%. This is typical for northern boreal forests [55,72].

Aphyllophoroid fungi with larger fruit bodies appeared during the last 60 years, replacing species with small fruit bodies. In the tundra and Arctic deserts, there are generally no species of class three (above 8 cm in height), and species belonging to class two are very rare. Six out of seven species forming large fruit bodies appeared in the third period. These were predominantly mycorrhizal *Ramaria* species which typically occur in hemiboreal forests. Litter saprobic species have usually smaller fruit bodies [73], and they decreased towards period three. An increase in the proportion of fungal species with large fruit bodies was also reported for clavarioid fungi of middle boreal forests [55].

Over the three studied periods, the proportions of the chorological groups of fungi significantly changed. Arcto-alpine fungi are species whose distributional peaks are in the arctic and alpine regions. Arcto-alpine fungi are usually widespread in the tundra and arctic deserts [56,71]. A reduction in the range of Arcto-alpine species became evident during the last 60 years and is widely discussed [14]. The winners were typical forest species, which significantly increased in diversity. This consolidates our observation that the mycobiota transformed from a forest-tundra into a northern taiga type during the studied period [58], thus reflecting changes in the vegetation cover and in soil- or litter conditions. However, several typical boreal fungi have not yet been reported from the region. Among them are ordinary inhabitants of coniferous forest like *Fomitopsis rosea* (Alb. & Schwein.) P. Karst., *F. cajanderi* (P. Karst.) Kotl. & Pouzar, *Ganoderma lucidum* (Curtis) P. Karst., *Leptoporus mollis* (Pers.) Quél., *Gloeoporus dichrous* (Fr.) Bres., *Amylocystis lapponica* (Romell) Bondartsev & Singer, *Phaeolus schweinitzii* (Fr.) Pat., and *Dichomitus squalens* (P. Karst.) D.A. Reid. Also, some usual boreal species growing on large birches and willows have not yet been detected: *Datronia mollis* (Sommerf.) Donk, *Hericium coralloides* (Scop.) Pers., *Trametes trogii* Berk., *Oxyporus corticola* (Fr.) Ryvarden, *O. populinus* (Schumach.) Donk, and *Trichaptum biforme* (Fr.) Ryvarden. (Table S5). It is not clear, whether the absence of these species is due to environmental conditions (e.g., substrate volume), or rather due to insufficient sampling.

Boreal aphyllophoroid fungi forming large fruiting bodies can be considered as indicators for the ongoing borealization in the Arctic, such species developing on large trunks in the Polar Urals are *Laricifomes officinalis, Laurilia sulcata, Datronia scutellata, Skeletocutis stellae,* and *Trametes suaveolens.* Litter saprobial species capable of forming fruiting bodies only on deeply thawed soil (*Clavariadelphus pistillaris, Coltricia perennis*), or mycorrhiza-forming species (*Cantharellus cibarius, Hydnum umbilicatum, Ramaria botrytis*) could also be considered as indicators. The reduced range and abundance of many Arcto-alpine fungi further confirms an ongoing borealization, because these species are indicators for tundra-alpine communities.

There is no economic activity in the study area. However, the Yamal-Nenets Autonomous District is currently the region with the most intensive development of infrastructure in the Russian Arctic. A railway passing through the Sob River valley connects the district with the European part of the country. The railway line is to be expanded, to increase freight traffic. This will be associated with deforestation and destruction of swamps close to the railway, and adversely affect the fragile high-latitudinal ecosystem in the region. Unfortunately, the Sob River valley and the slope of Mountain Slantsevaya are not within protected natural areas, although one of the northernmost forests of the Urals is located here. The slope of Mountain Rai-Iz is encircling the Sob River valley from the west and is already located inside the Polyarnouralsky Natural Park. We therefore recommend that the entire valley of the Sob River, including Mountain Slantsevaya, should be included in the boundaries of Polyarnouralsky Natural Park.

5. Conclusions

Over the past 60 years, noticeable rapid changes in the main bioclimatic parameters have occurred on the eastern macroscope of the Polar Urals, in the Sob River valley, as well as on Mountains Slantsevaya and Rai-Iz. Summer air temperatures increased by 2 °C and winter air temperatures by 1.8 °C. Summer precipitation increased by 30 mm and winter precipitation by 49 mm, and the growing season lasted 6–7 days longer. The permafrost thawed 1–2 m deeper, and cellulose decomposition increased 2–5 times. All these factors favoured tree growth, as detected by rising of the timberline (35–41 m) and increase in total forest area (30%),in radial growth of trees (4 times), in crown density (15–20%) and in phytomass (tenfold for all trees). The diversity and proportion of boreal herbaceous mesophilic plants increased in the river valley and on mountain slopes. These data are confirmed by an average increase in NDVI by 15%, in spruce and shrub vegetation by even 15–25%. This proves ongoing borealization and greening of the Polar Ural's vegetation.

Over 60 years, the number of species of aphyllophoroid fungi doubled. Fungi depend on the availability of suitable substrate, which is now available due to climate warming. The trees became older, and large-sized deadwood appeared, now enabling the growth of species forming large fruit bodies on such a substrate (e.g., Poroids which increased by 12%). The increase in phytomass of needles, small twigs, and the emergence of new mesophilic boreal plants led to a sharp increase in the number of litter saprobes, and to the appearance of boreal fungal species. Furthermore, a deeper soil thaw contributed to the emergence of humus saprobs and mycorrhiza forming species, and their proportion doubled. Large-body fungal species generally increased, thus clearly reflecting a transition of the in mycobiota from forest-tundra to taiga type. Many of the large-bodied wood and humus saprobial, as well as mycorrhiza-forming fungi, can be considered as indicators of the ongoing borealization of the Arctic.

The flip side of warming and greening is the weakening of native Arcto-alpine bryophilic species, which now try to survive in the state of neo-relicts in a decreasing number of available habitats, such as in *Sphagnum* bogs and Alpine vegetation on mountaintops. Basidio-lichen species are also sharply reduced. These species retain their position only above treeline.

New parasitic fungi appeared in the region. Grass parasites causing enormous damage in agriculture (snow mold disease) in the taiga zone have appeared over the past 10 years. Growing of various crops becomes now possible due to milder climate, and it can be assumed that corresponding parasitic fungi will soon appear. Also, wood-destroying fungi are expected to cause economical loss in the near future. The advancement of most species of fungi to the north is not limited by climatic conditions, but by the lack of the necessary substrate [71]. Aphyllophoroids are not only indicators of bioclimatic changes at high latitudes, but also for substrate availability. In this regard, they will take an active part in the transformation of the economic infrastructure of the region.

Supplementary Materials: The following are available online at http://www.mdpi.com/1999-4907/10/12/1112/s1. Table S1. Presence-absence data for 263 aphyllophoroid fungal species found at the Polar Urals (in the Sob River valley and slopes of the Mountains Slantsevaya and Rai-Is) during 3 study periods (1960-1979, 1980-1999, 2000-2019) and characteristics of the species (ecological strategy, fruit body type, height groups and chorological groups). Table S2. Time of trees establishment (A) and age structure (B) on the studied area. Table S3. Stock of aboveground biomass (t/ha) of *Larix* and *Picea* and species ratio (%) in the Sob River valley and on the slopes of Mountain Slansevaya. Table S4. Changes of biolimatic parameters in the Sob River valley and on the slopes of Mountain Slansevaya. Table S5. Large-bodies species of Aphyllophoroid fungi appeared in the region and what still absent. Figure S1. Rare and interesting Aphyllophoroid fungi found on the eastern slope of the Polar Urals: in the Sob River valley, on the slopes of Mountains Slantsevaya and Rai-Iz (1960-2019).

Author Contributions: A.G.S. conceived and co-ordinated the overall project. A.G.S., P.A.M., U.P. designed the sampling experiment. V.V.E. worked with satellite data and created NDVI map. A.G.S., P.A.M., N.M.D., V.V.K. and V.V.E. analyzed the data. A.G.S., P.A.M., U.P., and N.M.D. wrote the manuscript.

Funding: The study was carried out with financial support of the Russian Foundation for Basic Research (grant no. 18-05-00398) and of the Russian Science Foundation (grant no. 17-14-01112) for investigation of stands biomass and treeline position changes (P.A.M., N.M.D., V.V.K.).

Acknowledgments: We would like to thank for anonymous reviewers for their comments on an earlier version of the manuscript. We very much appreciate the support provided by the administration of the Arctic Research Station of the Institute of Plant and Animal Ecology UrB RAS (Labytnangi) for the miscellaneous assistance during this research. We are thankful to Stanislav Obolevski (Omsk) for allowing us to publish his data on rate of litter decomposition and soil thaw depth. Our team thanks Denis Kosolapov (Syktyvkar) and Juhani Paivarinta (Helsinki) for help in conducting expeditionary work on Slantsevaya Mountain.

Conflicts of Interest: The authors declare no conflict of interest.

References

1. IPCC Working Group II. Climate Change 2014: Impacts, Adaptation, and Vulnerability. 2014. Available online: https://www.ipcc.ch/working-group/wg2/ (accessed on 20 August 2019).

2. IPCC. Summary for Policymakers. In *IPCC Special Report on the Ocean and Cryosphere in a Changing Climate*; Pörtneretal, H.-O., Ed.; IPCC: Monaco, 2019; Available online: https://report.ipcc.ch/srocc/pdf/SROCC_FinalDraft_FullReport.pdf (accessed on 3 October 2019).

3. Forbes, B.C.; Fauria, M.M.; Zetterberg, P. No change without a cause-why climate change remains the most plausible reason for shrub growth dynamics in Scandinavia. *New Phytol.* **2011**, *190*, 805.

4. Körner, C.; Paulsen, J.A. World-wide study of high altitude treeline temperatures. *J. Biogeogr.* **2004**, *31*, 713–732. [CrossRef]

5. Walker, D.A.; Epstein, H.E.; Raynolds, M.K.; Kuss, P.; Kopecky, M.A.; Frost, G.V.; Daniels, F.J.A.; Leibman, M.O.; Moskalenko, N.G.; Matyshak, G.V.; et al. Environment, vegetation and greenness (NDVI) along the North America and Eurasia Arctic transects. *Environ. Res. Lett.* **2012**, *7*, 015504. [CrossRef]

6. Bhatt, U.S.; Walker, D.A.; Raynolds, M.K.; Bieniek, P.A.; Epstein, H.E.; Comiso, J.C.; Pinzon, J.E.; Tucker, C.J.; Steele, M.; Ermold, W.; et al. Changing seasonality of panarctic tundra vegetation in relationship to climatic variables. *Environ. Res. Lett.* **2017**, *12*, 055003. [CrossRef]

7. Xu, L.; Myneni, R.B.; Chapin, F.S.; Callaghan, T.V.; Pinzon, J.E.; Tucker, C.J.; Zhu, Z.; Bi, J.; Ciais, P.; Tømmervik, H.; et al. Temperature and vegetation seasonality diminishment over northern lands. *Nat. Clim. Chang.* **2013**, *3*, 581–586. [CrossRef]

8. Epstein, H.E.; Beringer, J.; Gould, W.A.; Lloyd, A.; Thompson, C.D.; Chapin, F.S., III; Michaelson, G.J.; Ping, C.; Rupp, T.S.; Walker, D. The nature of spatial transitions in the Arctic. *J. Biogeogr.* **2004**, *31*, 1917–1933. [CrossRef]

9. Paulsen, J.; Körner, C. A climate-based model to predict potential treeline position around the globe. *Alp. Bot.* **2014**, *124*, 1–12. [CrossRef]

10. Shiyatov, S.G.; Terent'ev, M.M.; Fomin, V.V.; Zimmermann, N.E. Altitudinal and horizontal shifts of the upper boundaries of open and closed forests in the Polar Urals in the 20th century. *Russ. J. Ecol.* **2007**, *38*, 223–227. [CrossRef]

11. Forbes, B.C.; Fauria, M.M.; Zetterberg, P. Russian Arctic warming and 'greening' are closely tracked by tundra shrub willows. *Glob. Chang. Biol.* **2010**, *16*, 1542–1554. [CrossRef]

12. Vaganov, E.A.; Shiyatov, S.G.; Mazepa, V.S. *Dendroclimatic Studies in the Ural-Siberian Subarctic*; Nauka: Novosibirsk, Russia, 1996. (In Russian)

13. Vinogradova, G.M.; Zavalishin, N.N.; Kuzin, V.I. Variability of seasonal characteristics of the climate of Siberia during the XX century. *Opt. Atmos. Ocean* **2000**, *13*, 604–607. (In Russian)

14. Tishkov, A.A.; Belonovskaya, E.A.; Vaisfeld, M.A.; Glazov, P.M.; Krenke, A.N.; Tertitsky, G.M. Greening of the tundra as a driver of current dynamic for Arctic biota. *Arct. Ecol. Econ.* **2018**, *2*, 31–44.

15. Bazilevich, N.I.; Grebenschikov, O.S.; Tishkov, A.A. *Geographical Principles of the Structure and Function of Ecosystems*; Institute of Geography: Moscow, Russia, 1986.

16. Biskaborn, B.K.; Smith, S.L.; Noetzli, J.; Matthes, H.; Vieira, G.; Streletskiy, D.A.; Schoeneich, P.; Romanovsky, V.E.; Lewkowicz, A.G.; Abramov, A.; et al. Permafrost is warming at a global scale. *Nat. Commun.* **2019**, *10*, 1–11. [CrossRef] [PubMed]

17. Yu, Q.; Epstein, H.E.; Walker, D.A.; Frost, G.V.; Forbes, B.C. Modeling dynamics of tundra plant communities on the Yamal Peninsula, Russia, in response to climate change and grazing pressure. *Environ. Res. Lett.* **2011**, *6*, 045505. [CrossRef]

18. Diez, J.M.; James, T.Y.; McMunn, M.; Ibáñez, I. Predicting species-specific responses of fungi to climatic variation using historical records. *Glob. Chang. Biol.* **2013**, *19*, 3145–3154. [CrossRef]

19. Ohenoja, E.; Kaukonen, M.; Ruotsalainen, A.L. Sarcosoma globosum – an indicator of climate change? *Acta Mycol.* **2013**, *48*, 81–88. [CrossRef]

20. Osono, T.; Takeda, H. Roles of diverse fungi in larch needle-litter decomposition. *Mycologia* **2003**, *95*, 820–826. [CrossRef]

21. Kauserud, H.; Stige, L.C.; Vik, J.O.; Økland, R.H.; Høiland, K.; Stenseth, N.C. Mushroom fruiting and climate change. *Proc. Nat. Acad. Sci. USA* **2008**, *105*, 3811–3814. [CrossRef]

22. Boddy, L.; Büntgen, U.; Egli, S.; Gange, A.C.; Heegaard, E.; Kirk, P.M.; Mohammad, A.; Kauserud, H. Climate variation effects on fungal fruiting. *Fungal Ecol.* **2014**, *10*, 20–33. [CrossRef]

23. Gange, A.C.; Gange, E.G.; Sparks, T.H.; Boddy, L. Rapid and recent changes in fungal fruiting patterns. *Science* **2007**, *316*, 71. [CrossRef]

24. Andrew, C.; Heegaard, E.; Halvorsen, R.; Martinez-Peña, F.; Egli, S.; Kirk, P.M.; Bässler, C.; Büntgen, U.; Høiland, K.; Boddy, L.; et al. Climate impacts on fungal community and trait dynamics. *Fungal Ecol.* **2016**, *22*, 17–25. [CrossRef]

25. Bässler, C.; Müller, J.; Dziock, F.; Brandl, R. Effects of resource availability and climate on the diversity of wood-decaying fungi. *J. Ecol.* **2010**, *98*, 822–832. [CrossRef]

26. Grau, O.; Geml, J.; Perez-Haase, A.; Ninot, J.; Semenova-Nelsen, T.A.; Penuelas, J. Abrupt changes in the composition and function of fungal communities along an environmental gradient in the high Arctic. *Mol. Ecol.* **2017**. [CrossRef] [PubMed]

27. Parmasto, E. *Aphyllophoroid fungi of Estonia*; Tartu Univ.: Tartu, Estonia, 1969. (In Russian)

28. Kotiranta, H.; Shiryaev, A.G. Notes on Aphyllophoroid fungi (Basidiomycota) in Kevo, collectedin 2009. *Kevo Notes* **2013**, *14*, 1–23.

29. Ordynets, A.; Heilmann-Clausen, J.; Savchenko, A.; Bässler, C.; Volobuev, S.; Akulov, O.; Karadelev, M.; Kotiranta, H.; Saitta, A.; Langer, E.; et al. Do plant-based biogeographical regions shape aphyllophoroid fungal communities in Europe? *J. Biogeogr.* **2018**, *45*, 1182–1195. [CrossRef]

30. Shiryaev, A.G. Spatial diversity of clavarioid mycota (Basidiomycota) at the forest-tundra ecotone. *Mycoscience* **2018**, *59*, 310–318. [CrossRef]

31. Geml, J.; Morgado, L.N.; Semenova, T.A.; Welker, J.M.; Walker, M.D.; Smets, E. Long-term warming alters richness and composition of taxonomic and functional groups of arctic fungi. *FEMS Microbiol. Ecol.* **2015**, *91*, fiv095. [CrossRef]

32. Geml, J.; Semenova, T.A.; Morgado, L.N.; Welker, J.M. Changes in composition and abundance of functional groups of arctic fungi in response to long-term summer warming. *Biol. Lett.* **2016**, *12*, 20160503. [CrossRef]

33. Shiryaev, A.G. Longitudianl changes of clavarioid funga (Basidiomycota) diversity in the tundra zone of Eurasia. *Mycology* **2017**, *8*, 135–146. [CrossRef]

34. Karatygin, I.V.; Nezdoiminogo, E.L.; Novozhilov, Y.; Zhurbenko, M.P. *Russian Arctic Fungi. Check-list*; SpbKPA: St. Petersburg, Russia, 1999. (In Russian)

35. Shiyatov, S.G.; Mazepa, V.S. The modern expansion of Siberian larch in to the mountain tundra of the Polar Urals. *Russ. J. Ecol.* **2015**, *6*, 403–410. (In Russian)

36. Shiyatov, S.G. *Dynamics of Woody and Shrubby Vegetation in the Mountains of the Polar Urals under the Influence of Modern Climate Changes*; Yekaterinburg Publish: Yekaterinburg, Russia, 2009. (In Russian)

37. Andreyashkina, N.I. Changes in principle characteristics of phytocenoses with participation of Larix sibirica Lebed. in the uppear treeline ecotone in the Polar Urals. *Vest. Tomsk. State Univ. Biol.* **2014**, *3*, 53–67. [CrossRef]

38. Walker, D.A.; Leibman, M.O.; Epstein, H.E.; Forbes, B.C.; Bhatt, U.S.; Raynolds, M.K.; Comiso, J.C.; Gubarkov, A.A.; Khomutov, A.V.; Jia, G.J.; et al. Spatial and temporal patterns of greenness on the Yamal Peninsula, Russia: Interactions of ecological and social factors affecting the Arctic normalized difference vegetation index. *Environ. Res. Lett.* **2009**, *4*, 045004. [CrossRef]

39. Kulay, G.A. Microbiological characteristics of some soil types in the tundra (Kharp station). *Ann. Rep. For. Soil Lab. for 1967.* **1968**, 30–33. (In Russian)

40. Kulay, G.A.; Ischenko, N.F. *Comparative characteristics of microbiological activity in some forest-tundra soils of Kharp station surrounds. Productivity of Subarctic Biocoenosis*; Firsova, V.P., Ed.; UrDAS: Sverdlovsk, Russia, 1970; pp. 221–223. (In Russian)

41. Solly, E.; Djukic, I.; Moiseev, P.; Andreyashkina, N.; Devi, N.; Goransson, H.; Mazepa, V.; Shiyatov, S.; Trubina, M.; Schweingruber, F.; et al. Treeline advances and associated shifts in the ground vegetation alter fine root dynamics and mycelia production in the South and Polar Urals. *Oecologia* **2017**, *183*, 571–586. [CrossRef] [PubMed]

42. Egorov, V.A.; Bartalev, S.A. The method of radiometric correction of distortions of the reflective characteristics of the earth cover in satellite measurement data caused by the influence of the terrain. *Mod. Probl. Remote Sens. Earth Space* **2016**, *13*, 192–201.

43. Schinner, F.; Öhlinger, R.; Kandeler, E.; Margesin, R. (Eds.) *Methods in Soil Biology*; Springer: Berlin/Heidelberg, Germany, 1996.

44. Elsakov, V.V.; Marushchak, I.O. Spectrozonal satellite images in identifying trends in climatic changes in forest phytocenoses of the western slopes of the Subpolar Urals. *Comput. Opt.* **2011**, *35*, 281–286.

45. Elsakov, V.V. Spatial and inter annual heterogeneity of changes in the vegetation cover of the tundra zone of Eurasia according to the MODIS 2000-2016 survey data. *Mod. Probl. Remote Sens. Earth Space* **2017**, *14*, 56–72. (In Russian)

46. Elsakov, V.V.; Kulygina, E.E. Vegetation of Yugorsky peninsula at climate change of the last decades. *Earth Obs. Remote Sens.* **2014**, *3*, 65–77. (In Russian)

47. Kapralov, D.S.; Fomin, V.V.; Shiyatov, S.G.; Moiseev, P.A. Changes in the composition, structure and altitudinal distribution of low forest at the upper limit of their growth in the Northern Urals mountains. *Russ. J. Ecol.* **2006**, *37*, 367–372. [CrossRef]

48. Hagedorn, F.; Shiyatov, S.G.; Mazepa, V.S.; Devi, N.M.; Grigor'ev, A.A.; Bartysh, A.A.; Fomin, V.V.; Kapralov, D.S.; Terent'ev, M.; Bugman, H.; et al. Treeline advances along the Urals mountain range-driven by improved winter conditions? *Glob. Chang. Biol.* **2014**, *20*, 3530–3543. [CrossRef]

49. Rinn, F.; Tsap, V. *3.6 Reference Manual: Computer Program for Tree-Ring Analysis and Presentation*; Bierhelderweg 20; Frank Rinn Distribution: Heidelberg, Germany, 1996.

50. Cook, E.R.; Kairiukstis, L.A. (Eds.) *Methods of Dendrochronology*; Springer: Dordrecht, The Netherlands, 1990.

51. Holmes, R.L. *Dendrochronological Program Library (Computer Program)*; Laboratory of Tree Ring Research, The University of Arizona: Tucson, AZ, USA, 1995.

52. Bakker, J.D. A new, proportional method for reconstructing historical tree diameters. *Can. J. For. Res.* **2005**, *35*, 2515–2520. [CrossRef]

53. Kazantseva, L.K. Mycoflora of the eastern slope of Polar Ural. *Zap. Sverdl. Dep. Bot. Soc. UrFAN* **1966**, *4*, 162–166.

54. Kotiranta, H.; Penzina, T.A. Notes on the North Ural Aphyllophorales (Basidiomycetes). In *Arctic and Alpine Mycology 5*; Yekaterinburg Publish: Yekaterinburg, Russia, 1998; pp. 67–81.

55. Shiryaev, A.G.; Mukhin, V.A.; Kotiranta, H.; Arefyev, S.P.; Safonov, M.A.; Stavishenko, I.V.; Kosolapov, D.A. Biodiversity of aphyllophoroid fungi of the Urals. In *Biological Diversity of Plant World of the Urals and Adjacent Territories*; Goschitsky Publish: Ekaterinburg, Russia, 2012; pp. 311–313.

56. Mukhin, V.A.; Knudsen, H. (Eds.) *Arctic and Alpine Mycology 5*; Yekaterinburg Publish: Yekaterinburg, Russia, 1998.

57. Shiryaev, A.G.; Khimich, Y.R.; Volobuev, S.V.; Morozova, O.V.; Koroleva, N.E.; Kosolapov, D.A.; Sokovnina, S.Y.; Shiryaeva, O.S.; Peintner, U. Greening of the Arctic and climate-driving dynamics of mycobiota at high latitudes. In *Ecological Problems of the Northern Regions and Ways to Their Solution*; Kola Science Center: Apatity, Russia, 2019; pp. 198–200. (In Russian)

58. Shiryaev, A.G. *Spatial Differentiation of Clavarioid Mycobiota of RUSSIA: Eco-Geographical Aspect*; Moscow University: Moscow, Russia, 2014

59. Djukic, I.; Kepfer-Rojas, S.; Schmidt, I.K.; Larsen, K.S.; Beier, C.; Berg, B.; Verheyen, K.; Luo, W. Early stage litter decomposition across biomes. *Sci. Total Environ.* **2018**, *626*, 1369–1394. [CrossRef] [PubMed]

60. Valdayskikh, V.V.; Nekrasova, O.A. Spatial variation and dynamics of seasonal thawing depth in Yamal forest-tundra soils. In *Soil Resources of Siberia: Challenges of the 21st Century*; Syso, A.I., Ed.; Tomsk State University Publish: Tomsk, Russia, 2017; pp. 34–37. (In Russian)

61. Kaverin, D.A.; Pastukhov, A.V.; Mazhitova, G.G. Temperature regime of tundra soils and underlying permafrost (Northeastern European Russia). *Earth's Cryosph* **2014**, *18*, 23–32. (In Russian)

62. Voigt, C.; Marushchak, M.E.; Lamprecht, R.E.; Jackowicz-Korczyński, M.; Lindgren, A.; Mastepanov, M.; Granlund, L.; Christensen, T.R.; Tahvanainen, T.; Martikainen, P.J.; et al. Increased nitrous oxide emissions from Arctic peatlands after permafrost thaw. *PNAS* **2017**, *114*, 6238–6243. [CrossRef] [PubMed]

63. Nikrad, M.P.; Kerkhof, L.J.; Häggblom, M.M. The subzero microbiome: Microbial activity in frozen and thawing soils. *FEMS Microbiol. Ecol.* **2016**, *92*, fiw081. [CrossRef] [PubMed]

64. Mazepa, V.S. Stand density in the last millennium at the upper treeline ecotone in the Polar Ural Mountains. *Can. J. For. Res.* **2005**, *35*, 2082–2091. [CrossRef]

65. Kapitonov, V.I. The finds of new (for Udmurtia) species of macromycetes. *Bull. Udmurt State Univ. Ser. Biol. Earth Sci.* **2013**, *4*, 9–24. (In Russian)

66. Potapov, K.O. Preliminary report on macromycetes in the dendrarium of Volzhsko-Kamsky Reserve. *Proc. Volzhsko-Kamsky Nat. Biosph. Reserve* **2016**, *7*, 167–172. (In Russian)

67. Bolshakov, V.N. (Ed.) *Red Data Book of Chelyabinsk Oblast. Animals, Plants, Fungi*; Reart: Moscow, Russia, 2017.

68. Stepanova-Kartavenko, N.T. *Aphyllophoroid Fungi of the Urals*; UrDAN: Sverdlovsk, Russia, 1969.

69. Miles, M.W.; Miles, V.V.; Esau, I. Varying climate response across the tundra, forest-tundra and boreal forest biomes in northern West Siberia. *Environ. Res. Lett.* **2019**, *14*, 075008. [CrossRef]

70. Khimich, Y.R.; Isaeva, L.G.; Borovichev, E.A. New findings of rare aphyllophoroid fungi from the Murmansk region (North-East Russia). *Folia Cryptogam. Est.* **2017**, *54*, 37–41. [CrossRef]

71. Shiryaev, A.G.; Zmitrovich, I.V.; Ezhov, O.N. Taxonomic and ecological structure of basidial macromycetes biota in polar deserts of the Northern Hemisphere. *Contemp. Probl. Ecol.* **2018**, *11*, 458–471. [CrossRef]

72. Shiryaev, A.G. Clavarioid fungi of the Urals. III. Arctic zone. *Mycol. Phytopatol.* **2006**, *40*, 294–307.
73. Bässler, C.; Heilmann-Clausen, J.; Karasch, P.; Brandl, R.; Halbwachs, H. Ectomycorrhizal fungi have larger fruit bodies than saprotrophic fungi. *Fungal Ecol.* **2015**, *17*, 205–212. [CrossRef]

 *forests*

Article

Microsite Effects on Physiological Performance of *Betula ermanii* at and Beyond an Alpine Treeline Site on Changbai Mountain in Northeast China

Dapao Yu [1], Qingwei Wang [1], Xiaoyu Wang [2,*], Limin Dai [1] and Maihe Li [3]

[1]   Key Laboratory of Forest Ecology and Management, Institute of Applied Ecology, Chinese Academy of Sciences, Shenyang 110016, China; yudp2003@iae.ac.cn (D.Y.); wangqw08@gmail.com (Q.W.); lmdai@iae.ac.cn (L.D.)
[2]   Jiyang College, Zhejiang A & F University, Zhuji 311800, China
[3]   Swiss Federal Research Institute WSL, CH-8903 Birmensdorf, Switzerland; maihe.li@wsl.ch
*   Correspondence: wangxiaoyu12@mails.ucas.ac.cn; Tel.: +86-248-397-0329

Received: 21 April 2019; Accepted: 6 May 2019; Published: 9 May 2019

Abstract: The alpine treeline demarcates the temperature-limited upper elevational boundary of the tree life form. However, this treeline does not always occur exclusively as a sharp "line", outposts of tree groups (OTG) with a height of at least 3 m are often observed in microsites up to several hundred meters beyond the line of continuous forest on some mountains. This suggests that other factors such as microenvironment may play a significant role in compensating for the alpine tree facing growth-limiting low temperature conditions. To test the microenvironment effects, this study compared the differences in growing conditions (climate and soil properties) and ecophysiological performance of Erman's birch (*Betula ermanii* Cham.) trees growing in a continuous treeline site (CTL, ~1950 m above sea level, a.s.l.) and OTGs (~2050 m a.s.l.) on Changbai Mountain in northeastern China. The results show the average 2-m air temperature for OTG was slightly lower in the non-growing season than which at the CTL (−10.2 °C < −8.4 °C), there was no difference in growing season air temperature and soil temperature at 10 cm depth between CTL and OTG. The contents of focal soil nutrients in CTL and OTG were similar. Difference in K and Mn contents between sites were detected in leaves, difference in K, Mn, and Zn in shoots. However, comparing similarity of ecophysiological performances at an individual level, trees at CTL and OTG show no significant difference. Our study reveals that mature trees at the CTL and OTG experience generally similar environmental conditions (climate and soil properties) and exhibit similar overall ecophysiological performance (reflected in carbon reserves and nutrients). This might provide insight into how mature trees might be able to survive in areas higher than the continuous treeline, as well as the importance of microclimatic amelioration provided by protective microsites and the trees themselves.

Keywords: Changbai Mountain; Erman's birch; microsite; alpine treeline; non-structural carbohydrates (NSCs)

1. Introduction

The alpine treeline marks the upper limit of the elevational distribution of trees (generally defined as 3m or greater in height) [1,2]. The climate at the alpine treeline is commonly characterized by low temperature, high radiation, strong winds, and a short growing season [2]. Of all these adverse factors, low temperature has long been recognized as the fundamental limit to tree growth [1]. The average temperature in the warmest month was initially suggested to explain the elevational position of the alpine treeline in the Alps (calculated as 10 °C), as reviewed by Gehrig-Fasel, et al. [3]. Other investigations around the world have found this temperature to range from 6 °C to 13 °C, depending on the geographical location of mountainous terrain [4]. According to Körner and Paulsen [5], the position of the global alpine treeline is commonly determined by soil temperature of 6.7 ± 0.8 °C at 10 cm-depth in the growing season. However, the importance of air temperature has recently been re-emphasized, as, for example, by Li, et al. [6] who reported a critical minimum air temperature as 0.7 ± 0.4 °C for xylem growth of Smith fir. In general, below these temperature thresholds, trees cannot grow to a mature size.

Despite the aforementioned limits, clonal tree groups and tree clusters with a height of > 3 m are quite often observed up to a few hundred meters beyond the alpine treeline on high mountains worldwide [7]. Such tree outposts distributed above the treeline, or outpost of tree groups (OTGs) are generally associated with depressions microsites and the microenvironment created by trees themselves. Microsites have been regarded as an integrated proxy for many environmental factors, including air and soil temperature [8,9], soil water and nutrient availability [10,11], geomorphologic processes [12], and snow redistribution and persistence [13,14]. Microsites modify environmental factors to influence the survival, growth, and development of trees beyond the alpine treeline [15,16]. For example, wind-blown ridge sites are commonly characterized by shallow soil and snow-free ground in winter. Leeward slopes and depressions tend to accumulate snow, which can protect seedlings from frost damage [17], desiccating winds [18,19], and xylem embolism [20]. On the other hand, thick snow cover and delays in onset of snowmelt will shorten the growing season and may increase fungal infections [21].

To date, most studies related to microsite effects have focused on seedlings and saplings [22–25]. Establishment, survival and growth rate of saplings in extreme mountain environment is considered to be highly connected to shelter availability and shelter type [23]. But microsite effects, especially in the canopy climate aspect, might weaken as trees grow to a height exceeding 3 m, since taller crowns decouple trees from the near-ground surface climate and more closely immerse them within the prevailing atmosphere, this was summaries as "boundary effect" by previous studies [24,26,27]. Root expansion and litter accumulation has also been shown to eliminate initial habitat influences on trees [28]. Thus, microsite effects may weaken to some degree as trees grow. Also, as the lapse rate of temperature, air temperature will drop by 0.6 °C as altitude increase per 100 m. That means the large trees at the OTGs might face severe low temperature limitation.

Treeline dynamics is a slow and long-term process, key characteristics of which can be identified through investigating environmental conditions and monitoring trees in long-term fixed plots. At the same time, comparing the differences in growing conditions (climate and soil properties) and ecophysiological performance of trees growing near the treeline (i.e., closed forest) vs. OTG trees may also lead to a better understanding of how trees respond to extreme environments at high elevations, and of possible trends in treeline dynamics from another perspective. Considering that OTG trees can grow to tree height (i.e., > 3 m), and the temperature conditions they might face as we discussed above, we would expect that: (1) there might be differences in 2 m air temperatures between CTL and OTGs, but similar soil temperature might be detected for CTL and OTG considering that soil temperature change might lag behind air temperatures (because of soil physics, e.g., heat capacity, heat conductance, also buffer against diurnal); (2) even though many studies have reported that carbon supply is not limited for trees at the treeline [29–32], in a recent study covering multiple tree species across a broad geographical scale, treeline trees had significantly lower non-structural carbohydrates (NSC) concentration in their roots in winter than those growing at low elevation sites [33], suggesting that carbon supply may still be limited for trees at the treeline. Considering that fact, we will also test if trees growing at CTL and OTG exhibit similar carbon reserves and other ecophysiological performance (e.g., nutrient supply), which can ensure that trees can grow to a large size. For different nutrients, we expect the macroelements (TC, TN, TP K) in leaves and shoots are not significantly different for CTL and OTG trees, based on the fact that these nutrients are required in relatively large quantities for the normal physiological processes.

A birch (*Betula*) species was selected as the focal object of study, given that as an important and widespread broadleaf alpine timberline genus, it has not been as thoroughly studied as have conifer treeline species [34]. Therefore, this study, by comparing growing environments of birches at (CTL) and beyond a treeline (OTG) on Changbai Mountain in Northeast China, and their corresponding ecophysiological performance (as reflected in NSCs and nutrient contents), may help us to better understand why large trees (>3 m in height) can be observed in microsites up to hundred meters beyond the continuous treeline, if microsites effects still works for these outposts trees.

2. Materials and Methods

2.1. Study Site Description

This study was conducted on Changbai Mountain (41°43′–42°26′N, 127°42′–128°17′E), in northeastern China. On the north slope, a pure Erman's birch (*Betula ermanii* Cham.) forest is distributed from 1700 m above sea level (a.s.l.) upward and forms a continuous and abrupt alpine treeline (CTL) at 1950 m a.s.l. Above the CTL, outposts of tree groups (OTG)—trees reaching 3m height—occur in depression microsites up to 2050 m a.s.l., accompanied by both krummholz birch (shrubby) and perennial shrub species *Rhododendron chrysanthum* Pall. and *Vacciniumu liginosum* L. [35].

The study area has a temperate continental montane climate, with cold windy winters and rainy summers. As the peak summit station recorded (Changbai Mountain Meteorological Service Bureau, during the past 30 years), average annual temperature ranges from –2.3 °C to –3.8 °C, and annual precipitation varies from 1000 mm to 1100 mm, approximately 80% of which occurs between June and September. Average annual wind speed ranges from 6 to 10 m s^{-1}, with gales of more than 200-day duration. The growing season for treeline trees generally extends from the end of May to the first severe frost, normally in late September. Soils originated from volcanic float stone, with depth ranging from 5 to 30 cm, have extremely low water-holding capacity [36,37], with soil pH of 5.2. Previous studies confirmed that global warming relieves the temperature limitation on the alpine treeline, contributing to an intense treeline upward shift (33 m per 1 °C increase) since 1985 [38].

2.2. Plant Sampling and Chemical Analyses

In 2012, along the contour of the abrupt, continuous alpine treeline (CTL), three plots (20 m × 30 m) spaced at a distance of at least 50 m were established. From the irregularly shaped outposts of tree groups (OTG), three plots of largest size (approximately 60 m^2) were selected that contained more than 5 mature birches at least 3 m of height, these plots were approximately 80 m higher in elevation than the CTL (Figure 1). In each CTL and OTG plot, 5 healthy, mature birch at least 3 m of height from each were selected for sampling. Non-shaded mature leaves and 1-year-old shoots (without bark) on leading branches were collected from selected trees. In each plot, leaves and shoots from the five selected trees were separately pooled, forming a mixed (i.e., 5-tree) sample for each tissue.

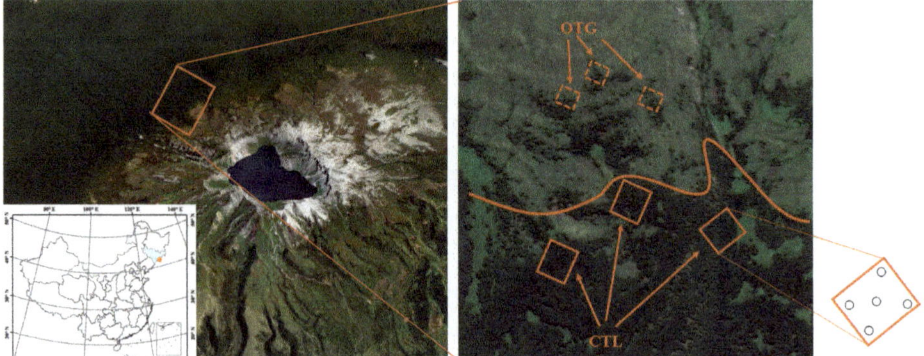

Figure 1. Study sites and plots located on the north-facing slope of Changbai Mountain, Northeast China. Rectangles with solid lines represent investigated study plots close to the continuous treeline (CTL, ~1950 m above sea level (a.s.l.)), rectangles with dotted lines represent outposts of tree groups (OTG, ~2050 m a.s.l.) (Adapted from Google Earth image), 5 birches in each plot were selected for sampling.

The NSC reserves in trees fluctuate during the growing season. In the early stage of the growing season, leaves and shoots act as carbon sinks, utilizing sugars transported from storage organs—i.e., roots and stems—that were stored in the past year. In the middle and late growing season, when the leaf development stage is finished, then carbohydrates are transported to stored organs [39]. With this in mind, sampling work in this study was conducted in the early, middle and late stages of the growing season (June 19, July 30 and September 7, respectively).

All plant samples were kept in a cool-box immediately after sampling. Samples were heated in a microwave oven (40 s at 600 W) within a maximum of 5 hours after collection to minimize biological activity and then were oven-dried to constant mass at 65 °C for approximately 48 hours. The oven-dried material was ground to fine powder and stored at 4 °C for further analysis.

The non-structural carbohydrates (NSC) concentration in plant samples is defined here as the sum of the mobile sugars plus starch. Anthrone method was used to determine total mobile sugars and starch concentrations [40]. Briefly, powdered material was extracted using 80% ethanol, incubated at 80 °C in a water bath shaker for 30 min, and then centrifuged at 5000 rpm for 5 min. After repeating this procedure twice, the combined supernatants for soluble sugar was spectrophotometrically measured at 620 nm using anthrone reagent and calculated the concentrations of soluble carbohydrate from standard regression equations using glucose as a standard. As for starch releases, the residue was boiled in distilled water for 15 min and then 9.2 M $HClO_4$ solution was added and kept for 15 min to hydrolyze the starch. Then an additional 4 mL of distilled water was added to the tube and centrifuged the mixture at 5000 rpm for 10 min, following which we extracted the pellet again with 2 mL 4.6 M $HClO_4$ solution. We then spectrophotometrically analyzed the combined supernatants for starch at 620 nm using anthrone reagent with glucose as a standard. We calculated starch concentration by multiplying glucose concentrations by a conversion factor of 0.9 [40,41]. The concentrations of soluble sugar and starch were all expressed on a dry matter (d.m.) basis (%).

Total carbon (TC, %) and total nitrogen (TN, mg/g, includes NO_3-N, NH_4-N, and microbe biomass forms) concentrations were determined with an elemental analyser (Elementar Analysensysteme Hanau, Germany). The total phosphorus (TP, mg/g) concentration was determined using an Auto Analyzer (AA3, Bran + Luebb GmbH, Germany). The nutrient concentrations of Potassium (K, mg/g), Calcium (Ca, mg/g), Magnesium (Mg, mg/g), Manganese (Mn, mg/g), Iron (Fe, μg/g), and Zinc (Zn, μg/g) in plant were measured through atomic absorption (PE Analyst 800, USA).

2.3. Soil Sampling and Chemical Analyses

Along with plant sampling, soil was sampled at a 10-cm depth at all selected trees. Using previous data as a reference, it was found that soil nutrients at the studied sites do not change substantially across the growing season (<10% change between seasons). Thus, soil samples collected from the same plot were mixed to create one sample for chemical measurements. After air-drying the soil, soil nutrients (TC, TN, TP, K, Ca, Mg, Mn, Fe, and Zn) were measured via the same methods used for plant tissues.

2.4. Climate Data Collection

At each site (CTL and OTG), micro-climatic loggers ($-30\ ^\circ$C~$50\ ^\circ$C, HOBO H8 Pro Temperature Logger, USA) were installed 2 m above the ground (on one tree at two selected plots, $n = 2$) avoiding direct sunlight to monitor 2-m air temperature. Loggers were also placed at a soil depth of 10 cm to monitor soil temperature ($n = 2$). All loggers recorded the temperature at 30 min intervals from May 20, 2009 to May 20, 2010. Growing season was defined to be when the average daily soil temperatures at 10-cm depth exceeded $3.2\ ^\circ$C [5].

2.5. Data Analyses

The average air and soil temperatures recorded from the same site (CTL or OTG) was used to calibrate each other (in case data miss for long time or large difference between the two loggers). The average daily air or soil temperature from each site was then calculated for further analysis. Paired-*t* test was used to test if air or soil temperature differs in growing and non-growing season between sites. Student's t-test was used to assess whether soil nutrients, differed significantly between CTL and OTG.

Two-way analysis of variance (ANOVA) was used to test if sites (CTL and OTG), growing season stages had significant effects on nutrients (TC, TN, TP, K, Ca, Mg, Mn, Fe, and Zn) or NSC and its components (total sugars and starch) in shoots and leaves. The Tukey–Kramer Honest Significant Difference method was used to examine differences between each responsive factor that differs between sites. Factor Analysis of Mixed Data (FAMD) [42] was then employed to compare the similarity between individuals by taking into account a mixed types of variables—both quantitative (plant tissue nutrients, NSC and its components) and qualitative (sites, tissues, growing season stages)—to test the extent to which ecophysiological performance varied between CTL and OTG sites, tissues, and growing stages. All statistical tests were conducted and all figures drawn with R 3.5.1 [43].

3. Results

3.1. Climate Conditions and Soil Properties

The growing season at OTG (123 days) was 14 days shorter than the growing season at the CTL (137 days) (Figure 2a). The average 2-m air temperature for OTG was slightly lower in the non-growing season than the temperature at the CTL ($-10.2\ ^\circ$C $< -8.4\ ^\circ$C). The air temperature in the growing season and soil temperature at 10-cm-depth for the entire year showed no differences between sites ($p > 0.05$) (Figure 2a,b). The concentrations of focal soil nutrients in CTL and OTG were similar (Figure 3).

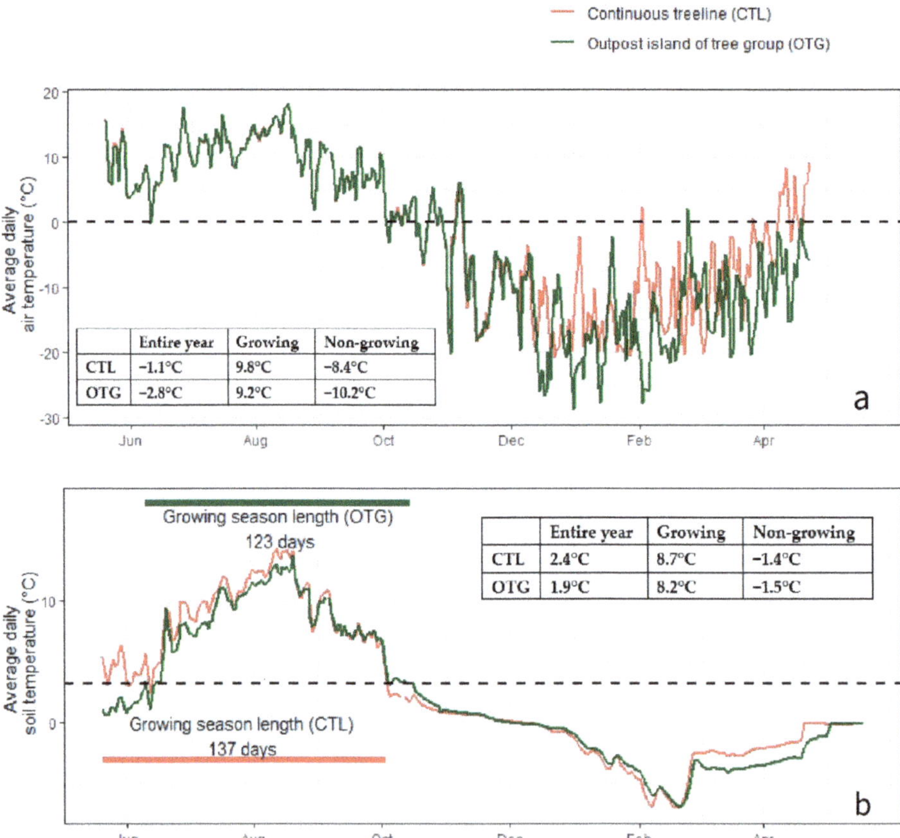

	Entire year	Growing	Non-growing
CTL	−1.1°C	9.8°C	−8.4°C
OTG	−2.8°C	9.2°C	−10.2°C

	Entire year	Growing	Non-growing
CTL	2.4°C	8.7°C	−1.4°C
OTG	1.9°C	8.2°C	−1.5°C

Figure 2. Average daily 2-m air temperature (a) and average daily 10-cm depth soil temperature (b) at continuous treeline site (CTL) and outpost-islands of tree group (OTG) on the north-facing slope of Changbai Mountain. Data was recorded from May 20, 2009 to May 20, 2010 (during the last 30 days, air temperature data was missing because of devices problem). The mean air or soil temperature in growing, non-growing, and the entire year was calculated and shown on the figure.

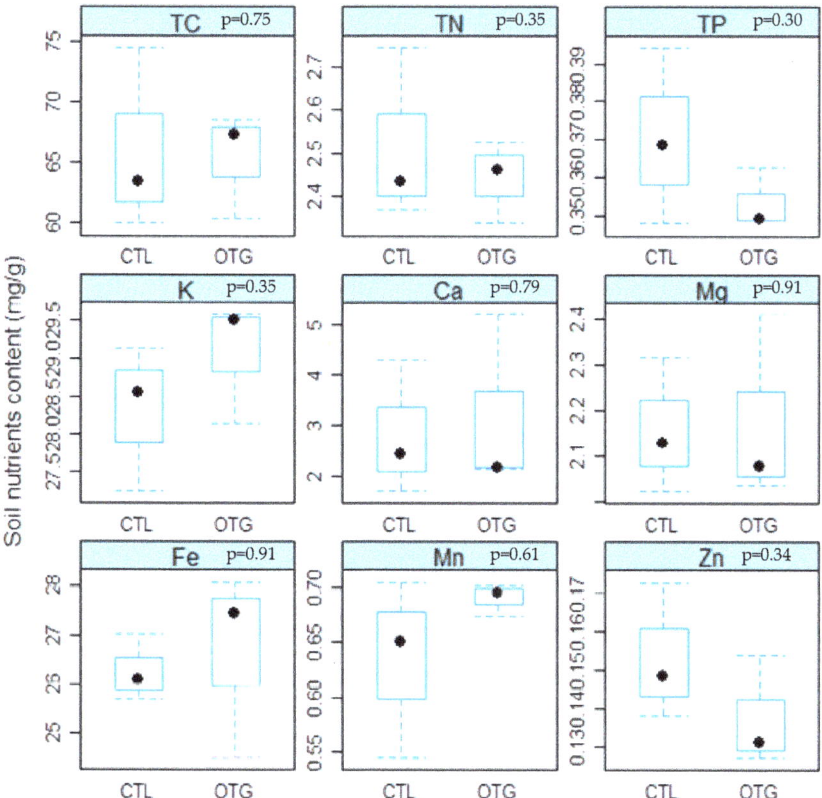

Figure 3. Soil nutrient contents (TC, TN, TP, K, Ca, Mg, Fe, Mn, and Zn) at the continuous treeline (CTL) and outposts of tree groups (OTG) sites on the north facing-slope of Changbai Mountain. The black points represent median values, the p values in panel titles show difference of variables between CTL and OTG analyzed by t-test. All the soil nutrients content showed no difference between CTL and OTG sites.

3.2. Plant Nutrients

Those measured nutrients in leaves other than Zn changed significantly across the growing season (all *p* < 0.01, Figure 4, Table S1). Leaf K and Mn contents differed significantly between CTL and OTG sites (*p* < 0.01), in which, leaves from OTG trees had significantly higher K but lower Mn than leaves from CTL trees (Figure 4a).

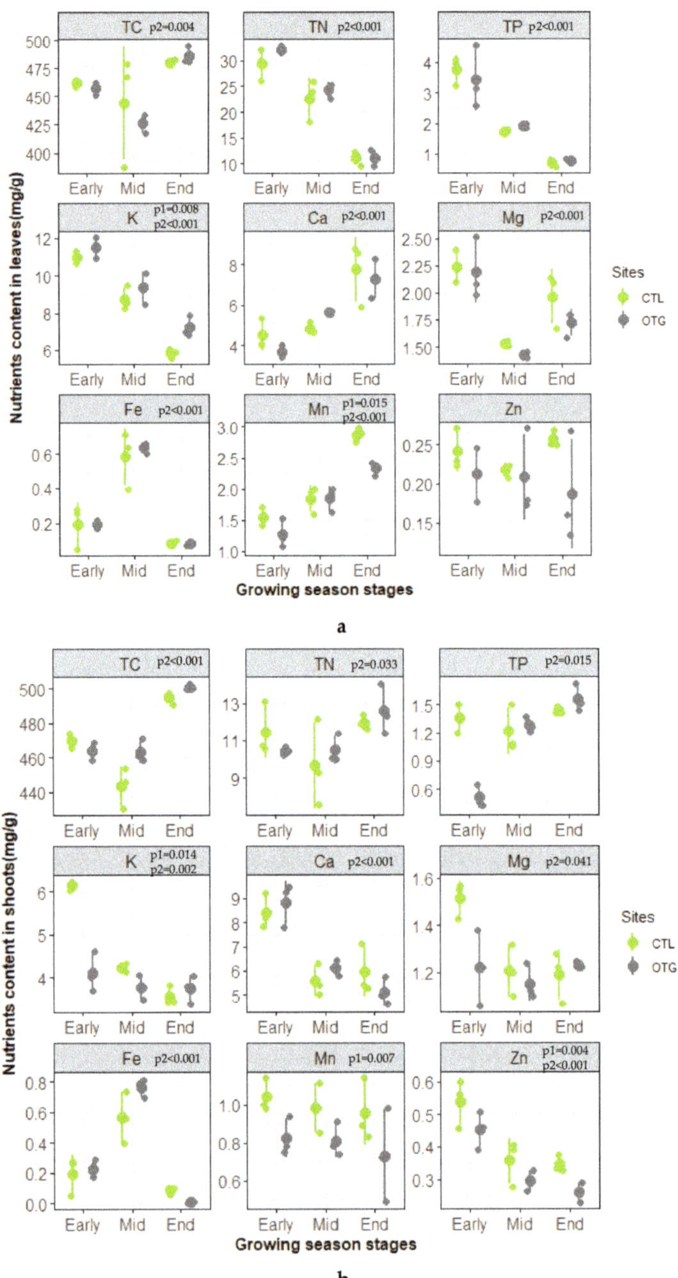

Figure 4. Nutrient contents in leaves (a) and shoots (b) of Erman's birch trees growing at the continuous treeline (CTL) and outposts beyond the treeline (OTG) on Changbai Mountain. The full data (scatter points) is shown on the figures, with large points show Mean ± SD. The p values from two-way ANOVA were shown in the figures, p1 represents difference between sites, while p2 represents difference between growing stages.

With the exception of Mn, shoot nutrients were significantly affected by growing season stage ($p < 0.05$). Site significantly affected K, Mn, and Zn in shoots ($p < 0.05$). Shoots of OTG trees had significantly lower K, Mn, and Zn compared to those of CTL trees (Figure 4b, Table S2). Other nutrients measured in shoots did not differ significantly between OTG and CTL trees.

3.3. Non-Structural Carbohydrates (NSCs)

In both leaves and shoots, the concentration of NSC, sugars, and starch all changed significantly with growing season stages ($p < 0.001$, Figure 5). Site had significant effects on NSC content in shoots: trees in OTG had significantly lower NSC ($p < 0.05$) in early season compared to those at the CTL (Figure 5; Tables S3 and S4).

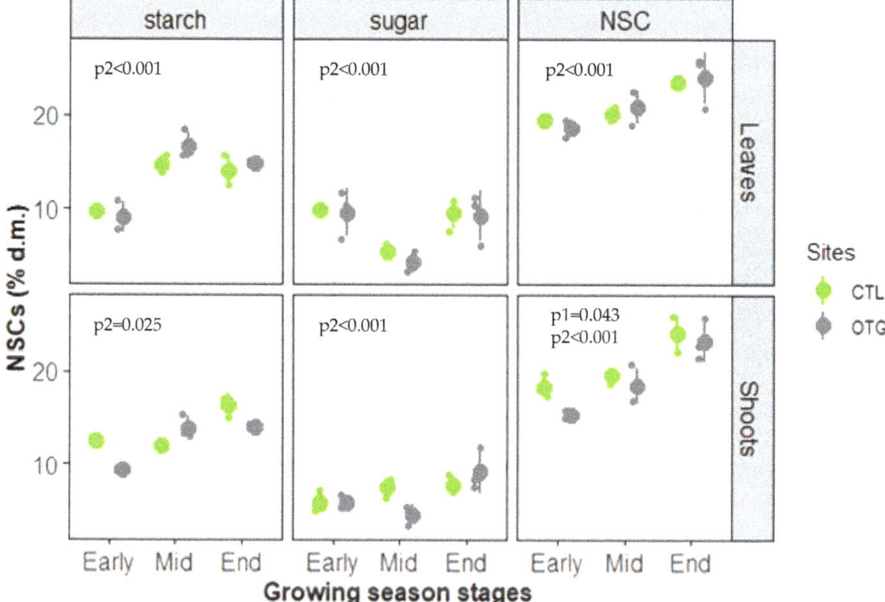

Figure 5. Non-structural carbohydrates (NSC and its components—total sugar and starch—dry mass (d.m.) fraction) (%) in leaves and shoots of Erman's birch grown at CTL and OTG. The full data is shown on the figures, with large points show Mean ± SD. The p values from two-way ANOVA were shown in the figure, p1 represents difference between sites, while p2 represents difference between sampling stages.

3.4. Relationships between Nutrients and NSC

Together, differences in organs, growing season stages, and sites (CTL vs. OTG) explained 54.8% of the variance in nutrients and NSC. The differences in overall ecophysiological performance of trees between CTL and OTG were far less than those between organs and stages (Figure 6a,b).

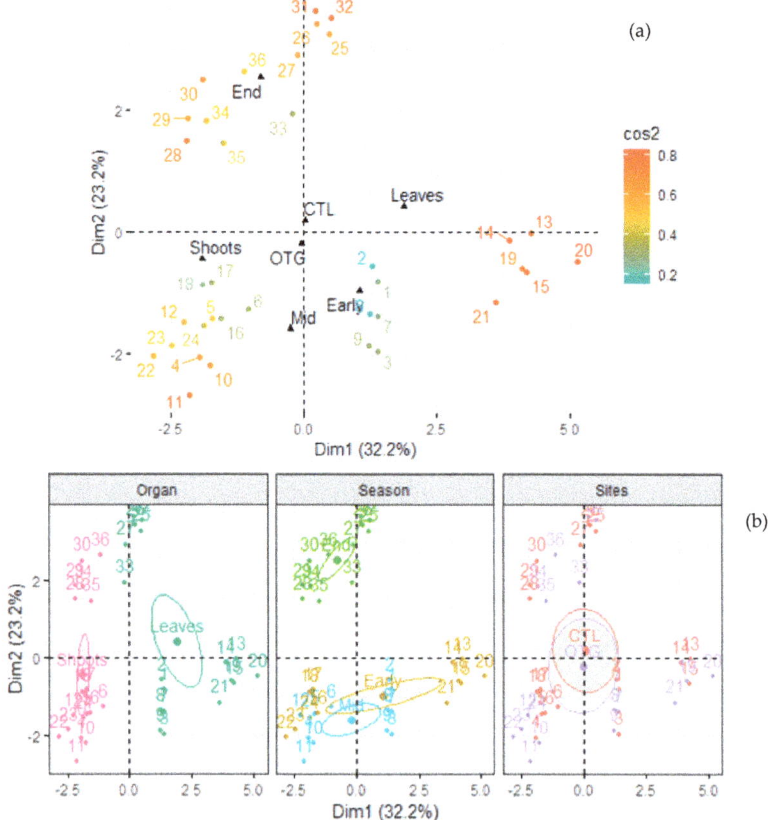

Figure 6. Factor analysis of mixed data (FAMD) used to show associations between quantitative variables (NSCs and nutrients) and qualitative variables (microsites, tissues and sampling seasons). On the factor map (a), 36 points represents characteristics of samples from 3 samples × 3 seasons × 2 organs × 2 sites. Cos2—the square of cosine value, represents the importance of the variable, a higher cos2 means this variable has a relatively larger contribution to the principal components, will distribute near the edge of the circle in the graph. Individual points with similar profiles are close to each other. On the factor map (b), multiple categorical variables were separately colored.

4. Discussion

4.1. Microsite-Modulated Climate in Growing and Non-Growing Seasons

Results showed that CTL and OTG sites with approximately 80 m difference in altitude had similar aboveground (2 m) air temperature during the growing season, but that OTG had colder air temperature than the CTL in the non-growing season (Figure 2a). This finding is partly consistent with our initial expectation that microsite effects may become negligible as trees grow to become taller than ~ 3 m, since taller crowns decouple trees from the near-ground surface climate and are more deeply immersed within the prevailing atmosphere [24,26,27]. Also, the altitude of OTGs is 80~100 m higher than which of CTL sites, this will also cause a temperature difference in some degree. But the inconsistency trend in growing and non-growing season shows that tree height not being the only factor that determines the microclimate of trees. As the previous study summarized, compact canopy with clustered branchlets and foliage will bring warmth to the trees [44]. The trees growing in OTG site adopting "compact" life forms might help them to create their own microclimate, decouple from the cold ambient air, and achieve higher heat accumulation in their leaf canopy than isolated trees. Thus, air temperature in the growing season at CTL and OTG were similar. However, in the non-growing season birch trees (especially isolated OTG individuals) lose most of their heat-capturing ability due to the loss of their deciduous leaves. That could explain why OTG trees had lower non-growing season air temperature.

In the present study, CTL and OTG had similar soil temperatures at 10 cm depth in both growing and non-growing seasons (Figure 2b). Soil temperature at the treeline is thought to be affected by the combined effects of topographic features and snow cover (if in winter) [44]. The mature trees at and beyond the treeline were all observed to be situated in depressions. Trees growing in protective depressions in both CTL and OTG sites receive protection from snow cover during the winter. This finding might partly explain why these two sites also had similar non-growing season soil temperatures.

4.2. Similar Soil Nutrient Concentrations

No significant differences were detected between the CTL and OTG for any measured soil nutrients (Figure 3). This may be because the distance between CTL and OTG is not enough to produce large climatic conditions and soil nutrient mineralization differences. In winter, mineralization in unfrozen soil with snow, nutrient release after soil thawing and meltwater input will further reduce soil nutrient differences between individuals on fine scales [44]. At present, there have been no studies directly showing treeline trees are enduring nutrient deficits [44], e.g., Li, et al. [45] studied trees at Himalayan treelines, and other studies on *Betula pubescence* ssp. tortuosa and *Betula ermanii* Cham. [46,47]. The knowledge of nutrients supply status in OTGs requires consideration of nutrient conditions in plant tissues, considering that even many studies reported highest N content in the needles or leaves, while their growth rate was still the lowest [44].

4.3. Generally Similar Ecophysiological Performances

Leaves sampled at OTG displayed significantly higher concentrations of K and lower concentrations of Mn than did CTL. K plays an important role in improving plant survival under cold stress; higher K in OTG leaves thus enhanced their tolerance for cold [48]. The range of Mn concentrations suitable for plant growth is thought to be 0.03–0.10 mg/g, but in this study, leaf Mn accumulated throughout the growing stage and reached toxic levels for trees at both sites [49]. This accumulation might be related to the fact that Mn has low mobility and always accumulates in old leaves. Mn exits the tree with deciduous leaf fall [50]. For shoots, concentrations of K, Mn, and Zn were all significantly lower in OTG than at CTL (Figure 4b). In addition to improving cold tolerance, K is also thought to function in the long-distance transport of photosynthetic carbon (C) in the phloem [51]. The shortage of K in shoots at OTG in the early growing season stage is also in accordance with the supply status of C.

In leaves, no difference was detected for NSC and its components between CTL and OTG for the entire growing season (Figure 5). This indicates that C supply is not more limited for trees growing in OTG than at the CTL. For shoots, OTG had lower NSC content (Figure 5). Early growing season starch in shoots acts as the main C source for bud break and leaf development; 2–6 weeks after leaf expansion, the carbon reserves will recover [52]. On Changbai Mountain, bud break occurs from mid-May to June, while the samples in this study in which NSC was measured were taken just after leaf expansion, when starch in shoots might not have fully recovered. The phenological variation and difference in growing season length might also affect carbon reserves in trees. By mid-season, the concentration of starch in shoots recovered to a higher level. In general, however, NSC and its components did not differ between sites, indicating that C supply of OTG trees is not more limited relative to that of CTL trees during the entire growing season. In line with nutrients status and carbon reserves results, the ecophysiological performances of trees growing at the CTL and OTG sites were found to differ much less with respect to site than with respect to organs and influence of sampling time (Figure 6).

As discussed in the previous studies [27,44], the reason why mature trees cannot rise to higher elevations like other seedlings or shrubs is because they cannot maintain enough heat like the latter plants, that is, heat level is still limiting the growth of mature trees. For seedlings, the microsites (special microtopography type such as depressions) has long been considered to be the precondition for their establishment [53]. Successfully established seedlings can grow to adult size after several warm winters. In this study, the adult trees observed at the OTG sites were evidence that trees could grow above the treeline. Evaluating the climatic conditions of CTL and OTG trees, it was found that in the growing season there was no difference between trees in CTL and OTG (for both air and soil), but in the non-growing season OTG was ~2 °C lower in air temperature than which of CTL, their soil temperature was still similar. The result obtained may come from the combined effects of canopy climate and microtopography, but because of the limited data, we cannot clearly distinguish their effect separately.

5. Conclusions

On the north slope of Changbai Mountain, Erman's birch is the single tree species forming a continuous treeline at 1950 m a.s.l. However, outposts of tree groups reaching a height of 3 m can also be observed in microsites up to 80 meters beyond the continuous treeline. Trees growing at and beyond the alpine treeline displayed similar thermal and edaphic conditions as well as ecophysiological performances (as reflected in carbon reserves and nutrients). This might help to provide insight into how mature trees might be able to survive in areas higher than the continuous treeline, as well as the importance of microclimatic amelioration provided by protective microsites and the trees themselves.

Supplementary Materials: The following are available online at http://www.mdpi.com/1999-4907/10/5/400/s1, Table S1: Multiple comparisons among average nutrients contents in leaves at different growing season stages and sites; Table S2: Multiple comparisons among average nutrients contents in shoots at different growing season stages and sites; Table S3: Multiple comparisons among average NSCs concentration in leaves at different growing season stages and sites; Table S4: Multiple comparisons among average NSCs concentration in shoots at different growing season stages and sites.

Author Contributions: Under the direction of M.L. and D.Y., X.W. contributed primarily to the writing of the manuscript. Q.W. worked with investigative work. L.D. offered valuable advice on the study.

Funding: This research was funded by the National Natural Science Foundation of China (41571197, 41371076), and in part supported by the Open Fund from the Academy of Science, Changbai Mountain (201503) and Startup Fund for Young Scholar (Jiyang College, JY2018RC05).

Acknowledgments: We sincerely appreciate suggestions for manuscript revisions from Jiaqing Liu and Hua Chen.

Conflicts of Interest: The authors declare no conflict of interest.

References

1. Wieser, G.; Holtmeier, F.-K.; Smith, W.K. Treelines in a changing global environment. In *Trees in a Changing Environment*; Springer: Dordrecht, The Netherlands, 2014; pp. 221–263.

2. Wieser, G.; Tausz, M. Current concepts for treelife limitation at the upper timberline. In *Trees at Their Upper Limit*; Springer: Dordrecht, The Netherlands, 2007; pp. 1–18.

3. Gehrig-Fasel, J.; Guisan, A.; Zimmermann, N.E. Evaluating thermal treeline indicators based on air and soil temperature using an air-to-soil temperature transfer model. *Ecol. Model.* **2008**, *213*, 345–355. [CrossRef]

4. Li, M.-H.; NK, K. The state of knowledge on alpine treeline and suggestions for future research. *J. Sichuan For. Sci. Technol.* **2005**, *26*, 36–42.

5. Körner, C.; Paulsen, J. A world-wide study of high altitude treeline temperatures. *J. Biogeogr.* **2004**, *31*, 713–732. [CrossRef]

6. Li, X.; Liang, E.; Gricar, J.; Rossi, S.; Cufar, K.; Ellison, A. Critical minimum temperature limits xylogenesis and maintains treelines on the Tibetan Plateau. Available online: https://www.biorxiv.org/content/early/2016/12/13/093781.full.pdf (accessed on 4 May 2019).

7. Friedrich-Karl, H.; Gabriele, B. Feedback effects of clonal groups and tree clusters on site conditions at the treeline: Implications for treeline dynamics. *Clim. Res.* **2017**, *73*, 85–96.

8. Scherrer, D.; Körner, C. Topographically controlled thermal-habitat differentiation buffers alpine plant diversity against climate warming. *J. Biogeogr.* **2011**, *38*, 406–416. [CrossRef]

9. Drik, W.; Roland, P.; Jörg, L. Alpine Soil Temperature Variability at Multiple Scales. *Arct. Antarct. Alp. Res.* **2010**, *42*, 117–128.

10. Schirmer, M.; Lehning, M. Persistence in intra-annual snow depth distribution: 2. Fractal analysis of snow depth development. *Water Resour. Res.* **2011**, *47*. [CrossRef]

11. Stoeckel, D.M.; Miller-Goodman, M.S. Seasonal nutrient dynamics of forested floodplain soil influenced by microtopography and depth. *Soil Sci. Soc. Am. J.* **2001**, *65*, 922–931. [CrossRef]

12. Derbyshire, E.; Hails, J.R.; Gregory, K.J. *Geomorphological Processes: Studies in Physical Geography*; Elsevier: London, UK, 2013.

13. Erickson, T.A.; Williams, M.W.; Winstral, A. Persistence of topographic controls on the spatial distribution of snow in rugged mountain terrain, Colorado, United States. *Water Resour. Res.* **2005**, *41*. [CrossRef]

14. Litaor, M.I.; Williams, M.; Seastedt, T.R. Topographic controls on snow distribution, soil moisture, and species diversity of herbaceous alpine vegetation, Niwot Ridge, Colorado. *J. Geophys. Res. Biogeosci.* **2008**, *113*. [CrossRef]

15. Hughes, N.M.; Johnson, D.M.; Akhalkatsi, M.; Abdaladze, O. Characterizing Betula Litwinowii Seedling Microsites at the Alpine-Treeline Ecotone, Central Greater Caucasus Mountains, Georgia. *Arct. Antarct. Alp. Res.* **2009**, *41*, 112–118. [CrossRef]

16. Johnson, A.C.; Yeakley, J.A. Seedling Regeneration in the Alpine Treeline Ecotone: Comparison of Wood Microsites and Adjacent Soil Substrates. *Mt. Res. Dev.* **2016**, *36*, 443–451. [CrossRef]

17. Mayr, S.; Hacke, U.; Schmid, P.; Schwienbacher, F.; Gruber, A. Frost drought in conifers at the alpine timberline: Xylem dysfunction and adaptations. *Ecology* **2006**, *87*, 3175–3185. [CrossRef]

18. Malanson, G.P.; Butler, D.R.; Fagre, D.B.; Walsh, S.J.; Tomback, D.F.; Daniels, L.D.; Resler, L.M.; Smith, W.K.; Weiss, D.J.; Peterson, D.L.; et al. Alpine treeline of Western North America: Linking organism-to-landscape dynamics. *Phys. Geogr.* **2007**, *28*, 378–396. [CrossRef]

19. Renard, S.M.; McIntire, E.J.; Fajardo, A. Winter conditions—Not summer temperature -influence establishment of seedlings atwhite spruce alpine treeline in Eastern Quebec. *J. Veg. Sci.* **2016**, *27*, 29–39. [CrossRef]

20. Mamet, S.D.; Kershaw, G.P. Age-dependency, climate, and environmental controls of recent tree growth trends at subarctic and alpine treelines. *Dendrochronologia* **2013**, *31*, 75–87. [CrossRef]

21. Barbeito, I.; Brucker, R.L.; Rixen, C.; Bebi, P. Snow Fungi-Induced Mortality of Pinus cembra at the Alpine Treeline: Evidence from Plantations. *Arct. Antarct. Alp. Res.* **2013**, *45*, 455–470. [CrossRef]

22. Hofgaard, A.; Dalen, L.; Hytteborn, H. Tree recruitment above the treeline and potential for climate-driven treeline change. *J. Veg. Sci.* **2009**, *20*, 1133–1144. [CrossRef]

23. Resler, L.M.; Butler, D.R.; Malanson, G.P. Topographic shelter and conifer establishment and mortality in an alpine environment, Glacier National Park, Montana. *Phys. Geogr.* **2005**, *26*, 112–125. [CrossRef]

24. Li, M.H.; Yang, J. Effects of microsite on growth of Pinus cembra in the subalpine zone of the Austrian Alps. *Ann. For. Sci.* **2004**, *61*, 319–325. [CrossRef]

25. Li, M.H.; Yang, J.; Kräuchi, N. Growth responses of Picea abies and Larix decidua to elevation in subalpine areas of Tyrol, Austria. *Can. J. For. Res.* **2003**, *33*, 653–662.

26. Greenwood, S.; Chen, J.C.; Chen, C.T.; Jump, A.S. Temperature and sheltering determine patterns of seedling establishment in an advancing subtropical treeline. *J. Veg. Sci.* **2015**, *26*, 711–721. [CrossRef]

27. Scherrer, D.; Koerner, C. Infra-red thermometry of alpine landscapes challenges climatic warming projections. *Glob. Chang. Biol.* **2010**, *16*, 2602–2613. [CrossRef]

28. Boateng, J.O.; Heineman, J.L.; McClarnon, J.; Bedford, L. Twenty year responses of white spruce to mechanical site preparation and early chemical release in the boreal region of northeastern British Columbia. *Can. J. For. Res.* **2006**, *36*, 2386–2399. [CrossRef]

29. Hoch, G.; Körner, C. The carbon charging of pines at the climatic treeline: A global comparison. *Oecologia* **2003**, *135*, 10–21. [CrossRef]

30. Hoch, G.; Körner, C. Growth, demography and carbon relations of Polylepis trees at the world's highest treeline. *Funct. Ecol.* **2005**, *19*, 941–951. [CrossRef]

31. Peili, S.; Christian, K.; Hoch, G. A test of the growth-limitation theory for alpine tree line formation in evergreen and deciduous taxa of the eastern Himalayas. *Funct. Ecol.* **2008**, *22*, 213–220.

32. Günter, H.; Christian, K. Global patterns of mobile carbon stores in trees at the high-elevation tree line. *Glob. Ecol. Biogeogr.* **2011**, *21*, 861–871.
33. Li, M.H.; Jiang, Y.; Wang, A.; Li, X.; Zhu, W.; Yan, C.F.; Du, Z.; Shi, Z.; Lei, J.; Schönbeck, L.; et al. Active summer carbon storage for winter persistence in trees at the cold alpine treeline. *Tree Physiol.* **2018**, *9*, 1345–1355. [CrossRef]
34. Liang, E.; Dawadi, B.; Pederson, N.; Eckstein, D. Is the growth of birch at the upper timberline in the Himalayas limited by moisture or by temperature? *Ecology* **2014**, *95*, 2453–2465. [CrossRef]
35. Yu, D.P.; Wang, Q.W.; Liu, J.Q.; Zhou, W.M.; Qi, L.; Wang, X.Y.; Zhou, L.; Dai, L.M. Formation mechanisms of the alpine Erman's birch (*Betula ermanii*) treeline on Changbai Mountain in Northeast China. *Trees* **2014**, *28*, 935–947. [CrossRef]
36. Chen, L.Z.; Bao, X.C.; Li, C.G. Structural characteristics of certain dominant species on the northern slope of Changbai mountain in Jilin province. *J. Plant Ecol.* **1964**, *2*, 207–225.
37. Yu, D.P.; Wang, G.G.; Dai, L.M.; Wang, Q.L. Dendroclimatic analysis of Betula ermanii forests at their upper limit of distribution in Changbai Mountain, Northeast China. *For. Ecol. Manag.* **2007**, *240*, 105–113. [CrossRef]
38. Du, H.; Liu, J.; Li, M.H.; Büntgen, U.; Yang, Y.; Wang, L.; Wu, Z.; He, H.S. Warming-induced upward migration of the alpine treeline in the Changbai Mountains, northeast China. *Glob. Chang. Biol.* **2018**, *24*, 1256–1266. [CrossRef] [PubMed]
39. Tamir, K.; Yann, V.; Günter, H. Coordination between growth, phenology and carbon storage in three coexisting deciduous tree species in a temperate forest. *Tree Physiol.* **2016**, *36*, 847–855.
40. Mitsuru, O.; Takuro, S.; Toshiaki, T. Redistribution of carbon and nitrogen compounds from the shoot to the harvesting organs during maturation in field crops. *Soil Sci. Plant Nutr.* **1991**, *37*, 117–128.
41. Wang, Q.-W.; Qi, L.; Zhou, W.; Liu, C.-G.; Yu, D.; Dai, L. Carbon dynamics in the deciduous broadleaf tree Erman's birch (*Betula ermanii*) at the subalpine treeline on Changbai Mountain, Northeast China. *Am. J. Bot.* **2018**, *105*, 42–49. [CrossRef] [PubMed]
42. Pagès, J. Analyse Factorielle de Donnees Mixtes. *Rev. Stat. Appl.* **2004**, *4*, 93–111.
43. R Core Team. *R: A Language and Environment for Statistical Computing*; R Foundation for Statistical Computing: Vienna, Austria, 2015.
44. Körner, C. *Alpine Treelines*; Springer: Basel, Switzerland, 2012.
45. Li, M.H.; Xiao, W.F.; Shi, P.; Wang, S.G.; Zhong, Y.D.; Liu, X.L.; Wang, X.D.; Cai, X.H.; Shi, Z.M. Nitrogen and carbon source–sink relationships in trees at the Himalayan treelines compared with lower elevations. *Plant Cell Environ.* **2008**, *31*, 1377–1387. [CrossRef]
46. Gaku, K. Intraspecific variation of leaf traits in several deciduous species in relation to length of growing season. *Ecoscience* **1996**, *3*, 483–489.
47. Karlsson, P.S.; Nordell, K.O. Intraspecific variation in nitrogen status and photosynthetic capacity within mountain birch populations. *Ecography* **1988**, *11*, 293–297. [CrossRef]
48. Manning, D.A. Mineral sources of potassium for plant nutrition. A review. *Agron. Sustain. Dev.* **2010**, *30*, 281–294. [CrossRef]
49. Limin, R.; Peng, L. Review of manganese toxicity and the mechanisms of plant tolerance. *Acta Ecol. Sin.* **2007**, *27*, 357–367.
50. Xu, W.B.; Shao, X.Q.; Wang, Y.T.; Wang, K. Advances in research on physiological effects of manganese on plants and manganese poisoning. *Grassl. Lawns* **2011**, *3*, 5–14.
51. Epron, D.; Cabral, O.M.; Laclau, J.P.; Dannoura, M.; Packer, A.P.; Plain, C.; Battie-Laclau, P.; Moreira, M.Z.; Trivelin, P.C.; Bouillet, J.P.; et al. In situ 13CO2 pulse labelling of field-grown eucalypt trees revealed the effects of potassium nutrition and throughfall exclusion on phloem transport of photosynthetic carbon. *Tree Physiol.* **2016**, *36*, 6–21. [CrossRef] [PubMed]

52. Klein, T.; Bader, M.K.F.; Leuzinger, S.; Mildner, M.; Schleppi, P.; Siegwolf, R.T.W.; Korner, C. Growth and carbon relations of mature Picea abies trees under 5years of free-air CO_2 enrichment. *J. Ecol.* **2016**, *104*, 1720–1733. [CrossRef]
53. Friedrich-Karl, H.; Gabriele, B. Landform Influences on Treeline Patchiness and Dynamics in a Changing Climate. *Phys. Geogr.* **2012**, *33*, 403–437.

Article

Leaf and Soil δ^{15}N Patterns Along Elevational Gradients at Both Treelines and Shrublines in Three Different Climate Zones

Xue Wang [1,*], Yong Jiang [2], Haiyan Ren [3], Fei-Hai Yu [1] and Mai-He Li [4]

1   Institute of Wetland Ecology & Clone Ecology/Zhejiang Provincial Key Laboratory of Plant Evolutionary Ecology and Conservation, Taizhou University, Taizhou 318000, China
2   Institute of Applied Ecology, Chinese Academy of Sciences, Shenyang 110016, China
3   College of Grassland, Resources and Environment, Key Laboratory of Grassland Resources of the Ministry of Education, Key Laboratory of Forage Cultivation, Processing and High Efficient Utilization of the Ministry of Agriculture, and Key Laboratory of Grassland Management and Utilization of Inner Mongolia Autonomous Region, Inner Mongolia Agricultural University, Hohhot 010011, China
4   Forest Dynamics, Swiss Federal Research Institute WSL, Zuercherstrasse 111, CH-8903 Birmensdorf, Switzerland
*   Correspondence: xuewang@tzc.edu.cn; Tel.: +86-136-2662-0017

Received: 16 May 2019; Accepted: 2 July 2019; Published: 3 July 2019

Abstract: The natural abundance of stable nitrogen (N) isotope (δ^{15}N) in plants and soils can reflect N cycling processes in ecosystems. However, we still do not fully understand patterns of plant and soil δ^{15}N at alpine treelines and shrublines in different climate zones. We measured δ^{15}N and N concentration in leaves of trees and shrubs and also in soils along elevational gradients from lower altitudes to the upper limits of treelines and shrublines in subtropical, dry- and wet-temperate regions in China. The patterns of leaf δ^{15}N in trees and shrubs in response to altitude changes were consistent, with lower values occurring at higher altitude in all three climate zones, but such patterns did not exist for leaf $\Delta\delta^{15}$N and soil δ^{15}N. Average δ^{15}N values of leaves ($-1.2‰$) and soils ($5.6‰$) in the subtropical region were significantly higher than those in the two temperate regions ($-3.4‰$ and $3.2‰$, respectively). Significant higher δ^{15}N values in subtro4pical forest compared with temperate forests prove that N cycles are more open in warm regions. The different responses of leaf and soil δ^{15}N to altitude indicate complex mechanisms of soil biogeochemical process and N sources uptake with environmental variations.

Keywords: ^{15}N natural abundance; nitrogen cycling; treeline; shrubline; altitude

---

1. Introduction

Variation in natural abundance of the stable isotope ^{15}N (δ^{15}N) in plants and soils can reflect N cycling in ecosystems [1] because it is related to the isotope compositions of N inputs and outputs and the internal N transformations [2]. Moreover, δ^{15}N can reflect the degree of the 'openness' of the N cycles, with higher values indicating greater N losses and a more open N cycling [3,4]. Thus, assessing patterns of δ^{15}N may help understand the process of N cycling in ecosystems [1,5].

Many studies have examined the variations of spatial patterns of δ^{15}N in plants and soils [6–9] and showed correlations of δ^{15}N and environmental factors. Elevation appears to be a major influence on leaf and soil δ^{15}N due to its natural environmental variations, such as soil water and temperature. At lower elevations where temperature tends to be higher, δ^{15}N values are likely to be higher [10,11]. Since litter decomposition and N mineralization can be accelerated by a direct increase in temperature at lower elevations [12]. However, an increasing trend or no change of δ^{15}N with elevational gradients has also been observed in some regions [4,13]. This is because not only the local microclimate but the primary physiological and biogeochemical processes regulate the N transformations in soils [2]. The soil processes including N mineralization, nitrification, denitrification and NH_3 volatilization all discriminate against ^{15}N and lead to different leaf and soil $\delta^{15}N$ signatures [7]. In addition, mycorrhizal fungi types and various N uptake sources can influence leaf $\delta^{15}N$ signatures [14–16]. Therefore, the mechanisms of elevational response for $\delta^{15}N$ could be complex.

In particular, at high elevations in alpine ecosystems, where are expected to be active in N dynamics with the increasing global warming [17]. The distributions of plant species on alpine ecosystems are constrained by upper altitude limits, resulting in markedly boundaries such as treelines and shrublines [18,19]. Although temperature has been suggested as a primary driver underlying the formation of such boundaries [20], low soil nutrient availability may also be responsible for reduced growth of trees and shrubs at their upper limits [21]. A recent study found that the growth of trees and shrubs in an alpine ecosystem increased with slightly increased nutrient availability [22], which in turn suggesting the potential increasing nutrient dynamics with the expansion of trees or shrubs. Thus, understanding changes of N cycling at alpine ecosystems is particularly important as both alpine treelines and shrublines may shift in the face of global climate change [23–25]. The signature of $\delta^{15}N$ at alpine ecosystems could serve as a powerful signal for potential effects of climate change on N cycling. For instance, experimental warming increased foliar $\delta^{15}N$ in the Swiss Alps [26], indicating an opening of N cycles with climate warming. However, we still do not fully understand the patterns of $\delta^{15}N$ in plants and soils at high altitudes of alpine ecosystems and whether plant and soil $\delta^{15}N$ at alpine treelines and shrublines show similar patterns.

We sampled leaves of trees and shrubs and soils at the upper limits of alpine treelines and shrublines and the lower altitudes in three different climate zones (subtropical, dry-temperate and wet-temperate) and measured their $\delta^{15}N$ values. Specifically, we address the following questions: (1) Do leaf and soil $\delta^{15}N$ patterns change with climatic zones? We hypothesized that leaf and soil $\delta^{15}N$ values are higher in subtropical zones than in temperate zones as temperature is higher and available N is richer in subtropical forests than in temperate forests. (2) Do leaf and soil $\delta^{15}N$ vary with altitude? We hypothesized that leaf and soil $\delta^{15}N$ would decrease with altitude as temperature decreases with increasing altitude. (3) Do leaf $\delta^{15}N$ values of trees and shrubs respond differently to changing altitude? We hypothesized that the response of leaf $\delta^{15}N$ to altitude was similar in trees and shrubs as the distribution of trees and shrubs are both restrained to the cold temperature.

2. Materials and Methods

2.1. Site Description and Sample Collection

The study sites were three mountains in China (Balang Mts., Qilian Mts. and Changbai Mts.), located in three climate zones (summarized in Table 1, Figure 1): Balang Mts. (102°52′–103°24′ E, 30°45′–31°25′ N) are located in Wolong Nature Reserve in southwestern China. The climate is subtropical, with the mean annual precipitation of about 995 mm and the annual mean temperature of 12.8 °C (measured at 1920 m a.s.l. according to the data from Wolong Nature Reserve Authority) [27]. Qilian Mts. (102°58′–103°01′ E, 37°14′–37°20′ N) are located in northwestern China. Its climate is dry-temperate, with the mean annual precipitation of about 435 mm and the annual mean temperature of 0.6 °C (measured at 2787 m a.s.l. by the Qilian weather station) [28]. Changbai Mts. (126°55′–129°00′ E, 41°23′–42′36° N) are located in northeastern China, with wet-temperate climate. For Changbai Mts., the mean annual precipitation increases from 1000 to 1100 mm and annual mean temperature decreases from −2.3 to −3.8 °C at the altitude from 1950 m to 2000 m a.s.l [29].

Table 1. The characteristics of sampling sites.

Mountain	Climate Zone	MAT (°C)	MAP (mm)	Longitude (E)	Latitude (N)	Species	Life Form	Elevation (m) a.s.l.) *	Height (m)
Balang	Sup-tropical	12.8	995	102°58′	30°51′	*Abies faxoniana* Rehder & E.H.Wilson	Tree	2860	14 ± 1.7
								3290	9.2 ± 0.9
								3670	11.5 ± 0.4
				102°45′	30°53′	*Quercus aquifolioides* Rehd. et Wils.	Shrub	2840	4.9 ± 0.1
								3160	2.8 ± 0.4
								3590	2.9 ± 0.2
Qilan	Dry-temperate	0.6	435	100°17′	38°34′	*Picea crassifolia* Kom.	Tree	2540	13.7 ± 0.9
								2870	15.5 ± 1.4
								3250	11.2 ± 0.5
				100°18′	38°31′	*Salix gilashanica* C. Wang, P.Y. Fu	Shrub	3000	1.4 ± 0.2
								3250	1.2 ± 0.0
								3550	1.5 ± 0.2
Changbai	Wet-temperate	−2.3—−3.8	1000	128°04′	42°03′	*Betula ermanii* Cham.	Tree	1700	18.9 ± 5.2
								1860	13 ± 0.8
								2030	6.4 ± 4.5
				128°04′	42°02′	*Vaccinium uliginosum* Linn.	Shrub	1430	0.5 ± 0.0
								2000	0.2 ± 0.0
								2380	0.1 ± 0.0

* The elevations showed from low to high for each tree and shrub are presented as "lower", "middle" and "upper".

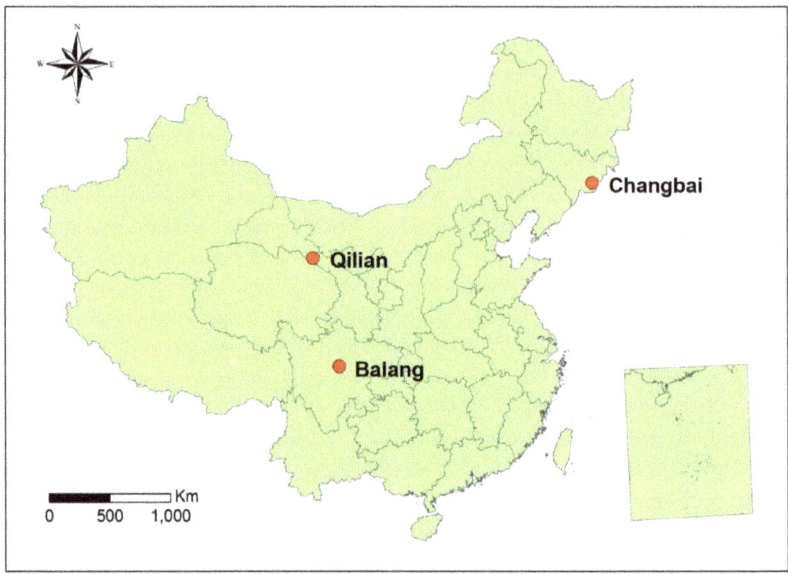

Figure 1. Location of three sampling sites of Balang Mt. (subtropical), Qilian Mt. (dry-temperate) and Changbai Mt. (wet-temperate).

In this study, the alpine treelines and shrublines were defined as the upper limits of individuals without visible disturbance and suppression. For Balang Mts., the treeline species was *Abies faxoniana* Rehder & E.H.Wilson and the shrubline species was *Quercus aquifolioides* Rehd. et Wils., and the altitude ranged from 2840 to 3670 m asl., (Table 1). For Qilian Mts., the treeline species was *Picea crassifolia* Kom. and the shrubline species was *Salix gilashanica* C. Wang, P.Y. Fu, and the altitude ranged from 2540 to 3540 m (Table 1). For Changbai Mts., the treeline species was *Betula ermanii* Cham. and the shrubline species was *Vaccinium uliginosum* Linn., and the altitude ranged from 1430 to 2380 m (Table 1). The soil was classified as Umbric Cryic Cambisols on Balang Mts., Calcaric Ustic Cambisols on Qilian Mts. and Andic Gelic Cambisols on Changbai Mts. [30].

At each site, leaf samples from the treeline and shrubline species were collected at three altitudes, i.e., the upper limit, the middle altitude and the lower altitude. At each altitude, soils (0–10 cm depth) under the canopy of the sampled trees or shrubs were collected after removing the layer of soil organic matter. They passed through a 2 mm sieve to remove stones and plant residues. At each altitude from each site, leaf samples were collected from six trees or shrubs, and a composite soil sample was collected from three randomly locations around corresponding trees or shrubs selected.

2.2. Chemical Analysis

Samples of leaves were ground into a fine powder after being oven-dried at 65 °C for 48 h, and then stored for the measurement of ^{15}N values and N concentration. For each soil sample, a fresh subsample was used for testing soil inorganic N (IN, $NO3^- + NH4^+$), and the remaining part was air dried and then milled to powder for the measurement of soil organic carbon (SOC), total carbon (TC), total N (TN) and ^{15}N abundance.

The ^{15}N abundance in leaves and soils, and concentrations of leaf N, soil TN and soil TC were analyzed in an elemental analyzer (Elementar Vario MICRO cube, Hanau, Germany) coupled with a stable isotope ratio mass spectrometer (Isoprime 100, Stockport, UK). Calibrated

DL-alanine (δ^{15}N = −1.7‰), glycine (δ^{15}N = 10.0‰) and histidine (δ^{15}N = −8.0‰) were used as the internal standards.

$$\delta^{15}\text{N (‰)} = [(R_{\text{sample}} - R_{\text{standard}})/R_{\text{standard}}] \times 10^3, \tag{1}$$

where R_{sample} is the ^{15}N/^{14}N ratio of the sample and R_{standard} is the ^{15}N/^{14}N ratio of the atmospheric N$_2$ [31]. Precision of duplicate measurements was <0.2%.

The soil IN was determined colorimetrically using an Auto Continuous Flow Analyzer (Bran and Luebbe, Norderstedt, Germany) after exacting fresh soil samples in 1 M KCL. For subtropical and wet temperate mountain zones, SOC was determined on ground soils using an elemental analyzer (Vario MACRO Cube, Elementar, Germany). For the dry temperate mountain, the ground soil samples were treated with 12 M HCl to remove inorganic C before organic C determination on the elemental analyzer [32]. Soil pH was determined in a 1:5 (soil:water) solution (*w/v*).

2.3. Statistical Analyses

The normality of the distribution and the homogeneity of the data were checked (Kolmogorov–Smirnov test) before statistical analyses. We used a three-way ANOVA to test the effects of life form (trees vs. shrubs), altitude (upper, middle and lower) and climatic zone (Balang Mts., Qilian Mts. and Changbai Mts.) on leaf and soil δ^{15}N, leaf $\Delta\delta^{15}$N (calculated as δ^{15}N$_{\text{leaf}}$ − δ^{15}N$_{\text{soil}}$), leaf N, soil inorganic N, soil total N and soil C:N. Multiple comparisons of means were conducted with Duncan's significant difference to test the different responses between altitudes. The correlations between variables were investigated with Pearson's correlation coefficient. All statistical analyses were performed using SPSS 22.0 (SPSS, Inc., Chicago, IL, USA).

3. Results

3.1. δ^{15}N in Leaves and Soils

δ^{15}N values varied significantly in different climate zones (sites; Table 2). Altitude and life form had significant effects on δ^{15}N in leaves and soils (Table 2). The two- and three-way interactions among variables also significantly influenced the δ^{15}N in leaves and soils (Table 2). Leaf δ^{15}N values of both trees and shrubs were lower at the upper limits than at lower altitudes in all three climate zones (Table 2, Figure 2a–c). There was no significant difference for leaf δ^{15}N between trees and shrubs in subtropical and dry-temperate forests, but leaf δ^{15}N of shrub (−1.3‰) was significantly higher than that of tree (−4.2‰) in wet-temperate forest (Table 2, Figure 2c). Leaf δ^{15}N was significantly different among climate zones ($p < 0.05$, Figure 2a–c), with that average δ^{15}N values of leaves in dry-temperate (−4.2‰) were lower than those in subtropical (−1.2‰) and wet-temperate (−2.8‰), but there was no difference of leaf δ^{15}N between subtropical and wet-temperate forest (Figure 2a–c).

Table 2. Three-way ANONAs of the effects of climate zones (site), altitude, life form and their interactions on Leaf $\Delta\delta^{15}$N (δ^{15}N$_{\text{leaf}}$ − δ^{15}N$_{\text{soil}}$), and δ^{15}N and N concentration in leaves and soils.

Source	df	Leaf δ^{15}N	Soil δ^{15}N	Leaf $\Delta\delta^{13}$N	Leaf N	Soil IN	Soil TN	Soil C:N
		F	F	F	F	F	F	F
Site (S)	2	83.4 ***	44.2 ***	26.0 ***	790.6 ***	27.7 ***	4.5 *	45.3 ***
Life form (L)	1	17.8 ***	7.7 **	0.2^{ns}	1776.8 ***	19.8 ***	0.0 ^{ns}	2.2 ^{ns}
Altitude (A)	2	26.7 ***	1.4 ^{ns}	8.2 ***	7.7 ***	0.8 ^{ns}	1.1 ^{ns}	4.1 *
S × L	2	4.7 *	2.5 ^{ns}	1.5 ^{ns}	1.0 ^{ns}	4.1 *	4.4 *	30.5 ***
S × A	4	4.6 **	5.7 ***	12.8 ***	7.3 ***	0.6 ^{ns}	5.8 ***	1.1 ^{ns}
L × A	2	32.3 ***	6.5 **	15.3 ***	450.9 ***	0.5 ^{ns}	0.8 ^{ns}	8.6 ***
S × L × A	4	6.7 ***	10.0 ***	11.0 ***	6.6 ***	3.8 **	3.6 **	12.7 ***

*** $p < 0.001$, ** $p < 0.01$, * $p < 0.05$, ^{ns} $p > 0.05$

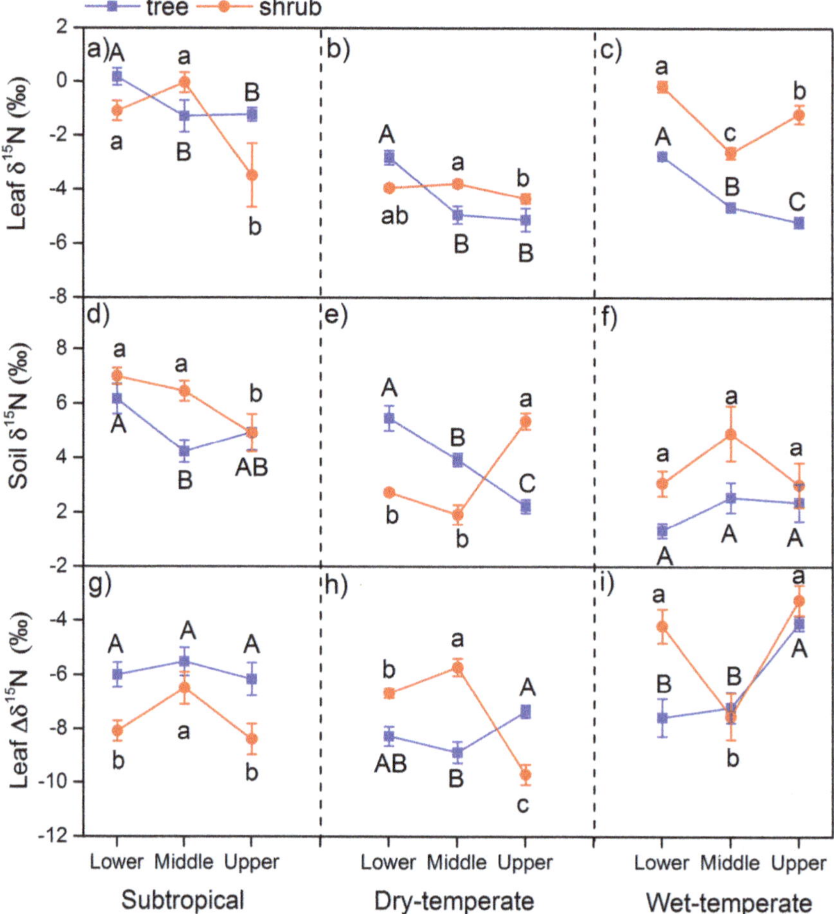

Figure 2. Leaf δ^{15}N, soil δ^{15}N and leaf $\Delta\delta^{15}$N (i.e., $\delta^{15}N_{leaf} - \delta^{15}N_{soil}$) of trees and shrubs along the altitude in subtropical (**a,d,g**), dry-temperate (**b,e,h**) and wet-temperate mountain forest (**c,f,i**), different lower case letters indicate the difference for shrub δ^{15}N between altitudes and upper case letters indicate the difference for tree δ^{15}N between altitudes.

The increasing altitude significantly decreased the value of soil δ^{15}N in subtropical forest but had no effect on that in wet-temperate forest (Table 2, Figure 2d–f). Soil δ^{15}N under the tree canopy decreased, however soil δ^{15}N under shrub canopy increased, with increasing altitude in dry-temperate forest because of the significant interaction between life form and altitude (Table 2, Figure 2e). In wet-temperate forest, soil δ^{15}N was significantly higher under shrub canopy than under tree canopy (Table 2, Figure 2f). The soil δ^{15}N in subtropical forest (5.6‰) was higher than those in dry-temperate (3.6‰) and wet-temperate forests (2.9‰) and there was no difference for soil δ^{15}N between dry- and wet-temperate forests (Figure 2d–f).

Altitude had a significant effect on leaf $\Delta\delta^{15}$N values (i.e., the difference of leaf δ^{15}N and soil δ^{15}N), however, this effect was changed by site and life form due to the significant two-way and three-way interactions (Table 2). For example, leaf $\Delta\delta^{15}$N values for shrubs were lower at the upper limits than at lower elevations, whereas $\Delta\delta^{15}$N values for trees tended to be higher at the upper limits in dry-temperate forest (Figure 2h). No significant difference for leaf $\Delta\delta^{15}$N values of trees and shrubs was found between the upper limits and lower elevations in subtropical forest, but leaf $\Delta\delta^{15}$N values for trees tended to be higher than those for shrubs in the subtropical forest (Figure 2g). Leaf $\Delta\delta^{15}N$ values were all negative for all trees and shrubs across the climate zones (Figure 2g–i).

3.2. N Concentrations in Leaves and Soils and Soil C:N Ratios

N concentrations in leaves varied in different climate zones (Table 2). Altitude had the significant effect on leaf N for dry- and wet-temperate forests but had no effect on that for subtropical forest (Table 2, Figure 3). Leaf N concentrations of shrubs were significantly higher than those of trees in both dry- and wet-temperate forests (Figure 3b–c). Average leaf N in wet-temperate forest was among the highest, and followed by dry-temperate forest and then subtropical had the least leaf N (Figure 3).

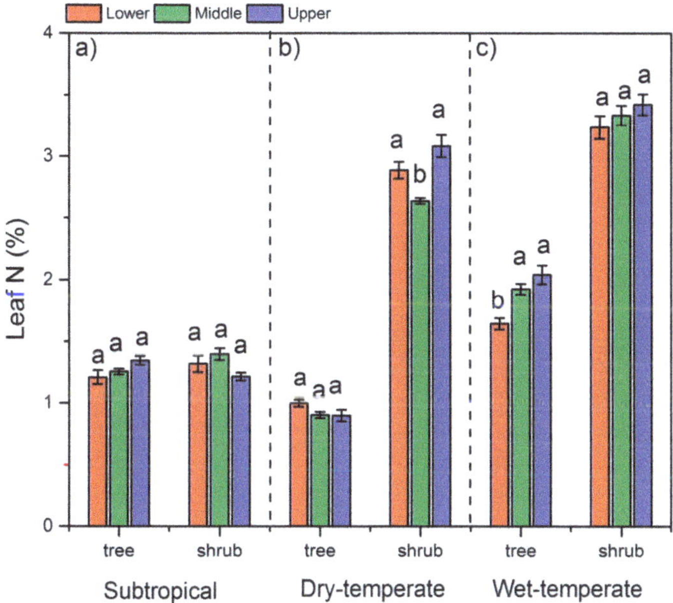

Figure 3. Leaf N of trees and shrubs along the altitude in subtropical (**a**), dry-temperate (**b**) and wet-temperate mountain forest (**c**), different lower case letters indicate the difference for leaf N concentration between altitudes.

Different climate zones had different soil total inorganic N (IN) and TN concentrations and soil C:N ratios (Table 2). There were significant two-way interactions of site and life form and three-way interactions of site, altitude and life form (Table 2). The response of soil available N to altitude was more sensitive in subtropical and dry-temperate forests than in wet-temperate where soil IN showed no change with increasing altitude (Figure 4a–c). Soil IN under tree canopy decreased but those under shrub canopy increased with increasing altitude in subtropical and dry-temperate forests, except that IN under shrub canopy in dry-temperate forest stabilized with increasing altitude (Figure 4a,b). The increasing altitude decreased soil TN under tree canopy but increased soil TN under shrub canopy in subtropical forest (Figure 4d). The increasing altitude increased soil TN under tree canopy but had no effect on soil TN under shrub canopy in dry-temperate forest, which was opposite in wet-temperate forest (Table 2, Figure 4e, f). Soil IN was lower but soil TN was higher in dry-temperate forest than those in subtropical and wet-temperate forests ($p < 0.05$, Figure 4). The soil C:N ratio under tree canopy increased but that under shrub canopy decreased with increasing altitude (Table 2; Figure 4h,i). Soil C:N ratio in wet-temperate forest (17) was the highest, followed by dry-temperate forest (15) and then subtropical forest (13; $p < 0.05$, Figure 4g–i).

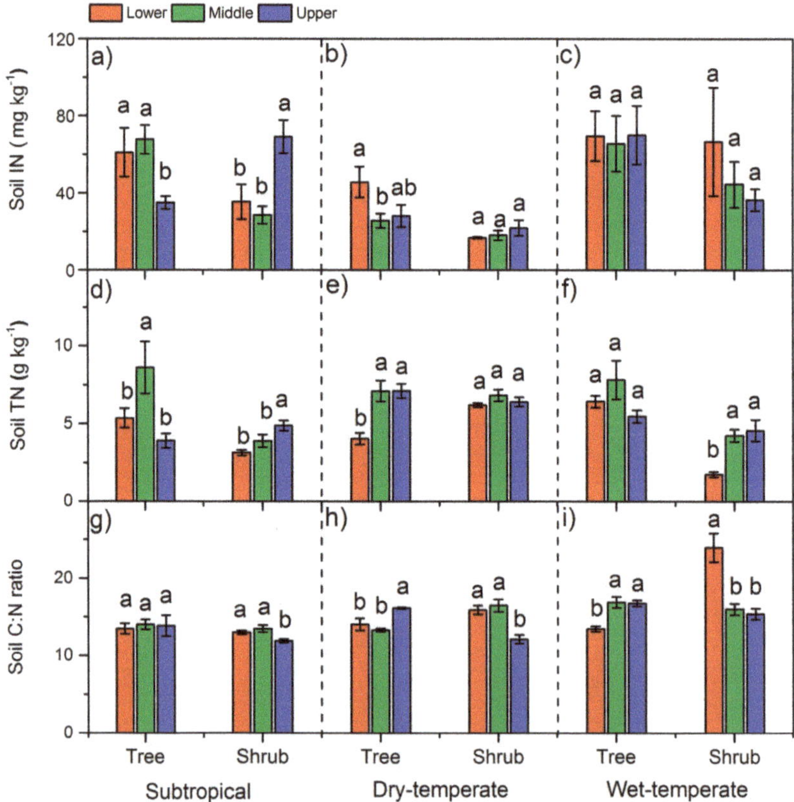

Figure 4. Soil IN, TN and C:N ratio under the canopy of trees and shrubs along the altitude in subtropical (**a,d,g**), dry-temperate (**b,e,h**) and wet-temperate mountain forest (**c,f,i**), lower case letters indicate the difference for the parameters between altitudes.

3.3. Correlations between $\delta^{15}N$ and Parameters in Leaves and Soils

Across all sampling mountains and plant types, both leaf $\delta^{15}N$ and soil $\delta^{15}N$ were negatively correlated with SOC and soil TN (Figure 5b–c); leaf $\delta^{15}N$ was negatively affected by pH (Figure 5a), however, soil $\delta^{15}N$ was negatively related to soil C:N (Figure 5d). Both leaf $\delta^{15}N$ and soil $\delta^{15}N$ were significantly correlated with leaf N in wet-temperate mountain forest (Table 3). Soil IN had a positive effect on leaf $\delta^{15}N$ and soil $\delta^{15}N$ in dry-temperate mountain forest (Table 3).

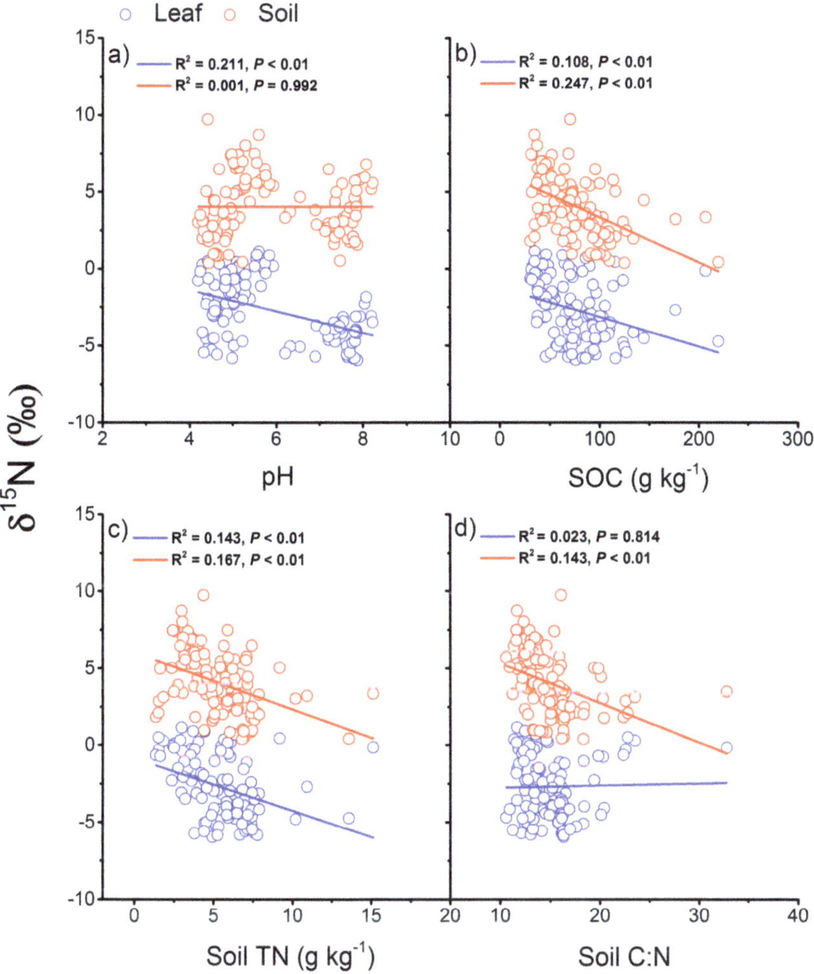

Figure 5. Relationships of δ^{15}N in leaves and soils and soil physicochemical parameters across three sampling sites.

Table 3. Correlation analyses (*R* values) of leaf δ^{15}N and soil δ^{15}N with leaf N and soil physicochemical parameters for each sampling site.

	Leaf N	Soil IN	Soil TN	Soil pH	SOC	Soil C:N
Subtropical						
Leaf δ^{15}N	0.249	−0.292	0.058	−0.439 **	0.036	−0.015
Soil δ^{15}N	0.072	−0.432 **	−0.508 **	−0.089	−0.555 **	−0.393 *
Dry-temperate						
Leaf δ^{15}N	0.172	0.496 **	−0.484 **	0.115	−0.340 *	0.006
Soil δ^{15}N	−0.05	0.476 **	−0.464 **	0.404 *	−0.578 **	−0.373 *
Wet-temperate						
Leaf δ^{15}N	0.682 **	−0.144	−0.629 **	0.123	−0.612 **	0.402 *
Soil δ^{15}N	0.459 **	−0.031	−0.433 **	−0.087	−0.388 *	0.099

** $p < 0.01$, * $p < 0.05$.

4. Discussion

4.1. The $\delta^{15}N$ and N in Different Climate Zones

The average $\delta^{15}N$ values in leaves and soils for subtropical forest was higher than the averages for both dry-and wet-temperate forests (Figure 2, $p < 0.05$). This finding is opposite to the synthesis of leaf $\delta^{15}N$ values in eastern Asian forests conducted by Fang et al. [33], in which no difference of leaf $\delta^{15}N$ values was found among tropical, subtropical and temperate forests. The opposite results may be due to the differences of the site size and species selected, since we only selected one mountain and two species in each climate zone. However, our result is consistent with the global compilation of Martinelli et al. [34], in which average leaf $\delta^{15}N$ and soil $\delta^{15}N$ values in tropical forests were respectively 6.5‰ and 8‰ higher than those in temperate forests. Similarly, another two global compilations have found the positive correlations of leaf and soil $\delta^{15}N$ with increasing MAT gradients when MAT was more than $-0.5\ ^\circ C$ and $9.8\ ^\circ C$, respectively [11,35].

The higher mean annual temperature in subtropical mountain forest might contribute to higher leaf $\delta^{15}N$ and soil $\delta^{15}N$ (Table 1). Since temperature has been suggested to be a critical factor to regulate the $\delta^{15}N$ by influencing the processes of soil mineralization, nitrification and denitrification [17,26]. These processes all fractionate nitrogen, in which more ^{15}N-depleted nitrogen can lose and consequently the nitrogen remaining is ^{15}N-enriched [2]. Previous studies also suggested that the $\delta^{15}N$ was related to soil N [36] because higher soil N concentrations can accelerate the rates of microbial N transformations [37], which probably resulted in greater potential for losses of ^{14}N and therefore higher ^{15}N retention in leaves and soils [36]. Subtropical forest has been suggested to have more available N in soils [34], however, both soil available N and total N in the subtropical forest in our study were not higher, and we even found the negative relationship between $\delta^{15}N$ and soil TN across the climate zones (Figure 5c), indicating that soil N is not the main cause for the higher $\delta^{15}N$. In addition, $\delta^{15}N$ has also suggested to be associated with N sources [2]. Atmospheric N deposition are the main sources for both plant and soil N pools. From the review of N deposition in China by Liu et al. [38], subtropical forest had a higher N deposition than temperate forest, which might cause plant $\delta^{15}N$ depletion because the deposited N is usually ^{15}N-depleted [39,40]. Therefore, more atmospheric N should lead to ^{15}N more depleted in leaves and soils. However, enhanced N deposition could also increase the soil N availability, which may further stimulate soil N biogeochemical process and thereafter lead to more enriched ^{15}N in soils.

4.2. Variation of $\delta^{15}N$ and N Along the Altitude

Although the relationship of leaf $\delta^{15}N$ and altitude was not linear in all the three climate zones, we found that leaf $\delta^{15}N$ of both trees and shrubs was consistently lower at the upper limit than at lower altitude. Unlike the patterns of leaf $\delta^{15}N$ along the altitude, the response of leaf $\Delta\delta^{15}N$ and soil $\delta^{15}N$ to altitude was climate zone-depended, showing no consistent trend for the three climate zones. Previous studies have observed a significant relationship between $\delta^{15}N$ and altitude. For example, Sah and Brume [41] found leaf $\delta^{15}N$ of *Pinus roxburghii* was negatively correlated with altitude (ranging from 1200 to 2200 m a.s.l.) in a pine forest in Nepal. Similarly, leaf $\delta^{15}N$ was more negative at higher altitude, which has been observed in Gongga Mountain in southwest China, ranging from 1100 to 4900 m a.s.l. [42]. However, leaf $\delta^{15}N$ decreased with altitude from 400 to 1350 m a.s.l. and then increased above 1350 m a.s.l. in Dongling mountain [43]. The previous study of Vitousek et al. [4] also found that there was no relationship between leaf $\delta^{15}N$ and altitude.

The altitudes in the present study ranged from 2840 to 3670 m a.s.l., 2540 to 3550 m a.s.l. and 1430 to 2380 m a.s.l. in subtropical, dry-temperate and wet-temperate mountain forests, respectively (Table 1). The lower leaf $\delta^{15}N$ at higher altitude might be attributed to decreased temperature induced by high altitude because cold temperature could weaken the activity of soil microbes and thereafter reduce the N uptake for plants and soil N loss from ammonia volatilization and other gas N losses (such as N_2O, NO), and thereby more ^{14}N-enriched retains in soils [41]. However, the response of leaf

$\Delta\delta^{15}N$ values to elevation was inconsistent in three climate zones. The negative $\Delta\delta^{15}N$ values for all trees and shrubs indicated the mineral N uptake during ^{15}N discrimination processes, and different plants discriminate against ^{15}N via associations with mycorrhizal fungus, which deliver ^{15}N-depleted N to plants [14,44]. For soil $\delta^{15}N$ along the altitude, however, not only temperature but also other factors tended to have effect on soil $\delta^{15}N$ in our study areas (Figure 2f). In general, soil $\delta^{15}N$ could index long-term dynamics of N cycling, and other factors, such as, soil pH, SOC and soil C:N ratio, might co-regulate soil $\delta^{15}N$ (Table 3, Figure 5).

4.3. Variation of $\delta^{15}N$ and N between Life Forms

Across the climate zones, leaf $\delta^{15}N$ of both shrubs and trees were more depleted at the upper limits than at lower altitudes. This result indicates that in spite of different species, the depleted $\delta^{15}N$ at higher altitude might be a general pattern.

However, the values of $\delta^{15}N$ varied between life forms in our study, especially the significant difference of $\delta^{15}N$ between shrub and tree in wet-temperate mountain forest. Different $\delta^{15}N$ values indicated different use of N sources in the same regions. The shrub *Vaccinium uliginosum* is a typical ericoid mycorrhizal plant, which could be more depleted in ^{15}N [11,15,45]. Since ericoid plants have been considered to be more reliant on mycorrhizal fungi and thereafter obtained more depleted ^{15}N [46]. However, *Vaccinium uliginosum* was more enriched than *Betula ermanii* (ectomycorrhizal plant) in wet-temperate forest and even other plants (ectomycorrhizal plants or arbuscular mycorrhizal plants) selected in both subtropical and dry-temperate forest in the present study. This is consistent with the study of $\delta^{15}N$ among life forms conducted by Schulze et al. [47], in which *Vaccinium* was more enriched than other plants at the northern treeline of Alaska. The authors attributed it to more organic N (normally ^{15}N enriched) uptake by ericoid mycorrhizae. In addition, leaf $\delta^{15}N$ in trees were not more enriched than in shrubs at their corresponding upper limits (i.e., treeline and shrubline), although the shrublines are higher than the treelines. It has been suggested that higher leaf N concentrations tend to higher leaf $\delta^{15}N$ [48,49]. For instance, from the synthesis of the global patterns of N isotope compositions, a large proportion of the variation in leaf $\delta^{15}N$ was explained by leaf N concentrations, although only occurring above a mean annual temperature (MAT) of -0.5 °C [11]. In the present study, higher leaf $\delta^{15}N$ in shrub might be attributed to its higher leaf N concentration (Table 3). However, higher leaf N in shrub in dry-temperate did not lead to its higher leaf $\delta^{15}N$. This suggests that $\delta^{15}N$ is species- and site-determined in our study.

5. Conclusions

Our results showed that leaf $\delta^{15}N$ and soil $\delta^{15}N$ were higher in subtropical forest than in dry- and wet-temperate forests. Leaf $\delta^{15}N$ of both treeline and shrubline species in three climate zones decreased with increasing altitude, whereas the response of leaf $\Delta\delta^{15}N$ and soil $\delta^{15}N$ to altitude varied in different climate zones. $\delta^{15}N$ values differed between trees and shrubs in different climate zones. Different responses of leaf and soil $\delta^{15}N$ to altitude indicate the complexity of soil biogeochemical process and N sources uptake along with environmental variations. Higher $\delta^{15}N$ values in subtropical forest indicate that N cycles are more open in warm regions. The nutrient-related effect can also explain the patterns of $\delta^{15}N$, but their effects are species- and site-dependent. Overall, the patterns of N isotopes give insights into understanding the potential climate and edaphic influence on N cycles in high-latitude and high-altitude ecosystems.

Author Contributions: Funding acquisition, X.W. and M.-H.L.; Methodology, F.-H.Y.; Supervision, Y.J. and M.-H.L.; Writing—original draft, X.W.; Writing—review & editing, H.R. and F.-H.Y.

Funding: This work was funded by Research and Innovation Initiatives of Taizhou University (2017PY033), the National Natural Science Foundation of China (41371076), and Sino-Swiss Science and Technology Cooperation (SSSTC) program (EG 06-032015).

Acknowledgments: We would like to thank Yun-Ting Fang for the comments on earlier versions of the manuscript. We appreciate the help of Xiao-bin Li on the field sampling.

Conflicts of Interest: The authors declare that they have no conflict of interest.

References

1. Robinson, D. δ^{15}N as an integrator of the nitrogen cycle. *Trends Ecol. Evol.* **2001**, *16*, 153–162. [CrossRef]
2. Högberg, P. 15N natural abundance in soil-plant systems. *New Phytol.* **1997**, *137*, 179–203. [CrossRef]
3. Austin, A.; Vitousek, P.M. Nutrient dynamics on a precipitation gradient in Hawai'i. *Oecologia* **1998**, *113*, 519–529. [CrossRef] [PubMed]
4. Vitousek, P.M.; Shearer, G.; Kohl, D.H. Foliar 15N natural abundance in Hawaiian rainforest: Patterns and possible mechanisms. *Oecologia* **1989**, *78*, 383–388. [CrossRef] [PubMed]
5. Handley, L.L.; Austin, A.T.; Stewart, G.R.; Robinson, D.; Scrimgeour, C.M.; Raven, J.A.; Schmidt, S. The 15-N natural abundance (δ^{15}N) of ecosystem samples reflects measures of water availability. *Aust. J. Plant. Phys.* **1999**, *26*, 185–199.
6. Mizutani, H.; Kabaya, Y.; Wada, E. Linear correlation netween latitude and soil ^{15}N enrichment at seabird rookeries. *Naturwissenschaften* **1991**, *78*, 34–36. [CrossRef]
7. Craine, J.M.; Brookshire, E.N.J.; Cramer, M.D.; Hasselquist, N.J.; Koba, K.; Marin-Spiotta, E.; Wang, L. Ecological interpretations of nitrogen isotope ratios of terrestrial plants and soils. *Plant Soil* **2015**, *396*, 1–26. [CrossRef]
8. Gavazov, K.; Hagedorn, F.; Buttler, A.; Siegwolf, R.; Bragazza, L. Environmental drivers of carbon and nitrogen isotopic signatures in peatland vascular plants along an altitude gradient. *Oecologia* **2016**, *180*, 257–264. [CrossRef] [PubMed]
9. Bai, E.; Boutton, T.W.; Liu, F.; Wu, X.B.; Archer, S.R.; Hallmark, C.T. Spatial variation of the stable nitrogen isotope ratio of woody plants along a topoedaphic gradient in a subtropical savanna. *Oecologia* **2009**, *159*, 493–503. [CrossRef] [PubMed]
10. Amundson, R.; Austin, A.T.; Schuur, E.A.G.; Yoo, K.; Matzek, V.; Kendall, C.; Uebersax, A.; Brenner, D.L.; Baisden, W.T. Global patterns of theisotopic composition of soil and plant nitrogen. *Glob. Biogeochem. Cycles* **2003**, *17*, 1031. [CrossRef]
11. Craine, J.M.; Elmore, A.J.; Aidar, M.P.M.; Bustamante, M.; Dawson, T.E.; Hobbie, E.A.; Kahmen, A.; Mack, M.C.; McLauchlan, K.K.; Michelsen, A.; et al. Global patterns of foliar nitrogen isotopes and their relationships with climate, mycorrhizal fungi, foliar nutrient concentrations, and nitrogen availability. *New Phytol.* **2009**, *183*, 980–992. [CrossRef] [PubMed]
12. Rustad, L.; Campbell JLMarion, G.; Norby, R.; Mitchell, M.; Hartley, A. A meta-analysis of the response of soil respiration, net nitrogenmineralization, and aboveground plant growth to experimental ecosystemwarming. *Oecologia* **2001**, *126*, 543–562. [CrossRef] [PubMed]
13. Yi, X.F.; Yang, Y.Q. Enrichment of stable carbon and nitrogen isotopes of plant populations and plateau pikas along altitudes. *J. Anim. Feed Sci.* **2006**, *15*, 661–667. [CrossRef]
14. Hobbie, E.A.; Peter, H. Nitrogen isotopes link mycorrhizal fungi and plants to nitrogen dynamics. *New Phytol.* **2012**, *196*, 367–382. [CrossRef] [PubMed]
15. He, X.H.; Xu, M.G.; Qiu, G.Y.; Zhou, J.B. Use of 15N stable isotope to quantify nitrogen transfer between mycorrhizal plants. *J. Plant Ecol.* **2009**, *2*, 107–118. [CrossRef]
16. Michelsen, A.; Schmidt, I.K.; Jonasson, S.; Quarmby, C.; Sleep, D. Leaf 15 N abundance of subarctic plants provides field evidence that ericoid, ectomycorrhizal and non-and arbuscular mycorrhizal species access different sources of soil nitrogen. *Oecologia* **1996**, *105*, 53–63. [CrossRef] [PubMed]
17. Chang, R.; Wang, G.; Yang, Y.; Chen, X. Experimental warming increased soil nitrogen sink in the Tibetan permafrost. *J. Geophys. Res. Biogeosci.* **2017**, *122*, 1870–1879. [CrossRef]
18. Dial, R.J.; Smeltz, T.S.; Sullivan, P.F.; Rinas, C.L.; Timm, K.; Geck, J.E.; Tobin, S.C.; Golden, T.S.; Berg, E.C. Shrubline but not treeline advance matches climate velocity in montane ecosystems of south-central Alaska. *Glob. Chang. Biol.* **2016**, *22*, 1841–1856. [CrossRef]
19. Myers-Smith, I.H.; Hik, D.S. Climate warming as a driver of tundra shrubline advance. *J. Ecol.* **2018**, *106*, 547–560. [CrossRef]
20. Körner, C. A re-assessment of high elevation treeline positions and their explanation. *Oecologia* **1998**, *115*, 445–459. [CrossRef]

21. Sullivan, P.F.; Ellison, S.B.Z.; McNown, R.W.; Brownlee, A.H.; Sveinbjoernsson, B. Evidence of soil nutrient availability as the proximate constraint on growth of treeline trees in northwest Alaska. *Ecology* **2015**, *96*, 716–727. [CrossRef] [PubMed]

22. Mohl, P.; Morsdorf, M.A.; Dawes, M.A.; Hagedorn, F.; Bebi, P.; Viglietti, D.; Freppaz, M.; Wipf, S.; Korner, C.; Thomas, F.M.; et al. Twelve years of low nutrient input stimulates growth of trees and dwarf shrubs in the treeline ecotone. *J. Ecol.* **2019**, *107*, 768–780. [CrossRef]

23. Myers-Smith, I. Shrub Line Advance in Alpine Tundra of the Kluane Region: Mechanisms of Expansion and Ecosystem Impacts. *Arctic* **2007**, *60*, 447–451. [CrossRef]

24. Harsch, M.A.; Hulme, P.E.; McGlone, M.S.; Duncan, R.P. Are treelines advancing? A global meta-analysis of treeline response to climate warming. *Ecol. Lett.* **2009**, *12*, 1040–1049. [CrossRef] [PubMed]

25. Hagedorn, F.; Shiyatov, S.G.; Mazepa, V.S.; Devi, N.M.; Grigor'ev, A.A.; Bartysh, A.A.; Fomin, V.V.; Kapralov, D.S.; Terent'ev, M.; Bugman, H.; et al. Treeline advances along the Urals mountain range—Driven by improved winter conditions? *Glob. Chang. Biol.* **2014**, *20*, 3530–3543. [CrossRef] [PubMed]

26. Dawes, M.A.; Schleppi, P.; Hattenschwiler, S.; Rixen, C.; Hagedorn, F. Soil warming opens the nitrogen cycle at the alpine treeline. *Glob. Chang. Biol.* **2017**, *23*, 421–434. [CrossRef] [PubMed]

27. Wang, A.; Wang, X.; Tognetti, R.; Lei, J.P.; Pan, H.L.; Liu, X.L.; Jiang, Y.; Wang, X.Y.; He, P.; Yu, F.H.; et al. Elevation alters carbon and nutrient concentrations and stoichiometry in *Quercus aquifolioides* in southwestern China. *Sci. Total Environ.* **2018**, *622*, 1463–1475. [CrossRef] [PubMed]

28. Qiang, W.Y.; Wang, X.L.; Chen, T.; Feng, H.Y.; An, L.Z.; He, Y.Q.; Wang, G. Variations of stomatal density and carbon isotope values of Picea crassifolia at different altitudes in the Qilian Mountains. *Trees* **2003**, *17*, 258–262.

29. Yu, D.; Wang, Q.; Liu, J.; Zhou, W.; Qi, L.; Wang, X.; Zhou, L.; Dai, L. Formation mechanisms of the alpine Erman's birch (*Betula ermanii*) treeline on Changbai Mountain in Northeast China. *Trees* **2014**, *28*, 935–947. [CrossRef]

30. WRB, I.W.G. *World Reference Base for Soil Resource 2014, International Soil Classification System for Naming Soils and Creating Legends for Soil Maps*; World Soil Resources Reports No. 106; FAO: Rome, Italy, 2014.

31. Mariotti, A. Atmospheric nitrogen is a reliable standard for natural ^{15}N abundance measurements. *Nature* **1983**, *303*, 685–687. [CrossRef]

32. Wang, R.; Dungait, J.A.J.; Creamer, C.A.; Cai, J.; Bo, L.; Xu, Z.; Zhang, Y.; Ma, Y.; Yong, J. Carbon and nitrogen dynamics in soil aggregates under long-term nitrogen and water addition in a temperate steppe. *Soil Sci. Soc. Am. J.* **2015**, *79*, 527–535. [CrossRef]

33. Fang, Y.; Koba, K.; Yoh, M.; Makabe, A.; Liu, X. Patterns of foliar delta δ^{15}N and their control in Eastern Asian forests. *Ecol. Res.* **2013**, *28*, 735–748. [CrossRef]

34. Martinelli, L.A.; Piccolo, M.C.; Townsend, A.R.; Vitousek, P.M.; Cuevas, E.; Mcdowell, W.; Robertson, G.P.; Santos, O.C.; Treseder, K. Nitrogen stable isotopic composition of leaves and soil: Tropical versus temperate forests. *Biogeochemistry* **1999**, *46*, 45–65. [CrossRef]

35. Craine, J.M.; Elmore, A.J.; Wang, L.; Augusto, L.; Baisden, W.T.; Brookshire, E.N.J.; Cramer, M.D.; Hasselquist, N.J.; Hobbie, E.A.; Kahmen, A.; et al. Convergence of soil nitrogen isotopes across global climate gradients. *Sci. Rep.* **2015**, *5*, 8280. [CrossRef] [PubMed]

36. Pardo, L.H.; Templer, P.H.; Goodale, C.L.; Duke, S.; Groffman, P.M.; Adams, M.B.; Boeckx, P.; Boggs, J.; Campbell, J.; Colman, B. Regional Assessment of N Saturation using Foliar and Root δ^{15}N. *Biogeochemistry* **2006**, *80*, 143–171. [CrossRef]

37. Booth, M.; Stark, J.E. Controls on nitrogen cycling in terrestrial ecosystems: A synthetic analysis of literature data. *Ecol. Monogr.* **2005**, *75*, 139–157. [CrossRef]

38. Liu, X.; Duan, L.; Mo, J.; Du, E.; Shen, J.; Lu, X.; Zhang, Y.; Zhou, X.; He, C.; Zhang, F. Nitrogen deposition and its ecological impact in China: An overview. *Environ. Pollut.* **2011**, *159*, 2251–2264. [CrossRef] [PubMed]

39. Garten, C.T.J. Nitrogen isotope composition of ammonium and nitrate in bulk precipitation and forest throughfall. *Int. J. Environ. Anal. Chem.* **1992**, *47*, 33–45. [CrossRef]

40. Houlton, B.Z.; Edith, B. Imprint of denitrifying bacteria on the global terrestrial biosphere. *Proc. Natl. Acad. Sci. USA* **2009**, *106*, 21713–21716. [CrossRef] [PubMed]

41. Sah, S.; Brume, R. Altitudinal gradients of natural abundance of stable isotopes of nitrogen and carbon in the needles and soil of a pine forest in Nepal. *J. For. Sci.* **2003**, *49*, 19–26. [CrossRef]

42. Liu, X.-Z.; Wang, G. Measurements of nitrogen isotope composition of plants and surface soils along the altitudinal transect of the eastern slope of Mount Gongga in southwest China. *Rapid Commun. Mass Spectrom.* **2010**, *24*, 3063–3071. [CrossRef] [PubMed]
43. Liu, X.Z.; Wang, G.A. Nitrogen isotope composition characteristics of modern plants and their variations along an altitudinal gradient in Dongling Mountain in Beijing. *Sci. China* **2010**, *53*, 128–140. [CrossRef]
44. Hobbie, E.A.; Colpaert, J.V. Nitrogen Availability and Colonization by Mycorrhizal Fungi Correlate with Nitrogen Isotope Patterns in Plants. *New Phytol.* **2003**, *157*, 115–126. [CrossRef]
45. He, X.H.; Critchley, C.; Bledsoe, C. Nitrogen Transfer Within and Between Plants Through Common Mycorrhizal Networks (CMNs). *Crit. Rev. Plant Sci.* **2003**, *22*, 531–567. [CrossRef]
46. Hobbie, E.A.; Jumpponen, A.J. Foliar and fungal 15N:14N ratios reflect development of mycorrhizae and nitrogen supply during primary succession: Testing analytical models. *Oecologia* **2005**, *146*, 258–268. [CrossRef]
47. Schulze, E.D.; Chapin, F.S.; Gebauer, G. Nitrogen nutrition and isotope differences among life forms at the northern treeline of Alaska. *Oecologia* **1994**, *100*, 406–412. [CrossRef]
48. Pardo, L.H.; McNulty, S.G.; Boggs, J.L.; Duke, S. Regional patterns in foliar ^{15}N across a gradient of nitrogen deposition in the northeastern US. *Environ. Pollut.* **2007**, *149*, 293–302. [CrossRef]
49. Kang, H.; Liu, C.; Yu, W.; Wu, L.; Chen, D.; Sun, X.; Ma, X.; Hu, H.; Zhu, X. Variation in foliar δ ^{15}N among oriental oak (*Quercus variabilis*) stands over eastern China: Patterns and interactions. *J. Geochem. Explor.* **2011**, *110*, 8–14. [CrossRef]

 *forests* 

Article

Temperature, Wind, Cloud, and the Postglacial Tree Line History of Sub-Antarctic Campbell Island

Matt S. McGlone [1,*], **Janet M. Wilmshurst** [1,2], **Sarah J. Richardson** [1], **Chris S.M. Turney** [3] and **Jamie R. Wood** [1]

[1]   Manaaki Whenua-Landcare Research, Lincoln 7640, New Zealand;
      janet.wilmshurst@auckland.ac.nz (J.M.W.); richardsons@landcareresearch.co.nz (S.J.R.);
      woodj@landcareresearch.co.nz (J.R.W.)
[2]   School of Environment, The University of Auckland, Auckland 1142, New Zealand
[3]   Palaeontology, Geobiology and Earth Archives Research Centre, and ARC Centre of Excellence for
      Australian Biodiversity and Heritage, School of Biological, Earth and Environmental Sciences,
      University of New South Wales, New South Wales 2052, Australia; c.turney@unsw.edu.au
*   Correspondence: mcglonem@landcareresearch.co.nz; Tel.: +64-3-321-9733

Received: 3 October 2019; Accepted: 5 November 2019; Published: 7 November 2019

Abstract: Campbell Island, which is 600 km south of New Zealand, has the southernmost tree line in this ocean sector. Directly under the maximum of the westerlies, the island is sensitive to changes in wind strength and direction. Pollen records from three peat cores spanning the tree line ecotone provide a 17,000-year history of vegetation change, temperature, and site moisture. With postglacial warming, tundra was replaced by tussock grassland 12,500 years ago. A subsequent increase of shrubland was reversed at 10,500 years ago and wetland-grassland communities became dominant. Around 9000 years ago, trees spread, with maximum tree line elevation reached around 6500 to 3000 years ago. This sequence is out of step with Southern Ocean sea surface temperatures, which were warmer than 12,500 to 9000 years ago, and, subsequently, cooled. Campbell Island tree lines were decoupled from temperature trends in the adjacent ocean by weaker westerlies from 12,500 to 9000 years ago, which leads to the intrusion of warmer, cloudier northern airmasses. This reduced solar radiation and evapotranspiration while increasing atmospheric humidity and substrate wetness, which suppressed tree growth. Cooler, stronger westerlies in the Holocene brought clearer skies, drier air, increased evapotranspiration, and rising tree lines. Future global warming will not necessarily lead to rising tree lines in oceanic regions.

Keywords: tree line; sub-Antarctic; westerly winds; postglacial; Holocene; Southern Ocean; climate change; palynology; cloud; peat

---

1. Introduction

Tree lines are confidently predicted to rise as a result of global climate change in the near future, which will lead to a profound impact on alpine ecosystems [1]. Nevertheless, there are many uncertainties regarding the impact on the tree line ecotone [2] and even some doubt as to whether all tree lines will rise. For instance, a global survey of tree line response to climatic warming over the last 100 years showed that, while some were rapidly advancing, many had either responded sluggishly or not at all [3]. Oceanic tree lines, such as those that characterize New Zealand and southern South America, are among those that have failed to respond to recent warming. Why this may be so, and under what circumstances these tree lines may change, is, therefore, of great interest [4]. The sub-Antarctic islands lying several hundred kilometers to the south of New Zealand offer a unique opportunity to explore these questions.

Two island groups several hundred kilometers to the south of the New Zealand mainland (Auckland and Campbell) are at the southern limit to trees in the Australasian sector of the Southern Ocean (Figure 1). Lying directly under the maximum of the southern westerly wind belt, they have windy, moist, cool, cloudy, and highly oceanic climates and are nearly entirely blanketed with deep organic soils, but have low-growing forest on their leeward flanks [5]. Although local environmental factors can prevent the growth of trees, globally, the high elevation limit to the tree line is associated with a minimum growing season length and mean temperature [6,7]. Palaeoecological studies suggest that tree lines on these islands may have been decoupled from a regional temperature change and, thus, provide an exception to this global pattern [8]. Rise to dominance of the forest in these islands was delayed until the mid-Holocene [8], which is a lag inconsistent with the warm, early postglacial sea surface temperatures (SSTs) in the adjacent ocean [9]. The anomalous response of sub-Antarctic tree lines to oceanic warming may, therefore, have been driven by other factors linked to the westerlies such as humidity, sunshine hours, and wind strength. In this study, we explore this possibility.

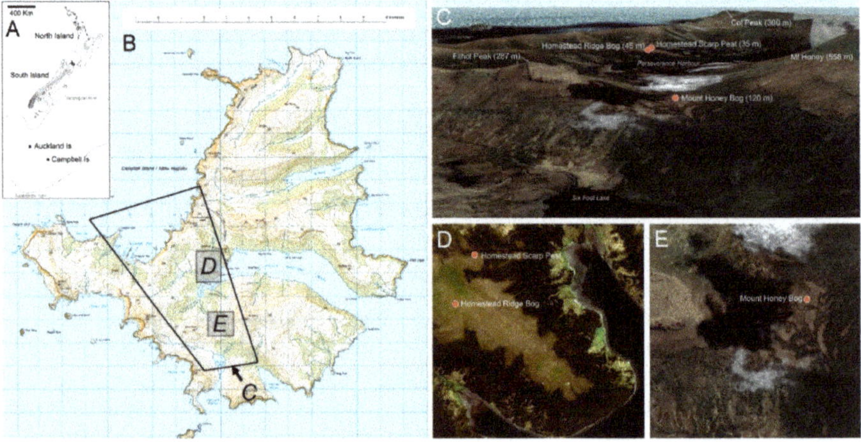

Figure 1. Campbell Island. Location in the southern part of the ocean (**A**), topographic map (**B**), oblique view, Perseverance Harbour (**C**), vertical view, Homestead Ridge bog (**D**), vertical view, and Mount Honey Saddle bog (**E**). Imagery from Google Earth.

In this study, we present the postglacial vegetation, peat, and climatic history of extensive peat bog complexes on the southernmost of these forested islands, including sub-Antarctic Campbell Island (52°34′ S, 169°09′ E) some 600 km south of the New Zealand mainland. Truncated pollen diagrams with pollen-based temperature estimates have been reported [8]. We now provide the full pollen results for these and an additional site, along with moisture proxies. We also present summaries of cloud levels, daily peat water table depth measurements, and peat vegetation studies, which were undertaken many years ago to support ecological studies and mapping, but not compiled until recently [10]. We use these vegetation and peat analyses from bogs with contrasting elevations, exposure, and hydrology to help determine the influence of temperature, wind, and cloudiness on the postglacial climatic regime and explore their effects on the vegetation and position of the tree line.

2. Physical Geography and Vegetation Cover of Campbell Island

2.1. Geology, Topography, Soils, and Climate

Campbell Island has a surface area of 113 km^2 and consists of the eroded eastern segment of a Pliocene volcanic complex [11]. Basaltic flows and intrusions make up two-thirds of the island and schists, conglomerates, mudstones, limestone tuffs, and breccias make up the rest. The island has a rugged topography, with some peaks over 500 m elevation and evidence of glacial action in the form of cirques, U-shaped valleys, and moraine-like landforms [5,12]. The west coast is exposed to the prevailing westerly winds and has steep cliffs eroded by strong wave action, while the eastern flanks have gentle to moderately steep slopes and broad valleys (Figure 1).

Campbell Island has a mild, humid, extremely windy climate with a muted annual cycle [13]. Mean annual temperature at sea level is 6.8 °C (monthly range 9.3–4.6 °C). The mean daily minimum of July (the coldest month) is 2.5 °C, and the mean daily maximum in January (warmest month) is 11.9 °C. Campbell Island technically has a year-long growing season, since, even in winter, daily mean minima are more than 2.5 °C, which is well above the 0.9 °C mean daily minimum that defines the tree line growing season limit [7]. The mean annual temperature of 6.5 °C at the tree line on Campbell Island is nearly identical to the global tree line average of 6.4 °C [7], which suggests that this relationship is valid for this location. Mean annual precipitation of 1404 mm falls on an average of 252 d with little seasonal variation. Sunshine hours are 664 per annum. Cool, moist conditions make decomposition slow, and, thus, highly organic peat soils cover nearly the entire landscape. Only fresh slips, cliffs, and steep slopes of ≥30°, as well as rocky outcrops and exposed tops above 400 m a.s.l. have mineral soils. An overview of the soils is provided in References [11,14] and their nutrient status in References [14–17].

2.2. Vegetation

The vegetation and plant ecology of Campbell Island is documented in References [16,18,19]. Key vegetation types are shown in Figure 2a–d. Well drained, sheltered lowland sites are in a low forest (≤5 m) or scrub dominated by *Dracophyllum scoparium* and *Dracophyllum longifolium* over the subdominant shrubs *Myrsine divaricata* and *Coprosma ciliata*, and a fern ground cover of *Polystichum vestitum*. The tree line is defined as the presence of groups of trees at least 3 m tall [6] and, by this standard, the Campbell Island tree line sits at 30 m a.s.l. (Figure 2a), even though *Dracophyllum* stands up to 2.5 m tall dominate sheltered sites up to 120 m elevation. We regard this elevational range as a tree line ecotone. *Myrsine* and *Coprosma* shrublands become more abundant with increasing elevation and are often the main cover from 100 m to 250 m on sheltered slopes where the last shrubland patches occur (Figure 2c). *Chionochloa antarctica* tussock grassland is the main vegetation above 150–200 m (Figure 2d) but, in turn, intergrades with a herb field of macrophyllous forbs (*Anisotome antipoda*, *A. latifolia*, *Pleurophyllum hookeri*, *P. speciosum*, and *Bulbinella rossii*), which become prominent above 250–300 m (Figure 2e). The open, exposed tops (above 300 m) are in a tundra of low-growing macrophyllous forbs, sedges, rushes, and stunted grasses.

Figure 2. *Cont.*

Figure 2. Main vegetation types on Campbell Island. Tree line *Dracophyllum* at 30 m a.s.l. near the head of Perseverance Inlet (**a**), *Oreobolus* cushion bog, c. 40 m (**b**), *Myrsine* and *Coprosma* shrubland, c. 160 m (**c**), grassland above tree line, c. 220 m (**d**), macrophyllous forb community, c. 400 m (**e**).

Wind, drainage, and exposure to sea spray alter this general pattern markedly. The exposed western cliffs and steep slopes are covered with macrophyllous forb communities with *Poa litorosa* tussocks, grassland extending from the sea level to the tundra, and woody vegetation being confined to low-growing wind-trimmed shrubland in sheltered locations. Ombrogenous (rain-fed) moorlands occur on low-angle valley bottoms and on gentle slopes, particularly on eastern headlands. In this region, extensive cushion bogs form dominated by the sedges *Oreobolus pectinatus* and *Isolepis* spp., together with *Centrolepis* spp., *Astelia subulata*, and stunted *Dracophyllum scoparium* (Figure 2b). Wet fens have a cover of *Carex appressa*, and the macrophyllous forb *Pleurophyllum criniferum* and *Blechnum montanum*.

3. Study Sites (Figure 1)

Previous investigations have shown that the raised bogs of Campbell Island provide the most complete record of past vegetation while, at the same time, sensitively recording fluctuations in soil surface moisture [12]. Two bogs were selected to provide maximum information on tree line fluctuations: one lying at the limit to *Dracophyllum* forest (Homestead Ridge—two core sites) and another within the upper *Dracophyllum* shrubland ecotone (Mt Honey Saddle).

3.1. Homestead Ridge Bog (HRB)

The Homestead Ridge lies at the head of Perseverance Harbor between two inlets, known as Camp Cove and Tucker Inlet. Its crest is mantled by a large, raised, oligotrophic cushion bog, about 1 km long and averaging 300 m wide and extending over c. 25 ha (Figure 1d). This raised bog is part of a larger bog complex, cut by steep-sided streams, extending inland to the west and covering much of the local drainage basin. The central dome is covered by a mosaic of wetland cushions of *Oreobolus*, *Isolepis*, *Centrolepis*, *Astelia*, moss, stunted *Dracophyllum scoparium*, and scattered *Chionochloa* tussocks. The flanks of the bog are mostly covered with a *Dracophyllum longifolium* low forest or shrubland with subdominant *Myrsine divaricata*, *Coprosma cuneata*, and *Coprosma ciliata*, as are the banks of the streams in the western bog complex. Gaps in the low canopy are occupied by *Polystichum vestitum* ferns, *Chionochloa antarctica* tussocks, and, close to the shore, where sea lions haul out, and swards of the macroforb, *Bubinella rossii*.

The bog forms an elongated, low-angle (1–5°) peat dome, averaging 8.6 m deep, with a maximum depth of 9.6 m at the crest (45 m a.s.l.). At the same time, the oligotrophic bog gives way to *Dracophyllum* shrub and a low forest. The slopes steepen to 10–15°. The peat underlying the oligotrophic bog is a wet, fibrous sedge peat with abundant wood in the top 6–7 m, fining in the bottom meter to a gyttja-like peat above the blue silt base. Under the scrub-covered and forest-covered margins is a dry, dense, coarsely fibrous, well humified peat with abundant wood throughout the top three quarters of the profile. It varies in depth according to the slope from 2.0–4.0 m. The core (laboratory i.d. X9901) was taken from the bog summit, which has a vegetation cover of cushion bog and scattered patches of low-growing *Dracophyllum scoparium*.

3.2. Homestead Scarp Peat (HSP)

Homestead Scarp (c. 35 m a.s.l.) lies on the north-eastern flank of Homestead Ridge bog complex (Figure 1d) in a steep-sided small valley covered in dense *Dracophyllum longifolium* 1.5–2 m high, with occasional *Myrsine divaricata* and *Coprosma* spp. In this region, recent erosion has exposed a 2 m peat bank. A pit was dug for a further 2 m into the base of the bank for a total sample exposure of 3.9 m of peat above the coarse gravel basement (laboratory i.d. X9911).

3.3. Mount Honey Saddle Bog (MHB)

At approximately 120 m a.s.l. in the saddle between Mount Honey and Filhol Peak, an oligotrophic bog complex occupies the headwaters of streams draining to the south (Figure 1d). The bog complex lies within the upper *Dracophyllum*-dominant scrub some 30 m vertically below the upper boundary of the closed scrub. *Dracophyllum* in the vicinity has a canopy height of c. 1 m on exposed ridges but up to 2.5 m in sheltered depressions. The bog has a hummocky microtopography, with a cover of approximately 30% stunted (0.1–0.6 m high) *Dracophyllum scoparium*, alongside occasional tufts of *Chionochloa* and prostrate *Myrsine* and *Coprosma cuneata*, over a ground cover dominated by *Astelia*, *Oreobolus*, and *Dicranoloma* moss. A single c. 7 m core of fibrous sedge peat was taken from near the highest point of the complex (laboratory i.d. X9903).

4. Methods

4.1. Water Table, Vegetation Cover, and Surface Wetness Index

In order to assess the relationship of oligotrophic bog vegetation cover to the water table, we located 21 1 × 1 m^2 plots across Homestead Ridge Bog in December 1998, including austral summer. The 11 plots were distributed at regular intervals along 2 transects running from shrubland to the bog dome summit. A further 10 were subjectively allocated to sample specific communities or soil moisture conditions by selecting markedly different vegetation compositions. Vegetation cover (%) was estimated visually. A 15 × 15 cm pit was dug adjacent to the plot down to the water table and its depth was measured as distance from the surface of the litter layer to the water surface after the level equilibrated (c. 20 minutes for most plots). Vegetation cover percentages and water table depth are provided in Table S2.

No unusually heavy rainfall occurred during the measurement period and the average water table depth is nearly identical to the long-term summer water table as recorded from a site within our survey area [10]. We tested whether vegetation composition was significantly related to the water table depth using permutational multivariate analysis of variance (PERMANOVA) [20] (implemented using the function *adonis* in the *vegan* library [21] in R v3.5.1). To determine how individual taxa responded to water table depth, the percent coverage of each taxa was correlated with the water table depth using Spearman's rank correlation. We removed all singleton taxa before analyses since they provide negligible information about how the plant community is related to the water table depth [22,23].

Results from a modern pollen survey of the Campbell vegetation communities spanning significant gradients of temperature and moisture [24] were used via ordination techniques (non-metric multidimensional scaling = NMS) [21,25,26] to develop a surface wetness index (SI: Derivation of the surface wetness index, Figures S4 and S5).

4.2. Peat Sampling

HRB and MHB were sampled with a 10-cm diameter and a 5-cm diameter 'D'-section corer, respectively. Core segments of 50-cm lengths were transferred to plastic half drain pipes, wrapped in plastic film, and stored at 4 °C for return to the mainland. HSP was sampled by cutting back an exposure and taking block samples ca. 25 cm × 10 cm. Subsamples were taken for loss-on-ignition (LOI), macrofossils, pollen analysis, and radiocarbon dating. LOI samples of 1 cm^3 were taken at 2-cm intervals, and combusted in a muffle furnace at 550 °C for 2 h to determine the dry weight percentage of organic matter [27].

4.3. Radiocarbon

Material for radiocarbon analysis was obtained from peat, leaf, or wood fragments. It was cleaned in distilled water and then dried at <70 °C. Ages have been calibrated using a Bayesian statistical approach (on the OxCal 3.5 calibration program with resolution = 4; OxCal, Oxford, UK, https://c14.arch.ox.ac.uk) against the INTCAL98 data set [28] and are reported as calibrated years before the present (= BP).

4.4. Peat Humification

Humification analysis was used to determine the degree of peat decomposition using a modified version of the semi-quantitative sodium hydroxide extraction technique [29]. In this technique, light transmission through the sodium hydroxide extract under standardized conditions is used as a measure of the humic acids present in the sample. The lower the values of light transmission is, the more decomposed the peat is. A 0.2 g dry weight peat from each depth was pulverised and extracted, as in the methods outlined in Reference [29]. The filtration stage of the procedure was replaced by centrifugation of the samples for three minutes at 3000 rpm, and retention of the supernatant, which is a technique that produces results identical to the standard procedure. The percentage transmission of light through the supernatant was measured on a spectrophotometer at 540 nm.

4.5. Pollen Analysis

Pollen samples (volume 1.2 cm^3) were prepared using standard procedures (hot 10% KOH, 40% HF, and acetolysis) [30]. Samples were spiked in the initial processing step with exotic *Lycopodium* spores for calculation of palynomorph concentrations. Pollen and spore counts were continued until a pollen sum of at least 250 grains was reached. Percentage calculations are based on island pollen and spore types but exclude all known extra-island pollen and spore types. Pollen and spore naming conventions follow [31] except that the new Nothofagaceae classification is accepted here: *Fuscospora* pollen type replaces the previous *Nothofagus fusca* pollen type and *Lophozonia menziesii* pollen type is the previous *Nothofagus menziesii* pollen type [32]. Pollen results were calculated and drafted using TILIA and TILIA.GRAPH [33].

5. Results

5.1. Chronostratigraphy

A total of 85 radiocarbon dates (HSP, 20, HRB, 44, MHB, 21) were obtained for the three sites (Table S1). Details of the Baysian derivation of calibrated ages for Figure 3 are given in Table S1. Despite the issue of downward intrusion of roots, which is a problem with peat bog dates, there are only a few age reversals and a high degree of internal consistency within and between peat profiles. An external validation of the accuracy of the dating is provided by the record of Podocarpacaeae long-distance pollen derived from the postglacial spread of the tall forest on the New Zealand mainland at c. 11,500 BP (SI Figures 1–3).

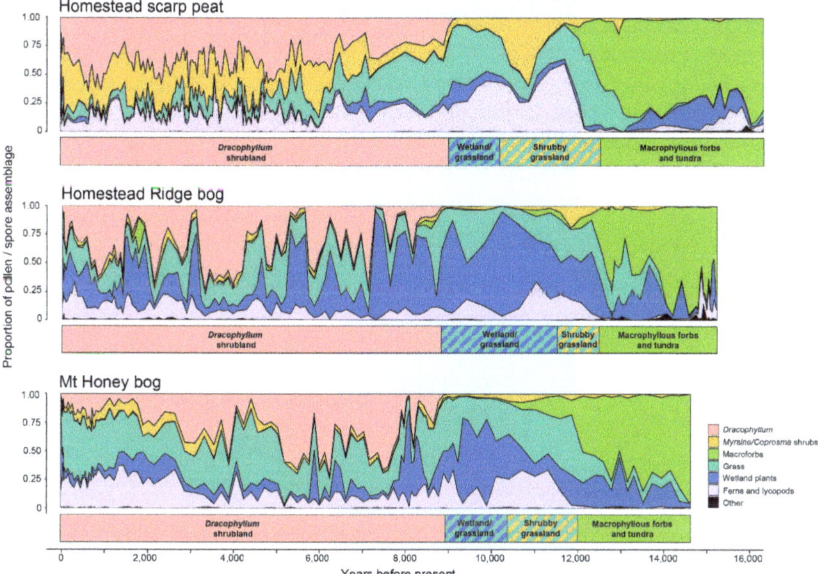

Figure 3. Summary percentage pollen and spore diagrams for Homestead Scarp Peat (HSP), Homestead Ridge Bog (HRB), and Mount Honey Bog (MHB). Blue vertical shading represents the period during the Holocene (12,000 to 6500 BP) when the tree line was repressed.

5.2. Vegetation Sequences Based on Pollen Stratigraphy

Pollen results from the three sites are summarized in Figure 3. Detailed results are shown in SI Figures 1–3.

Four vegetation zones are recognised based on the dominant pollen and spore types (Figure 3, Table 1, full results, Figures S1–S3).

1. **Macrophyllous forb**: The oldest sediments are dominated by macrophyllous forbs and include occurrences of the poorly dispersed *Stilbocarpa polaris*, and have relatively little wetland, low shrub, grassland, or fern representation. The small ground fern *Grammitis* is characteristic of the earliest sediments. While *Grammitis* spores cannot be identified to species, it is likely, given its association with tundra dominants, to be the dwarf alpine fern *G. poeppigiana*, which forms mats on rocks in the tundra zone [19]. A sparse macrophyllous forb tundra growing in a rock-covered and silt-covered landscape was, therefore, the first postglacial vegetation. The absence of abundant

grass or wetland pollen types points to the tundra zone above about 300–400 m as the closest modern analogue. In contrast to the modern tundra, few fern spores are recorded, aside from the alpine *Grammitis*. Since fern spores are among the few well dispersed types on the island [24], tundra-like conditions extended right to the contemporary sea level about 100 m lower than the present day.

2. **Grassland-shrub-wetland**: At all three sites, there is an abrupt to gradual transition toward less macrophyllous forbs and more grasses, which is followed by increased shrubs (*Coprosma* and *Myrsine divaricata*) and ferns. This sequence is best exemplified at well drained HSP where a *Myrsine/Coprosma* shrub-dominated zone (11,400-10,200 BP) not apparent at the other sites is recognised.

3. **Wetland-grassland:** Wetland types begin to predominate in *Astelia* and *Centrolepis*. HSP, being the best drained of the three sites, has only a weak wetland response, but wetland types are more abundant at the other two sites than elsewhere in the profile. *Hymenophyllum* ferns, which are abundant in the preceding zone, are also common in this region.

4. ***Dracophyllum*-scrub:** The island-wide simultaneous spread of *Dracophyllum* marks the onset of this zone, but grassland and ferns remain abundant, and wetland at the bog sites, for two to three thousand years longer. All sites have significant fluctuations in abundance of the major components, but most markedly in the raised bog, HRB, where *Dracophyllum* oscillates in response to wetter or drier conditions. This stage ended with the settlement of the island in the late 19th century and the changes it brought with fire and sheep grazing [34–37].

Table 1. Pollen vegetation zonation (periods calibrated years BP).

Vegetation Zones	Homestead Ridge Bog (HRB: X9901)	Homestead Scarp Peat (HSP: X9911)	Mt Honey Bog (MHB: X9903)
Macrophyllous forbs-tundra	15,300–12,500	16,800–12,500	14,700–12,000
Grassland-shrub	12,500–11,500	12,500–10,200	12,000–10,400
Wetland-grassland	11,500–8800	10,200–9000	10,400–8900
Dracophyllum	8900–55	9000–55	8900–55

These zones are only approximately equivalent in time and composition because of the different elevations and soil moisture conditions. For instance, the wetland-grassland phase was more pronounced and much longer (2700 years vs. 1200 years) in the poorly drained bog crest HRB compared with the close by, but well-drained scarp site, HSP.

5.3. Water Table Relationship with Bog Vegetation

A general pattern of vegetation cover and water table depth has been already documented for the island (Table S2). The detailed survey of the raised bog crest showed that vegetation composition was strongly predicted by water table depth (PERMANOVA $F_{1,19}' = 6.92$, $p < 0.001$, $R^2 = 0.27$). The percentage cover of the sedge *Oreobolus pectinatus* and the grass *Poa litorosa* have a significant positive correlation with the water table depth, while the forb *Astelia subulata* and graminoids *Isolepis aucklandica* and *Centrolepis* spp. have a significant negative correlation with the water table depth (Figure 4). The shrub species had positive relationships with water table depth but not as strong as might be expected from the general dominance of shrubs on well-drained peats (see Table 2). Because the well-drained shrub-dominant sites were not sampled, we suggest the results reflect the dynamics of stunted shrubs on the raised bog surface and its margins. *Myrsine divaricata* and *Coprosma* spp. are an insignificant component of the cushion bog vegetation whereas *Dracophyllum scoparium* is abundant and patches of various sizes occur across the entire surface, which averages 20%–25% of the ground cover.

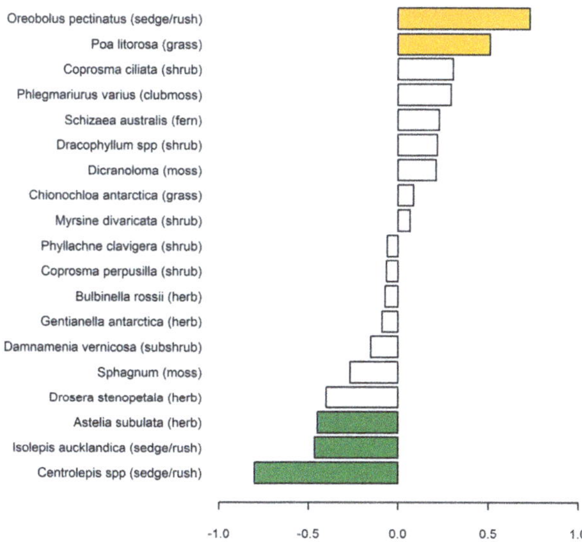

Figure 4. Homestead Ridge Bog. Vegetation cover percentage correlations with the water table. Green-filled bars are statistically significant negative correlations, yellow bars are statistically significant positive correlations, and empty filled bars are not statistically significant.

Table 2. Vegetation cover in relation to the water table depth. Two series of daily water table measurements (February 1976 to February 1977, and December 1983 to November 1984) were made at permanent sites on Homestead Ridge raised bog and near the Meteorological Station, Perseverance Harbour, Campbell Island. Data from Reference [10].

Site Vegetation Cover (Elevation)	Annual Mean Water Table Depth (cm)	Standard Deviation
Cushion vegetation, Homestead Ridge (45 m)	5.8	7.1
Sedge fen, Met. Station (10 m)	8.1	9.9
Macrophyllous forbs/cushion, Homestead Ridge, raised bog (45 m)	17.0	12.0
Tussock grassland, Met. Station (10 m)	34.2	19.0
Dracophyllum low forest slope marginal to Homestead Ridge, raised bog (35 m)	58.2	20.3
Dracophyllum/grassland, Met. Station (10 m)	78.1	13.2
Dracophyllum/*Coprosma* forest, Met. Station (20 m)	109.5	7.6

5.4. Peat Vertical Growth, Surface Wetness, and Humification

A surface wetness indicator was developed from an ordination of three sites (HRB, HSP, and MHB) and the modern surface sample data was presented in Reference [24] (Figure 5c). The second axis is well aligned with the results of the modern water table depth results, with *Dracophyllum* and ferns contrasted with *Astelia antarctica* and *Centrolepis* spp. (Figure S4). The two bog sites (HRB and MHB) show a decline surface wetness from 14,000 BP onwards. The maximum wetness is evident between 10,500 and 9000 BP while the forest site (HSP) has fluctuating but dry conditions along with markedly increased moisture between 10,500 and 9000 BP (Figure 5c). All three sites became drier shortly after 9000 BP and followed a fluctuating trend toward less surface moisture, which culminated between 6000 and 2000 BP.

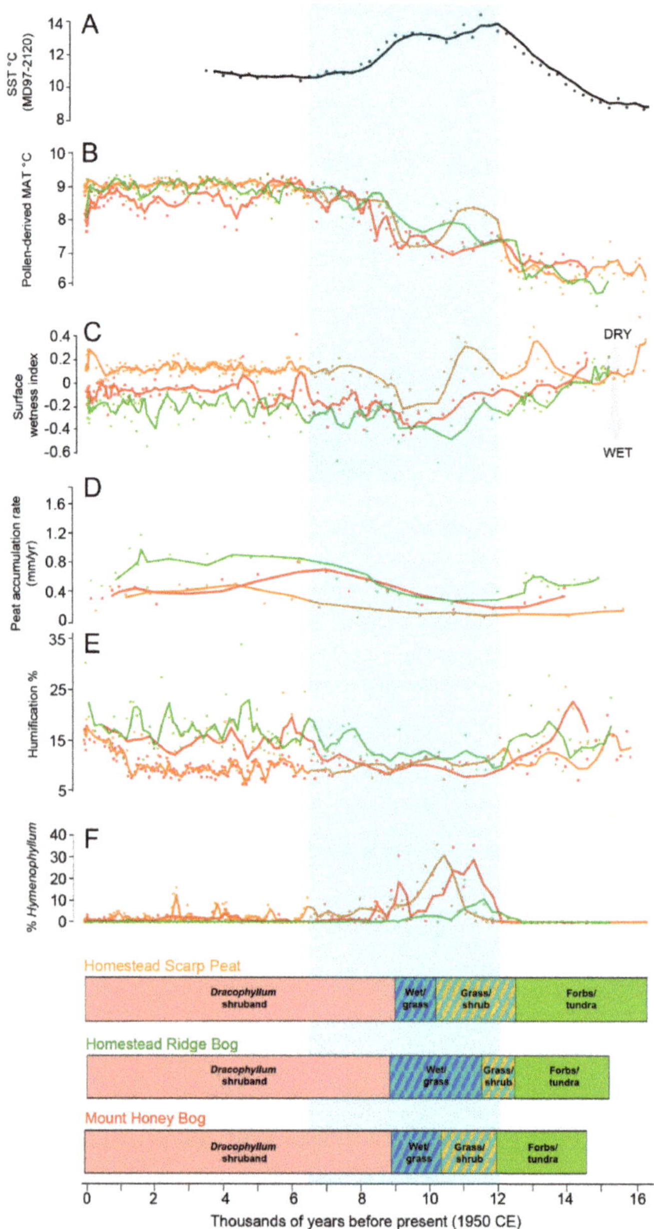

Figure 5. SST, deep sea core MD97-2120 off south-eastern South Island [38] (**a**). Panels (**b**) to (**f**) from analyses of peat sites HSP (yellow), HRB (green), and MHB (red). Equivalent mean January (summer) Campbell temperature [8] (**b**). Surface wetness proxy (NMS axis 2) (**c**). Vertical peat accumulation (**d**). Humification index (% transmission) (**e**). *Hymenophyllum* (filmy fern) spore percentages (**f**). Blue vertical shading represents the period during the Holocene (12,000 to 6500 BP) when the tree line was repressed.

Peat vertical growth (Figure 5d) is a proxy for net accumulation, the product of decay of organic material in the profile, and net plant productivity. The two bog sites (HRB and MHB) show a gradual decline in vertical growth over the early Holocene to reach a low value between 12,000 and 9000 BP. They recommend accelerated growth after 9000 BP. The high elevation MHB has an early Holocene peak in vertical growth centered on 6000 BP and then a steep decline. The low elevation HRB peaks at around the same time but continues to grow at about the same rate, which leaves aside rapid fluctuations in the late Holocene. The HSP peat soil grew slowly until 8000 BP, increased until 4500 BP, and then grew at a steady rate.

Peat humification matches the changes in vertical growth well (Figure 5e). The slow growing HSP peat soil is well-humified for the last 1000 years, which likely reflects undecomposed material in the surface that has had insufficient time to breakdown. HRB and MHB show a general increase in humification from 13,500 BP with sustained intense humification from 12,500 through 8000 BP, at which time, peat decomposition in these bog sites matched that of the well-drained HSP soil. By 6500 BP, humification at both sites was at relatively low levels that were sustained for the rest of the Holocene.

6. Discussion

6.1. Climatic Interpretation

The January (= summer) temperature estimates (Figure 5b) generated by transfer functions based on a modern pollen rain-temperature data set [8] need to be interpreted cautiously, since they are based on the current disposition of vegetation. We, therefore, refer to them as 'temperature equivalents' since they indicate the January temperature the pollen assemblages suggest if the same type of climatic regime prevailed.

The early (17,000 to c. 12,500 BP) pollen assemblages of the two lowland site profiles are typical of areas with abundant exposed rock and tundra vegetation. The only modern analogues are high elevation sites above 500 m where peat does not form and cold temperatures (January temperatures are ≤6 °C in the tundra zone), frost heave, and high winds induce stunted vegetation cover. The pollen transfer function reconstructions for the early part of the macrophyllous forb stage corroborate this estimate and show no pronounced shift in temperature until about 12,000 BP (Figure 5b). Tundra, therefore, covered the island surface from the sea level to at least 120 m elevation.

A transition from macrophyllous forb tundra to grassland-shrubland began c. 14,000–13,000 BP and was associated with a trend toward wetter conditions in the two bog sites, but not the well-drained HSP site. At HSP, first *Coprosma* shrubland and then *Myrsine divaricata* rose to dominance, with shrub cover peaking at c. 11,000 BP. This may be taken for the more general trend on well drained slopes. *Phlegmariurus varius*, which is a clubmoss common in low shrubland, and *Hymenophyllum*, which is a filmy fern more abundant in the current pollen assemblages above c. 100 m [24], were prominent at all sites over this period (Figure 5f). High percentages of *Myrsine* pollen on the present landscape are inevitably associated with high percentages of *Dracophyllum* pollen [24]. Therefore, the absence of the latter is significant, and indicates that forest climates have not been established. The major increase in shrubland noted at HSP was only weakly registered at the two bog sites. Equivalent January temperatures are estimated to have risen by c. 1 °C shortly after 12,000 BP. Between 11,000 and 10,500 BP, equivalent January temperatures were within 1 °C of the current temperature. The timing of this warming c. 12,000 BP is consistent with the end of the Antarctic Cold Reversal cooling event between 14,700 and 13,000 BP [39].

The grassland-wetland stage (present at all sites between 10,000 and 9000 BP) is characterised by the lowest percentages of shrub types in the post 12,500 BP period. The shrubland associate *Phlegmariurus varius* declined. *Hymenophyllum* ferns continued to be well represented. The reassertion of grassland cover and decline of shrubs is consistent with a lowering of elevational vegetation zones. The pollen transfer functions support such an interpretation since they show an equivalent temperature decline of c. 1 °C. The wetland index peaked, which could also be related to less evapotranspiration under cooler climates. However, vertical peat growth slowed at all three sites to a minimum and peat humification increased (Figure 5c–e) and this suggests an alternative explanation.

Vertical peat growth is determined by the balance between vegetation productivity and breakdown [40]. Most peat decay takes place in the few centimetres of the upper aerated acrotelm. If peat vertical growth was a faithful proxy for peat mass accumulation into the unaerated catotelm, steady peat accumulation would result in a linear relationship between peat height and age. However, compaction and slow anaerobic decay with increasing age and depth in the profile reduces the apparent vertical annual increment and upper peat layers. Therefore, they usually show an apparent acceleration of vertical growth [41]. Normally, a wetter peat will grow faster, all things being equal, because organic matter spends less time in the aerated acrotelm, which is thinner because of the higher water table. However, if vegetation productivity is low, even though the acrotelm is thin, the peat spends a greater amount of time within it, and, thus, decomposes more. Moreover, if winters and overnight temperatures are warm, peat breakdown increases as higher temperatures favour accelerated loss of organics [42]. Therefore, although the pollen sequences and temperature transfer functions developed from them, at first sight, seems to support a cooler summer climate, we prefer an interpretation based on changes in cloud cover and wind direction.

The peat surfaces became wetter after 13,000 BP, and much wetter between 10,500 and 9000 BP, and peat grew slowly (Figure 5c,d). Atmospheric humidity across the whole period (12,500 to 9000 BP) must have increased, as the filmy fern *Hymenophyllum* that is poorly represented in the modern pollen rain below 100 m [24], but common in the cloudier uplands, was abundant at all sites (Figure 5f). Rather than temperature decreases being the driver of these ecological changes, increased cloud immersion with accompanying low light, increased humidity and decreased transpiration, and wetter substrates [43] may have restricted shrub growth. Observations made on cloud bases over Campbell Island in the Perseverance Harbour basin [10] show that cloud immersion has a linear trend with altitude (Figure 6).

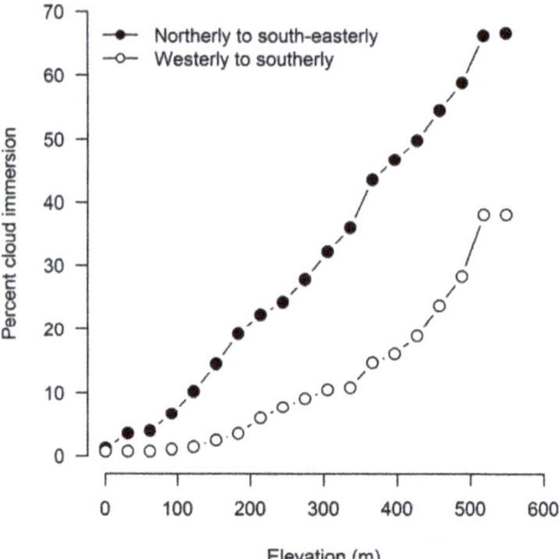

Figure 6. Percent of observations in which given elevations on the Mt Honey transect, Perseverance Harbour, Campbell Island were within the cloud base. A total of 576 twice daily (09:00 and 16:00 h) observations of cloud base were made from 1 March to 3 October 1991. Winds from northerly to south-easterly quarter (solid line) bring much lower cloud bases than those from westerly to southerly quarter (broken line). From Reference [10].

Wind direction has a strong effect on cloud base and mean daily temperature (Table 3). When prevailing winds were west to southwest, cloud bases were high (all peaks were clear in almost 60% of observations) and temperatures were cool (5.0 °C) (Table 3). Winds from the north-west quarter had cool temperatures but low cloud bases (peaks clear only in 21% of observations). Winds from the north and east led to cloud bases being low but temperatures were 3 °C warmer. Low cloud cover decreases warming during the day by reflecting solar radiation, but prevents heat loss overnight. The combination of northerly source air and prevention of overnight heat loss leads to markedly warmer temperatures. As plant growth depends on bright light and daytime warmth, the same daily mean temperature can have radically different consequences. Cloudiness at these latitudes, therefore, will not cool the land relative to the ocean but will suppress woody plant growth. Mild winters, in particular, are, paradoxically, a physiological stress factor for woody plants in oceanic environments since they break dormancy and encourage respiration and growth during a season when ground water levels are high and light levels are minimal [44]. For instance, *Dracophyllum* growth on Campbell Island is substantially reduced during warmer than normal winters [45]. In turn, loss of woody cover decreases evapotranspiration and increases soil moisture. This feedback cycle further reduces the suitability of soils for shrubs and encourages the spread of graminoids and cushion plants. Consistent with our interpretation is the lack of any resurgence of a cool climate and tundra indicators such as the macrophyllous forbs. If the early Holocene wetland-grassland phase represents a cloudiness and soil moisture induced depression of the limit to shrub-low forest on the island, it may have been substantial. The difference in the tree line altitude across the middle of the Southern Alps in New Zealand is c. 200 m, the cloudier, western slopes have lower elevation tree lines [46]. In terms of mean annual temperature, this is equivalent to more than a 1 °C depression, which shows that cloudiness could potentially offset a warming of this magnitude.

Table 3. Observations for which all peaks were clear of cloud, Perseverance Harbour Basin, 1 March to 3 October 1991. Data from Reference [10].

Wind Directions	Observations with all Peaks Clear (%)	Mean Daily Temperature (°C)	% Observations
All	39.5	6.1	100.0
North to east-southeast	21.0	8.1	27.4
North-northwest to west northwest	20.8	5.1	22.7
West to southeast	58.7	5.0	45.3
Calm	53.9	9.3	4.5

The near simultaneous expansion of *Dracophyllum* scrub across our sites argues for an abrupt change in the climate regime and a nearly 2 °C increase in equivalent January temperatures is reconstructed over the next 1000 years. However, the adjacent ocean was cooling from 9000 BP onward after an early Holocene peak [9,38,47]. Therefore, as the cooler equivalent temperatures reconstructed for the previous period were largely induced by cloudiness, it follows that the subsequent 'warming' reflects a reversion to clearer skies with increased southwesterly winds and, therefore, a real cooling of mean annual temperatures over the island. The establishment of *Dracophyllum* trees (as distinct from *Dracophyllum* shrubs) at the lower HSB site cannot have occurred before 9000 BP and we suggest not until 6200 BP when the grass and *Hymenophyllum* percentages fell to levels consistent with present day pollen rain from forested sites (Figure 5f). The timing is similar at the higher elevation MHB site. The Campbell Island fossil wood occurrences support this interpretation [48]. Buried *Dracophyllum* wood was collected along an elevational transect 65 to 100 m above the shrub line at c. 210 m in sheltered Southeast Harbour 3.5 km to the southeast of MHB, with no wood before 6000 BP.

The vegetation consequences of the interaction between wind and clear skies is complex. Although shrubs, and *Dracophyllum* in particular, become common at all three sites. They are not well aligned and each *Dracophyllum* curve has a different trajectory over the Holocene (Figure 4). The two lowland sites (HSP and HRB) achieve high *Dracophyllum* percentages at 6200 BP and, from then on, show marked fluctuations—especially the HRB—but no overall trend. In contrast, the upland MHB site after high values between 7500 and 5500 BP, undergoes a fluctuating decline in response to grassland expansion.

A sharp rise in monolete fern spores at 3000 BP at this site shows winds may have increased in strength in the late Holocene (Figure 4) since ground ferns are likely to be favoured by reduced competition from wind-stunted shrubs and grasses. A similar trend is seen in the late Holocene at Rocky Bay [12].

These differing Holocene long term trends in *Dracophyllum* between the upland and lowland sites are yet a further indication of how the separate components of the climate system can have markedly differing effects on sites depending on their altitude and exposure. The pollen-based temperature reconstructions (which, after 9000 BP, likely reflect actual January temperatures) do not show, in either case, a marked Holocene trend nor do sea surface temperatures in the adjacent ocean. *Dracophyllum* is, therefore, most likely responding directly to wind speed and bright sunshine hours. The upland MHB site lies in a saddle exposed to winds from the north-western and south-western sectors, while the lowland HSP site is sheltered. Any increase in southern sector winds is likely to subject MHB to strong, cold winds and reduced *Dracophyllum* growth, whereas HSP will experience clearer skies and increased growth.

Evidence for a direct wind effect comes from the absence or decline in the course of the Holocene of shrubs in the most wind-exposed sites on Campbell Island. West-facing Hooker Cliffs (90 m asl) north of the island, never had significant shrub representation. The south-facing Rocky Bay cliff-top (130 m a.s.l, c. 6 km west of MHB) lost its initial *Coprosma* shrub cover in the course of the Holocene [12]. The fossil wood transect from Southeast Harbour lacks wood between 2300 and 1000 BP, which coincides with an interval with wind-blown silt and stones in nearby cliffs [49] and a marked decline in *Dracophyllum* pollen at MHB, and is attributed to an episode of intense windiness [48]. Increased windiness is recorded in sediments from an Auckland Island fiord over the same interval [50]. We attribute this pattern to the increased prevalence of southwesterly winds bringing clearer skies and, thus, favouring woody growth in sheltered areas but stronger, cooler winds eliminate woody communities on exposed sites.

While, from c. 5000 BP, the general *Dracophyllum* trend is strongly negative at MHB and positive at HSP. There is more alignment from 2000 BP. Both sites show a *Dracophyllum* peak at or around 2000 BP, low centered on 1500 BP, and peak values between 1000 and 600 BP, and a decline thereafter. The periods of falling or low *Dracophyllum* values are well aligned with glacial advances in the Southern Alps of New Zealand, where there are two groups of closely spaced advances between 2000 and 1400 BP, and 700 and 200 BP [51]. These are broadly consistent with the timing of the Dark Ages Cool period (DAC: 1550–1250 BP), the Mediaeval Climate Anomaly (MCA: 1150–750 BP), and the Little Ice Age (LIA: 650–100 BP).

6.2. The Sea Surface Temperature versus Island Temperature Anomaly

Given that growing season mean temperature is closely correlated with tree line elevation [7], and, as SSTs in the adjacent ocean were 1 °C or more higher than present [8,9], the tree line during the early Holocene period (12,000 to 6500 BP) should have been at elevations of 150 m or more if current climatic patterns held. Summer mean temperatures on Campbell Island are presently closely aligned with the surrounding SST (as they are in coastal locations on the New Zealand mainland) and some substantial change would have had to occur to decouple this alignment.

McGlone et al. [8] argued that the cooling of the land relative to the ocean was a consequence of the weakening of the southern westerlies over the island and, another consequence was a decline in summer heat subsidies brought by cyclones embedded in that wind flow. While this is still a feasible explanation, increased cloudiness is more consistent with the evidence we have outlined above. Westerly cliff edge sites from sub-Antarctic Auckland Island and Campbell Island showed that wind-blown silt and stones were absent from their postglacial soils until c. 9000 BP [5,8]. This is consistent with results from an analysis of deep sea cores from the New Zealand sector, which suggested that the early Holocene (11,800 to 9800 BP) period was both less windy and warmer than currently south of the New Zealand mainland due to weakening of the westerly wind flow [9]. As we have shown, indicators of increased soil wetness, higher humidity, and expansion of wetland-grassland at this time can be most parsimoniously explained by increased north-easterly flow over the island with greater cloud immersion and less bright sunshine. Resumption of strong south-westerly flow after 9000 BP would have reduced the incidence of low cloud and led to the subsequent expansion of trees, tall woody shrubs, and drier peat surfaces peaking between 6500 and 3000 BP. The strong regionalisation of vegetation development after 9000 BP argues for the pre-eminent influence of wind in the Holocene. Exposed cliff sites and ridge tops appear to have lost what shrub cover that they had initially and, after 3000 BP, seem to have become even more wind-swept with open, stunted shrubs [12]. This is not to say that cooling of the ocean in the late Holocene had no effect. HRB and MHB clearly had reduced shrub cover during the DAC and the LIA.

6.3. Southern Ocean Context

There are only a few terrestrial peat records in the southern ocean region with which to compare those from Campbell Island. The Auckland Island group 150 km to the north lies in the same ocean waters and has a similar vegetation cover [5,52]. While the chronology is not as secure as in the Campbell Island peats, the general pattern is the same, with the major trees, *Dracophyllum longifolium* and *Metrosideros umbellata*, not abundant until after c. 9000 BP. Acceleration of vertical peat growth after 6000 BP at two sheltered sites at Dea's Head on the main island, and restriction of shrubs and small trees after ca. 9600 BP by intense windiness on the exposed Enderby Island site [5,53], supports the inferences made for Campbell Island. Evidence from wind-blown minerogenic input on Stewart Island immediately south of the New Zealand mainland points to stronger westerly winds after 5500 BP [54] as does isotopic and stratigraphic evidence from a lake in the far south and in the wind shadow east of the main ranges [55]. Further north on the New Zealand mainland, there are indications that tree lines were similarly inhibited by weaker westerly airflow. In the southeastern and central South Islands, tree lines did not reach current elevations until well after 9000 BP [56] and, in the northwestern South Island, not until 9500 BP [57].

A number of peat sites in Patagonia, Tierra del Fuego, and the Falkland Islands group between 55 and 51° S have vegetation records comparable to those of Campbell Island [58–67]. The initial timing of peat growth in the higher latitudes of Patagonia and Tierra del Fuego is similar to that of Campbell Island, commencing between 17,500 and 12,000 BP. The vegetation transitions from steppe-moorland to *Nothofagus* parkland by 12,000–11,000 BP, and then to closed *Nothofagus* forest during the early to mid-Holocene. Of particular relevance to the Campbell Island sequence is the late arrival of closed *Nothofagus* forest at upland sites close to the tree line. At Paso Garibaldi (54°43′ S, 500 m a.s.l), a closed forest was not present until after 8700 BP [64] and, at Las Contornas mire (54°41′ S, 420 m a.s.l), not until 6500 BP [63]. At Port Howard, Falkland Islands (51°20′S, 100–130 m a.s.l), the current vegetation of low shrubs did not replace herbaceous associations until 8000–7000 BP [66]. These patterns are confirmed in a recent review of vegetation change in Patagonia [68], which shows *Nothofagus* abundance not reaching current levels until 7000 BP in central Patagonia and 4000 BP in southern Patagonia.

The climatic explanation for the late establishment of forest or shrubland in southern South America differs from that given here for Campbell Island. In southern South America, SSTs were as warm or warmer than by 12,000 BP. However, weaker westerly flow brought more arid conditions and, thus, prevented the *Nothofagus* forest expanding, while, in the New Zealand sub-Antarctic regions, mist and low cloud associated with weaker westerlies suppressed the taller woody vegetation. Expansion of forest in the early to mid-Holocene in both regions was ultimately driven by increased windiness. However, in southern South America, rain-bearing westerlies delivered more precipitation and were pushed further inland, while, in the New Zealand sub-Antarctic regions, the south-westerly to westerly orientation of the wind flow brought higher cloud-bases and more spells of bright sunshine.

Recent tree line and scrub coverage on Campbell Island has been largely unresponsive to regional climate change. Around the turn of the 19th century and for several decades after, SSTs in the immediate region were as much as 0.5 °C below the 20th century average rising to 0.5 °C above 1970 to 1990 AD [69]. Although there has been vigorous regrowth of forest and shrubland across grassland induced by fire during the late 19th to early 20th century farming era [34], there has been no indication of elevational shifts in response to temperature [35,36]. This is consistent with our findings that past temperature shifts alone in these highly oceanic situations are insufficient to alter tree line dynamics, which has been noted for the mountains of mainland New Zealand [3,70,71].

7. Conclusions

While the postglacial spread of the forest on sub-Antarctic Campbell Island broadly follows the warming of the surrounding ocean, wind direction, wind strength, and cloudiness modified the timing and created gradients within the island, according to the degree of wind exposure. Decreased westerly to south-westerly wind flow permitted increased intrusion of north-easterly clouds between 12,500 and 9000 BP, which brings less sunshine, more cloud immersion, higher humidity, and weakened evapotranspiration that resulted in increased soil surface wetness. As a result, initial woody growth was inhibited despite rising ocean surface temperatures, and grassland and wetland spread. Establishment of forest and upland shrub coverage on the more sheltered leeward side of the island followed the switch to clearer skies as the current strong south-westerly airflow progressively established from 9000 BP onward. Woody vegetation reached maximum cover between 6500 and 3000 BP, but intensifying winds, at the same time, prevented forest or shrub spreading on the windward flanks. The warming of the southern part of the ocean in the early Holocene by 1–2 °C can be used as a proxy for anticipated warming in the near future. However, our results suggest temperature is only one of the factors that can affect the tree line position, and cloudiness and seasonality may be just as important. The failure of tree lines on Campbell Island and mainland New Zealand to respond to prolonged warming episodes during the 20th and 21st centuries supports this conclusion. Therefore, increasing mean annual temperatures will not necessarily lead to rising tree lines in highly oceanic regions unless these other factors are also in alignment.

Supplementary Materials: The following are available online at http://www.mdpi.com/1999-4907/10/11/998/s1. Table S1. Radiocarbon dates (old T1/2 and calibrated) for HRB, HSP, and MHB. Table S2. Homestead Ridge Bog vegetation cover and water table depths. Figure S1. Homestead Ridge Bog (X9901). Pollen diagram. Figure S2. Homestead Scarp (X9911). Pollen diagram. Figure S3. Mount Honey Saddle Bog(X9903). Pollen diagram. Derivation of the surface wetness index. Figure S4. NMS ordination of modern pollen and peat core samples at Campbell Island. Figure S5. NMS scores for modern pollen samples from 11 distinct Campbell Island plant communities.

Author Contributions: Conceptualization, M.S.M., J.M.W. and C.S.M.T. Data curation, J.M.W. Formal analysis, J.M.W., S.J.R., C.S.M.T., and J.R.W. Funding acquisition, M.S.M., J.M.W., and C.S.M.T. Investigation, M.S.M. and J.M.W. Project administration, J.M.W. Writing—original draft, M.S.M. Writing—review & editing, M.S.M., J.M.W., S.J.R., C.S.M.T., and J.R.W.

Acknowledgments: We thank the New Zealand Department of Conservation for permission to take samples from Campbell Island and for logistical support and Alison Watkins for technical support. This research was supported by Strategic Science Investment funding for Crown Research Institutes from the New Zealand Ministry of Business, Innovation and Employment's Science and Innovation Group.

Conflicts of Interest: The authors declare no conflict of interest.

References

1. IPCC. *Climate Change 2014: Impacts, Adaptation, and Vulnerability Part A: Global and Sectoral Aspects*; Cambridge University Press: Cambridge, UK; New York, NY, USA, 2014; p. 1132.
2. Halloy, S.R.P.; Mark, A.F. Climate-change effects on alpine plant biodiversity: A New Zealand perspective on quantifying the threat. *Arct. Antarct. Alp. Res.* **2003**, *35*, 248–254. [CrossRef]
3. Harsch, M.A.; Hulme, P.E.; McGlone, M.S.; Duncan, R.P. Are treelines advancing? A global meta-analysis of treeline response to climate warming. *Ecol. Lett.* **2009**, *12*, 1040–1049. [CrossRef] [PubMed]
4. Cieraad, E.; McGlone, M.S.; Huntley, B. Southern hemisphere temperate tree lines are not climatically depressed. *J. Biogeogr.* **2014**, *41*, 1456–1466. [CrossRef]
5. McGlone, M.S. The late Quaternary peat, vegetation and climate history of the southern oceanic islands of New Zealand. *Quat. Sci. Rev.* **2002**, *21*, 683–707. [CrossRef]

6.   Korner, C.; Paulsen, J. A world-wide study of high altitude treeline temperatures. *J. Biogeogr.* **2004**, *31*, 713–732. [CrossRef]

7.   Paulsen, J.; Körner, C. A climate-based model to predict potential treeline position around the globe. *Alp. Bot.* **2014**, *124*, 1–12. [CrossRef]

8.   McGlone, M.S.; Turney, C.S.M.; Wilmshurst, J.M.; Renwick, J.; Pahnke, K. Divergent trends in land and ocean temperature in the southern ocean over the past 18,000 years. *Nat. Geosci.* **2010**, *3*, 622–626. [CrossRef]

9.   Prebble, J.G.; Bostock, H.C.; Cortese, G.; Lorrey, A.M.; Hayward, B.W.; Calvo, E.; Northcote, L.C.; Scott, G.H.; Neil, H.L. Evidence for a Holocene climatic optimum in the southwest pacific: A multiproxy study. *Paleoceanography* **2017**, *32*, 763–779. [CrossRef]

10.   McGlone, M.S.; Meurk, C.D.; Crompton, M.B. *Climate and Water Table Depth Data from Campbell Island, New Zealand Subantarctic*; Manaaki Whenua-Landcare Research Contract Report; Landcare Research NZ Ltd.: Lincoln, New Zealand, 2019. [CrossRef]

11.   Campbell, I.B. Soil pattern of Campbell Island. *N. Z. J. Sci.* **1981**, *24*, 111–135.

12.   McGlone, M.S.; Moar, N.T.; Wardle, P.; Meurk, C.D. Late-glacial and Holocene vegetation and environment of Campbell Island, far southern new zealand. *Holocene* **1997**, *7*, 1–12. [CrossRef]

13.   De Lisle, J.F. Weather and climate of Campbell Island. *Pac. Insect Monogr.* **1964**, *7*, 34–44.

14.   Ross, D.J.; Campbell, I.B.; Bridger, B.A. Biochemical activities of organic soils from sub-antarctic tussock grasslands on Campbell Island. 1. Oxygen uptakes and nitrogen mineralization. *N. Z. J. Sci.* **1979**, *22*, 161–171.

15.   Foggo, M.N.; Meurk, C.D. A bioassay of some Campbell Island soils. *N. Z. J. Ecol.* **1983**, *6*, 121–124.

16.   Meurk, C.D.; Foggo, M.N.; Thomson, B.M.; Bathurst, E.T.J.; Crompton, M.B. Ion-rich precipitation and vegetation pattern on sub-antarctic Campbell Island. *Arct. Alp. Res.* **1994**, *26*, 281–289. [CrossRef]

17.   Meurk, C.D.; Blaschke, P.M. *How Representative Can Restored Islands Really Be?: An Analysis of Climo-Edaphic Environments in New Zealand*; Department of Conservation: Wellington, New Zealand, 1990.

18.   Meurk, C.D.; Given, D.R. *Vegetation Map of Campbell Island. Land Resources*; Department of Scientific and Industrial Research: Christchurch, New Zealand, 1990.

19.   Meurk, C.D.; Foggo, M.N.; Wilson, J.B. The vegetation of sub-antarctic Campbell Island. *N. Z. J. Ecol.* **1994**, *18*, 123–168.

20.   Anderson, M.J. A new method for non-parametric multivariate analysis of variance. *Austral Ecol.* **2001**, *26*, 32–46.

21.   Oksanen, J.; Blanchet, F.; Friendly, M.; Kindt, R.; Legendre, P.; McGlinn, D.; Minchin, P.; O'Hara, R.; Simpson, G.; Solymos, P. Vegan: Community Ecology Package. R Package v. 2.5-4. 2018. Available online: https://cran.r-project.org/web/packages/vegan/index.html (accessed on 3 October 2019).

22.   Warton, D.I.; Foster, S.D.; De'ath, G.; Stoklosa, J.; Dunstan, P.K. Model-based thinking for community ecology. *Plant Ecol.* **2015**, *216*, 669–682. [CrossRef]

23.   Wang, Y.; Naumann, U.; Wright, S.T.; Warton, D.I. Mvabund—An R package for model-based analysis of multivariate abundance data. *Methods Ecol. Evol.* **2012**, *3*, 471–474. [CrossRef]

24.   McGlone, M.S.; Meurk, C.D. Modern pollen rain, subantarctic Campbell Island, New Zealand. *N. Z. J. Ecol.* **2000**, *24*, 181–194.

25.   R Core Team. *R: A Language and Environment for Statistical Computing*; R Foundation for Statistical Computing: Vienna, Austria, 2018.

26.   Bruce, M.; Grace, J.; Urban, D. *Analysis of Ecological Communities*; MJM Software Design: Gleneden Beach, OR, USA, 2002.

27.   Bengtsson, L.; Enell, M.; Berglund, B. Handbook of Holocene palaeoecology and palaeohydrology. *Chem. Anal.* **1986**, 423–451.

28.   Stuiver, M.; Reimer, P.J.; Bard, E.; Beck, J.W.; Burr, G.S.; Hughen, K.A.; Kromer, B.; McCormac, G.; Van Der Plicht, J.; Spurk, M. Intcal98 radiocarbon age calibration, 24,000–0 cal bp. *Radiocarbon* **1998**, *40*, 1041–1083. [CrossRef]

29.   Blackford, J.; Chambers, F. Determining the degree of peat decomposition for peat-based palaeoclimatic studies. *Int. Peat J.* **1993**, *5*, 7–24.

30.   Moore, P.; Webb, J.; Collinson, M. *Pollen Analysis*; Blackwell Scientific: Oxford, UK, 1991.

31.   Moar, N.T.; Wilmshurst, J.M.; McGlone, M.S. Standardizing names applied to pollen and spores in New Zealand Quaternary palynology. *N. Z. J. Bot.* **2011**, *49*, 201–229. [CrossRef]

32. Heenan, P.B.; Smissen, R.D. Revised circumscription of *Nothofagus* and recognition of the segregate genera *Fuscospora*, *Lophozonia*, and *Trisyngyne* (Nothofagaceae). *Phytotaxa* **2013**, *142*, 1–31. [CrossRef]

33. Grimm, E.C. *TILIA Software Version 1.7.16*; Illinois State Museum, Research and Collection Center: Springfield, IL, USA, 2011. Available online: http://intra.museum.state.ilus/pub/grim?m/tilia/ (accessed on 3 October 2019).

34. McGlone, M.; Wilmshurst, J.; Meurk, C. Climate, fire, farming and the recent vegetation history of subantarctic Campbell Island. *Earth Environ. Sci. Trans. R. Soc. Edinb.* **2007**, *98*, 71–84. [CrossRef]

35. Bestic, K.L.; Duncan, R.P.; McGlone, M.S.; Wilmshurst, J.M.; Meurk, C.D. Population age structure and recent dracophyllum spread on subantarctic Campbell Island. *N. Z. J. Ecol.* **2005**, *29*, 291–297.

36. Wilmshurst, J.M.; Bestic, K.L.; Meurk, C.D.; McGlone, M.S. Recent spread of *Dracophyllum* scrub on subantarctic Campbell Island, New Zealand: Climatic or anthropogenic origins? *J. Biogeogr.* **2004**, *31*, 401–413. [CrossRef]

37. Bestic, K.L. *Dracophyllum Scrub Expansion on Subantarctic Campbell Island, New Zealand*; Lincoln University: Lincoln, New Zealand, 2002.

38. Pahnke, K.; Sachs, J.P. Sea surface temperatures of southern midlatitudes 0–160 kyr bp. *Paleoceanography* **2006**, *21*. [CrossRef]

39. Pedro, J.B.; Bostock, H.C.; Bitz, C.M.; He, F.; Vandergoes, M.J.; Steig, E.J.; Chase, B.M.; Krause, C.E.; Rasmussen, S.O.; Markle, B.R. The spatial extent and dynamics of the antarctic cold reversal. *Nat. Geosci.* **2016**, *9*, 51. [CrossRef]

40. Charman, D. *Peatlands and Environmental Change*; John Wiley & Sons: London, UK, 2002.

41. Rydin, H.; Jeglum, J.K. *The Biology of Peatlands*, 2nd ed.; Oxford University Press: Oxford, UK, 2013.

42. Ise, T.; Dunn, A.L.; Wofsy, S.C.; Moorcroft, P.R. High sensitivity of peat decomposition to climate change through water-table feedback. *Nat. Geosci.* **2008**, *1*, 763. [CrossRef]

43. Reinhardt, K.; Smith, W.K.; Carter, G.A. Clouds and cloud immersion alter photosynthetic light quality in a temperate mountain cloud forest. *Botany-Botanique* **2010**, *88*, 462–470. [CrossRef]

44. Crawford, R.M.M.; Jeffree, C.E.; Rees, W.G. Paludification and forest retreat in northern oceanic environments. *Ann. Bot.* **2003**, *91*, 213–226. [CrossRef] [PubMed]

45. Harsch, M.A.; McGlone, M.S.; Wilmshurst, J.M. Winter climate limits subantarctic low forest growth and establishment. *PLoS ONE* **2014**, *9*, e93241. [CrossRef] [PubMed]

46. Cieraad, E.; McGlone, M.S. Thermal environment of New Zealand's gradual and abrupt treeline ecotones. *N. Z. J. Ecol.* **2014**, *38*, 12–25.

47. Bostock, H.C.; Prebble, J.G.; Cortese, G.; Hayward, B.; Calvo, E.; Quiros-Collazos, L.; Kienast, M.; Kim, K. Paleoproductivity in the sw pacific ocean during the early Holocene climatic optimum. *Paleoceanogr. Paleoclimatol.* **2019**, *34*, 580–599. [CrossRef]

48. Turney, C.S.; M c Glone, M.; Palmer, J.; Fogwill, C.; Hogg, A.; Thomas, Z.A.; Lipson, M.; Wilmshurst, J.M.; Fenwick, P.; Jones, R.T. Intensification of southern hemisphere westerly winds 2000–1000 years ago: Evidence from the subantarctic Campbell and Auckland Islands (52–50° s). *J. Quat. Sci.* **2016**, *31*, 12–19. [CrossRef]

49. McGlone, M.S.; Moar, N.T. Pollen-vegetation relationships on the subantarctic Auckland Islands, New Zealand. *Rev. Palaeobot. Palynol.* **1997**, *96*, 317–338. [CrossRef]

50. Browne, I.M.; Moy, C.M.; Riesselman, C.R.; Neil, H.L.; Curtin, L.G.; Gorman, A.R.; Wilson, G.S. Late Holocene intensification of the westerly winds at the subantarctic Auckland Islands (51 degrees S), New Zealand. *Clim. Past* **2017**, *13*, 1301–1322. [CrossRef]

51. Schaefer, J.M.; Denton, G.H.; Kaplan, M.; Putnam, A.; Finkel, R.C.; Barrell, D.J.A.; Andersen, B.G.; Schwartz, R.; Mackintosh, A.; Chinn, T.; et al. High-frequency holocene glacier fluctuations in New Zealand differ from the northern signature. *Science* **2009**, *324*, 622–625. [CrossRef]

52. McGlone, M.S.; Wilmshurst, J.M.; Wiser, S.K. Lateglacial and Holocene vegetation and climatic change on Auckland Island, subantarctic New Zealand. *Holocene* **2000**, *10*, 719–728. [CrossRef]

53. Fleming, C.A.; Mildenhall, D.C.; Moar, N.T. Quaternary sediments and plant microfossils from Enderby Island, Auckland Islands. *J. R. Soc. N. Z.* **1976**, *6*, 433. [CrossRef]

54. Turney, C.; Wilmshurst, J.; Jones, R.; Wood, J.; Palmer, J.; Hogg, A.; Fenwick, P.; Crowley, S.; Privat, K.; Thomas, Z. Reconstructing atmospheric circulation over southern new zealand: Establishment of modern westerly airflow 5500 years ago and implications for southern hemisphere Holocene climate change. *Quat. Sci. Rev.* **2017**, *159*, 77–87. [CrossRef]

55. Anderson, H.J.; Moy, C.M.; Vandergoes, M.J.; Nichols, J.E.; Riesselman, C.R.; Van Hale, R. Southern hemisphere westerly wind influence on southern New Zealand hydrology during the lateglacial and Holocene. *J. Quat. Sci.* **2018**, *33*, 689–701. [CrossRef]

56. McGlone, M.S.; Basher, L. Holocene vegetation change at treeline, Cropp Valley, Southern Alps, New Zealand. In *Peopled Landscapes: Archaeological and Biogeographic Approaches to Landscape*; Haberle, S., David, B., Eds.; ANU E Press: Canberra, Australia, 2012; Volume 34, pp. 343–358.

57. Jara, I.A.; Newnham, R.M.; Vandergoes, M.J.; Foster, C.R.; Lowe, D.J.; Wilmshurst, J.M.; Moreno, P.I.; Renwick, J.A.; Homes, A.M. Pollen-climate reconstruction from northern South Island, New Zealand (41 degrees S), reveals varying high- and low-latitude teleconnections over the last 16,000 years. *J. Quat. Sci.* **2015**, *30*, 817–829. [CrossRef]

58. Heusser, C. Late Quaternary vegetation and climate of southern Tierra del Fuego. *Quatern. Res.* **1989**, *31*, 396–406. [CrossRef]

59. Heusser, C.J. Three late Quaternary pollen diagrams from southern Patagonia and their paleoecological implications. *Palaeogeogr. Palaeoclimatol. Palaeoecol.* **1995**, *118*, 1–24. [CrossRef]

60. McCulloch, R.D.; Davies, S.J. Late-glacial and Holocene palaeoenvironmental change in the central Strait of Magellan, southern Patagonia. *Palaeogeogr. Palaeoclimatol. Palaeoecol.* **2001**, *173*, 143–173. [CrossRef]

61. Fesq-Martin, M.; Friedmann, A.; Peters, M.; Behrmann, J.; Kilian, R. Late-glacial and Holocene vegetation history of the Magellanic rain forest in southwestern Patagonia, Chile. *Veg. Hist. Archaeobot.* **2004**, *13*, 249–255. [CrossRef]

62. Borromei, A.M.; Coronato, A.; Quattrocchio, M.; Rabassa, J.; Grill, S.; Roig, C. Late Pleistocene-Holocene environments in Valle Carbajal, Tierra del Fuego, Argentina. *J. S. Am. Earth Sci.* **2007**, *23*, 321–335. [CrossRef]

63. Borromei, A.M.; Coronato, A.; Franzen, L.G.; Ponce, J.F.; Saez, J.A.L.; Maidana, N.; Rabassa, J.; Candel, M.S. Multiproxy record of Holocene paleoenvironmental change, Tierra del Fuego, Argentina. *Palaeogeogr. Palaeoclimatol. Palaeoecol.* **2010**, *286*, 1–16. [CrossRef]

64. Markgraf, V. Paleoenvironments and paleoclimates in Tierra del Fuego and southernmost Patagonia, South America. *Palaeogeogr. Palaeoclimatol. Palaeoecol.* **1993**, *102*, 53–68. [CrossRef]

65. Pendall, E.; Markgraf, V.; White, J.W.C.; Dreier, M.; Kenny, R. Multiproxy record of late Pleistocene-Holocene climate and vegetation changes from a peat bog in Patagonia. *Quatern. Res.* **2001**, *55*, 168–178. [CrossRef]

66. Barrow, C.J. Postglacial pollen diagrams from South Georgia (sub-antarctic) and West Falkland Island (South Atlantic). *J. Biogeogr.* **1978**, *5*, 251–274. [CrossRef]

67. Smith, R.I.L.; Prince, P.A. The natural history of Beauchêne Island. *Biol. J. Linn. Soc.* **1985**, *24*, 233–283. [CrossRef]

68. Nanavati, W.P.; Whitlock, C.; Iglesias, V.; de Porras, M.E. Postglacial vegetation, fire, and climate history along the eastern Andes, Argentina and Chile (lat. 41–55 degrees s). *Quat. Sci. Rev.* **2019**, *207*, 145–160. [CrossRef]

69. Morrison, K.W.; Battley, P.F.; Sagar, P.M.; Thompson, D.R. Population dynamics of eastern rockhopper penguins on Campbell Island in relation to sea surface temperature 1942–2012: Current warming hiatus pauses a long-term decline. *Polar Biol.* **2015**, *38*, 163–177. [CrossRef]

70. Cullen, L.E.; Stewart, G.H.; Duncan, R.P.; Palmer, J.G. Disturbance and climate warming influences on New Zealand Nothofagus tree-line population dynamics. *J. Ecol.* **2001**, *89*, 1061–1071. [CrossRef]

71. Wardle, P.; Coleman, M.; Buxton, R.; Wilmshurst, J. Climatic warming and the upper forest limit. *Canterb. Bot. Soc. J.* **2005**, *39*, 90–98.

Article

Dendroclimatic Assessment of Ponderosa Pine Radial Growth along Elevational Transects in Western Montana, U.S.A.

Evan E. Montpellier [1,2,*], **Peter T. Soulé** [2], **Paul A. Knapp** [3] and **Justin T. Maxwell** [4]

[1] Department of Geography, Environment, and Society, University of Minnesota, 267 19th Ave S, Minneapolis, MN 55455, USA

[2] Department of Geography and Planning, Appalachian State University, P.O. Box 32066, Boone, NC 28608, USA; soulept@appstate.edu

[3] Department of Geography, Environment, and Sustainability, University of North Carolina at Greensboro, 237 Graham Building P.O. Box 26170, Greensboro, NC 27402, USA; paknapp@uncg.edu

[4] Department of Geography, Indiana University, 701 E Kirkwood Ave, Bloomington, IN 47405, USA; maxweljt@indiana.edu

* Correspondence: montp020@umn.edu; Tel.: +1-612-625-6080

Received: 11 November 2019; Accepted: 25 November 2019; Published: 2 December 2019

Abstract: Ponderosa pine (PP) is the most common and widely distributed pine species in the western United States, spanning from southern Canada to the United States–Mexico border. PP can be found growing between sea level and 3000 meters elevation making them an ideal species to assess the effects of changing climatic conditions at a variety of elevations. Here we compare PP standardized and raw growth responses to climate conditions along an elevational transect spanning 1000 meters in western Montana, U.S.A., a region that experienced a 20th century warming trend and is expected to incur much warmer (3.1–4.5 °C) and slightly drier summers (~0.3 cm decrease per month) by the end on the 21st century. Specifically, we assess if there are climate/growth differences based on relative (i.e., site-specific) and absolute (i.e., combined sites) elevation between groups of trees growing in different elevational classes. We find that values of the Palmer drought severity index (PDSI) in July are most strongly related to radial growth and that within-site elevation differences are a poor predictor of the response of PP to either wet or dry climatic conditions (i.e., years with above or below average July PDSI values). These results suggest that any generalization that stands of PP occurring at their elevational margins are most vulnerable to changing climatic may not be operative at these sites in western Montana. Our results show that when using standardized ring widths, PP growing at the lowest and highest elevations within western Montana exhibit differential growth during extreme climatological conditions with lower-elevation trees outperforming higher-elevation trees during dry years and vice versa during wet years.

Keywords: dendroclimatology; elevational gradients; drought; western Montana; Rocky Mountains

---

1. Introduction

Ponderosa pine (PP; *Pinus ponderosa* var. *ponderosa* Lawson and C. Lawson) is a geographically diverse, ecologically significant, and economically important tree species in western North America. In the Northern Rockies, U.S.A., PP occurs from approximately 700–2300 m elevation, making it not only a widespread species but one that also has an extensive altitudinal gradient [1,2]. Along the elevational gradient, significant climatic differences exist, with the warmest and driest conditions occurring at the lowest elevations and progressively cooler and wetter conditions with increasing elevation. Global circulation models for the Northern Rockies predict substantially warmer (3.1–4.5 °C) and slightly drier (~0.3 cm decrease per month) conditions in the next several decades [3–7] and these future conditions create an impetus to examine the historical responses of this dominant coniferous tree to changing climate and atmospheric composition.

If climatic change in the Northern Rockies [5–9] continues, regional PP forests likely will experience more frequent summer drought conditions. Prior work with PP in this region has demonstrated that the growth of PP is positively related to wetter and cooler conditions during the spring and summer months [10–13]. Thus, warmer conditions and more frequent droughts should negatively impact overall forest productivity. But the question remains as to what degree and, more importantly, which individuals will be most impacted? Are PP growing at the lowest elevations the most prone to growth declines or mortality because they grow along the climatic margin for survivability?

In a broad synthesis of research examining relationships between changing climate and tree mortality, Allen et al. [14] reported that the evidence linking recent increases in temperature and the frequency/severity of drought is associated with increasing tree senescence in multiple ecosystems worldwide. More specifically, McDowell et al. [15] (p. 399) reported that PP growing at lower elevations are more likely to exhibit "chronic water stress" and thus are more likely to die during periods of drought, with long-term implications for PP ecosystem modifications if the vectors for increasing aridity are manifest. Ganey et al. [16], hypothesized similar outcomes to McDowell et al. [15] in that tree senescence would be greatest at the low elevation sites as they are more frequently stressed by a combination of high temperature and limited moisture availability. Ganey et al. [16] examined mortality rates of multiple species within mixed PP forests that had experienced high rates of senescence in recent decades yet found that senescence was not closely linked with elevation. Tague et al. [17] examined rates of PP mortality and productivity along elevational transects to determine if models incorporating carbon allocation and hydroclimatic parameters can be used to understand spatial patterns of drought-induced tree senescence along elevational transects. Their elevation-specific results are mixed. They noted that net primary productivity (NPP) at high sites is greater largely due to higher precipitation rates and greater moisture storage at high elevations. However, the relationship between NPP and precipitation was similarly strong at both high and low elevation sites. They concluded that spatiotemporal patterns of temperature and moisture are strongly controlled by elevation and equally important in modeling the probability of tree senescence. While Lloret et al. [18] noted that climate change, which is manifest through warming and increasing drought frequency, is related to tree senescence globally. They found that rates of radial growth recovery for PP following drought are largely unchanged through time. They concluded that, long-term, climate change-induced senescence is not closely related to decreasing resilience but is rather controlled by tree reactions to specific events (e.g., an individual drought).

Previous work leaves the question: Could it be that many of these low-elevation PP populations already possess distinct characteristics that would confer climatically marginal populations an advantage over those PPs growing in the cooler and wetter conditions that exist with increasing elevation? Elevation has been identified as a key component regulating the growth of PP [19,20] with physiologic plasticity suggested as a means by which individuals adapt to elevational heterogeneity [20]. The degree to which the persistence of a species is influenced by local adaptation and/or phenotypic plasticity across populations could be critical to understanding the impact of climate change on overall populations [21].

Here, we address these climate/growth-response questions and discrepancies in the literature based on a dendrochronological sampling of PP along vertical transects in western Montana. We hypothesize there are differences in the climate-growth response of PP across elevational classes. Specifically, we posit that trees growing at lower elevations will exhibit a greater sensitivity to drought and an overall decline in radial growth. To investigate this hypothesis we: (1) assess drivers of PP radial growth at three study sites in western Montana; (2) compare standardized ring width of PP to climate along study-site specific elevational transects (i.e., relative elevation); and (3) investigate PP growth in response to varying extreme climatic conditions along an integrated elevational transect (i.e., absolute elevation).

2. Materials and Methods

2.1. Study Area

We collected samples from mature PP trees at three south- to southwest-facing sites in the northern Rockies of western Montana (Figure 1). All sites consisted of mixed open-canopy woodlands of PP and Douglas-fir (*Pseudotsuga menziesii* var. *glauca* (Mirb.) Franco)) and, when combined, allowed for sampling along an elevational transect of approximately 1020 m. Our sites were selected based on two principal criteria. First, PP growing on southerly slopes are more likely to experience summertime soil-moisture deficits than other aspects. Thus, the potential radial-growth effects of warmer, drier summers should be first apparent at these locations. Second, we sought to minimize the potential effects of human agency that could act as confounding factors in our analysis and selected sites with known histories of limited anthropogenic activity. Ferry Landing (FLF) is designated as a Research Natural Area in part because of a history of minimal disturbance (e.g., logging, grazing) [22], while our two other sites, Fish Creek (FCF) and The Grove (TGF), also have a history of minimal disturbance (personal communication, Steve Shelly—USFS). At each location, we began sampling at the slope base but above the height of permanent water sources and selected trees at approximately equally spaced intervals that allowed for at least 30 trees between the base and the ridgeline. Elevation ranged from ~800 m to ~1500 m at FLF with a mean elevation of ~1125 m, from ~1050 m to ~1500 m at FCF with a mean elevation of ~1250 m, and from ~1350m to ~1800 m at TGF with a mean elevation of ~1550 m. All sites experience Humid Continental climates on the Köppen system, and we present precipitation and temperature data from PRISM [23] to characterize climatic variability between sites. For each tree, we extracted two core samples at breast height (~1.3 m) using increment borers. We selected only trees with no canopy overlap to minimize potentially confounding growth effects from neighboring trees. We also avoided sampling any trees with large fire scars, broken tops, lightning strikes, dwarf mistletoe (*Arceuthobium* spp.), or other disturbances that could have impacted radial growth.

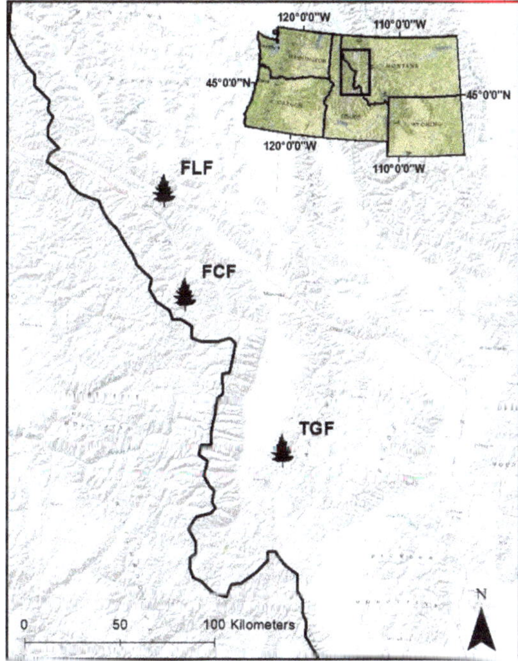

Figure 1. Study site locations in the Bitterroot National Wilderness of western Montana, USA.

2.2. Chronology Development

We used standard techniques to process the core samples [24]. We sanded the core samples until the cellular structure was clear under magnification and then used the list method for cross-dating [25]. We measured the individual ring-widths using WinDENDRO to 0.001 mm accuracy [26]. We used COFECHA [27] to confirm the accuracy of cross-dating.

We used a two-thirds smoothing spline to standardize our chronologies based on previous PP tree-ring studies [28,29] and because of potential forest disturbances (i.e., fire) in the region. While the biological growth in western trees is often modeled as negative exponential, stand-specific disturbances have the ability to under- or over-fit such curves. Cook and Peters [30] proposed the smoothing spline as an alternative method to negative exponential curves to accurately model tree growth in closed-canopy or disturbed forest types. In conjunction with our standardized chronologies, we retained raw radial growth rates to examine differences between biological growth at high and low elevations during wet and dry periods.

To assess climate response, we used three standardized chronologies (i.e., FLF_ALL, FCF_ALL, TGF_ALL), which contained as many core samples from the site that we were able to successfully cross-date (Table 1). At each site, we created two elevational classes (i.e., FLF_TOP, FLF_BOTTOM, etc.) that contained samples from the top and bottom ten trees (Table 1). We combined our trees into one elevational chronology using absolute elevation (ALL). From the ALL chronology, we divided the trees into two elevational classes, with 30 trees per group, and created combined chronologies (ALL_TOP and ALL_BOTTOM; Table 1).

Table 1. Descriptive chronology statistics.

Chronology	Mean DBH (cm)	Number of Samples	Mean Age (years)
FLF_ALL	73.5	57	182
FLF_TOP	73.5	20	208
FLF_BOTTOM	72.9	19	145
FCF_ALL	64.7	47	233
FCF_TOP	63.0	17	241
FCF_BOTTOM	65.2	15	230
TGF_ALL	75.8	50	223
TGF_TOP	78.5	18	171
TGF_BOTTOM	78.5	18	275
ALL	70.6	154	213
ALL_TOP	74.1	53	211
ALL_BOTTOM	69.0	55	193

2.3. Statistical Analysis

We searched for the primary climate drivers of standardized radial growth at the three sites using Spearman correlation between the ALL chronologies and monthly temperature, precipitation, and Palmer drought severity index (PDSI) data [31] from Montana Climate Division 1 from 1895–2011. The PDSI is a water balance-based measure of drought severity and considers antecedent moisture conditions for several months, with negative values representing dry conditions and positive values representing wet conditions [31]. We then compared the climate response for these primary climate drivers of temperature, precipitation, and PDSI between the TOP and BOTTOM (site-specific and ALL) chronologies using Spearman correlation and determined if there were significant differences in the overall strongest monthly relationships across the three sites using a Fisher's z-test. Specifically, we tested for significant differences in the climate/growth relationships between TOP and BOTTOM for temperature, precipitation, and PDSI using values from the month that produced the consistently highest Spearman r-values across the three sites (e.g., for PDSI this was the month of July).

We investigated if there was an elevational bias in PP growth (i.e., top trees had a modified growth rate when warm/dry or cool/wet conditions were present) during 1895–2011. To do this, we examined raw radial growth and standardized growth of individual trees during years when the July PDSI value was >2 (wet) and <−2 (dry). Although two cores were collected per tree when field sampling, the inability to cross-date a sample or visible core damage often precludes that core from being included in the final chronology. This creates a scenario in which some trees have one core sample while other trees have two. In instances when two cores were present, we calculated the average growth rate between the two samples during wet and dry periods. We then compared the average raw and standardized growth per tree for wet and dry years to elevational positioning using Spearman correlation. Lastly, we performed a Mann–Whitney U Test between trees growing at the top and bottom portions of our combined elevational transect (i.e., ALL_TOP and ALL_BOTTOM) chronologies during wet and dry periods in Montana Climate Division 1.

3. Results

Minimal differences exist in macroscale climatic conditions among the three study sites (Figure 2). As expected, the highest site (TGF) is the coldest, and the lowest site (FLF) is the warmest. Small differences in monthly precipitation totals are present with FCF receiving the greatest amount of annual precipitation (~63 cm), while TGF receives the least amount of annual precipitation (~51 cm).

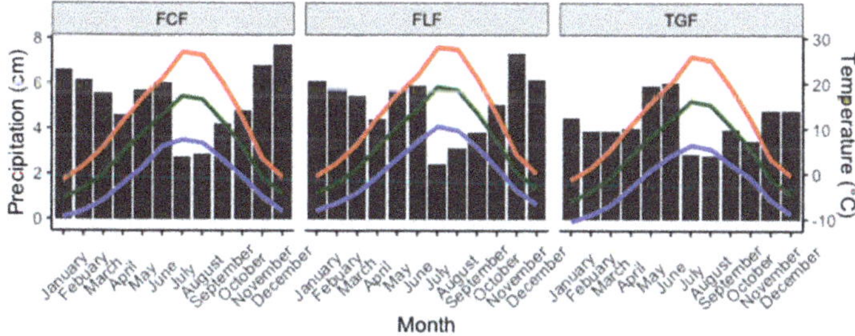

Figure 2. PRISM derived climographs for FCF, FLF, and TGF, respectively. Gray bars indicate average monthly precipitation amounts in centimeters while lines indicate average monthly maximum (red), mean (green), and minimum temperatures (blue) from 1895 to 2011 in Celsius.

Our investigation of the relationships between standardized growth and monthly measures of climate shows that moisture availability in mid-summer (i.e., July PDSI) is the principal driving force of PP growth. PP respond negatively to mid-summer temperatures and positively to precipitation (Table 2; Figure 3). For PDSI, the strongest relationships occur in the month of July. Since PDSI values in each month are partially dependent on moisture supply and demand in the preceding months [31], there is a cumulative component to the climate response for PP.

The overall (i.e., all trees in the site level chronology) climate response is comparable across the three study sites (Table 2). While small differences exist between sites when the analyses are divided into TOP and BOTTOM groupings of trees (Table 2), the Fisher z-test results showed there were no significant differences ($p < 0.05$) in the climate/growth relationships between TOP and BOTTOM trees for the strongest monthly relationship for each climate variable.

The ALL chronology (i.e., the chronology based on absolute elevation) reveals a significant and similar association with the same climatic variables discussed at the specific site level (Table 3). The Fisher z-test resulted in no significant difference in climate response between ALL_TOP and ALL_BOTTOM groupings along the combined elevational transect. When correlating (Spearman) raw radial growth of trees along the combined transect to elevation during wet and dry (PDSI > 2;

PDSI < −2) conditions, we found no significant association. However, for standardized radial growth values, we found that low elevation trees have significantly ($p < 0.001$) higher radial growth than higher elevation trees during dry periods, and high elevation trees have significantly ($p < 0.003$) greater radial growth than low elevation trees during wet periods (Figure 4). Lastly, our Mann–Whitney U Test used to compare the difference in means of high and low groupings resulted in significant differences between both dry ($p = 0.000$) and wet periods ($p = 0.014$).

Table 2. Spearman correlation *r*-Values between standardized radial growth values (1895–2011) for each elevational grouping at our three study locations and climate variables from Montana Climate Division 1.

	July Temperature	June Precipitation	July PSDI
FLF_Top	−0.410	0.258	0.542
p-Value	0.000	0.005	0.000
FLF_Bottom	−0.326	0.299	0.442
p-Value	0.000	0.001	0.000
FCF_Top	−0.297	0.286	0.493
p-Value	0.001	0.002	0.000
FCF_Bottom	−0.193	0.287	0.421
p-Value	0.037	0.002	0.000
TGF_Top	−0.410	0.328	0.461
p-Value	0.000	0.000	0.000
TGF_Bottom	−0.448	0.240	0.452
p-Value	0.000	0.009	0.000

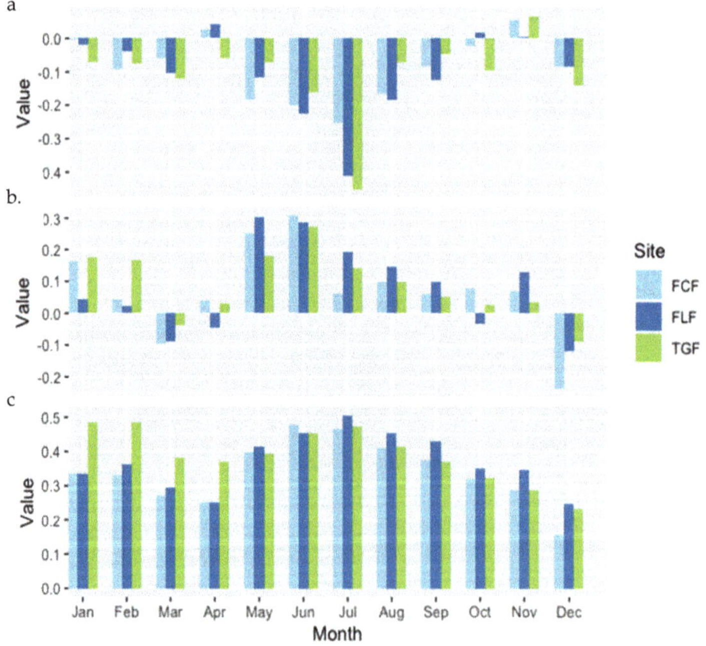

Figure 3. Monthly Spearman correlation *r*-Values between all standardized radial growth samples

at FCF (light blue), FLF (dark blue) and TGF (green), and: (**a**) temperature, (**b**) precipitation, and (**c**) Palmer drought severity index for Montana Climate Division 1.

Table 3. Spearman correlations between the standardized ALL chronologies and climate variables from Montana Climate Division 1.

	July Temperature	June Precipitation	July PSDI
ALL	−0.440	0.307	0.558
p-Value	0.000	0.001	0.000
ALL_Top	−0.477	0.257	0.533
p-Value	0.000	0.005	0.000
ALL_Bottom	−0.355	0.310	0.489
p-Value	0.000	0.001	0.000

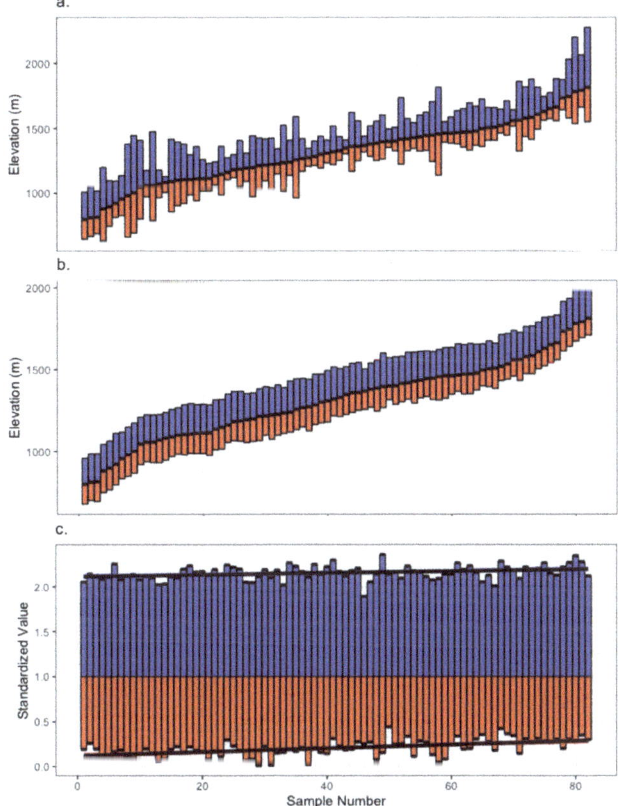

Figure 4. Raw and standardized tree growth along the ALL transect during wet and dry periods. The: (**a**) relative raw radial growth for wet (blue; PSDI >2) and dry (red; PSDI < −2) periods for each tree along the ALL elevational gradient. The solid black line represents individual tree elevation, and the magnitude of the vertical bars corresponds to the magnitude of average raw radial growth during wet and dry periods for that tree; (**b**) relative standardized radial growth for wet (blue; PSDI >2) and dry (red; PSDI < −2) periods for each tree along the ALL elevational gradient. The solid black line represents individual tree elevation, and the magnitude of the vertical bars corresponds to the magnitude of average standardized growth during wet and dry periods for that tree.; and (**c**) standardized growth

values during wet (blue; PSDI >2) and dry (red; PSDI < −2) periods for trees organized from lowest (left) to highest (right) elevation. Averaged growth during wet years (blue) have a value of one added to them and average growth for dry years (red) are subtracted by one (i.e., tree one (left) has an average growth value of ~1.05 for wet years and an average growth value ~0.81 for dry years).

4. Discussion

While slight differences exist in site-specific climatic conditions (Figure 2), the temperature patterns are logical (i.e., temperature decreased with elevational increase). We expected the highest annual precipitation at TGF due to the combination of higher elevation and orographic lifting processes, but the PRISM data suggest this location receives slightly less precipitation than FCF and FLF. This is likely a function of the location of TGF, east of the spine of the Bitterroot Mountain range, resulting in a rain shadow effect. (Figure 1). Although there are small site differences in precipitation and temperature, the overall climate response of trees is congruent (Figure 3; Tables 2 and 3). For change detection, a Mann–Kendall trend test showed that temperatures in Montana Climate Division 1 have a significant positive trend over the last 50-years (1862–2011; $p = 0.025$), but no trends were evident over both longer (1895–2011; $p = 0.439$) or shorter periods (1982–2011; $p = 0.080$). We also found no trends at 30-, 50- or the 117-year periods in precipitation or PDSI in Montana Climate Division 1, so the increasing temperatures have not translated into increasing aridity as measured via a water balance-based metric (i.e., the PDSI). However, summer temperature in the region is projected to continue to increase, and precipitation is projected to decrease [3–7]. If this occurs, PP in the region will experience increasing aridity during the growing season in future decades.

Raw radial growth patterns during dry and wet years in Montana Climate Division 1 have no directional relationship based on elevation (Figure 4a). For example, there is no propensity for lower (higher) elevation trees to grow faster than higher (lower) elevation trees during dry (wet) periods. The only discernible pattern is that trees with the highest rates of radial growth are growing faster than trees with the lowest rates during both wet and dry periods. Although all of our sampled trees were mature, this could be a function of declining radial growth with tree age. However, when assessing standardized radial growth based on absolute elevation during wet and dry years, distinct trends exist within the data (Figure 4b,c). The pattern reveals that, collectively, higher elevation trees outperform lower elevation trees during wet periods, and lower elevation trees outperform higher elevation trees during dry periods (Figure 4).

Both prior work (e.g., [13]) and our results (Figure 3) demonstrate that PP is positively related to wetter and cooler conditions during the spring and summer months. Thus, warmer, drier conditions and more frequent droughts should negatively impact overall forest productivity, and PP growing at the lowest elevations may be the most prone to growth declines or mortality because they grow along the climatic margin for survivability [15]. Conversely, these low-elevation populations may already possess distinct characteristics, such as exceptional water-use efficiency, that would confer climatically marginal populations an advantage over those PPs growing in the cooler and wetter conditions that exist with increasing elevation. Specifically, the ability to withstand xylem cavitation is often a characteristic of individuals growing in the more xeric portion of a species range [32], and individuals may experience decreasing growth response to drought with increasing elevation [33]. Elevation has been identified as a key component regulating the growth of PP, with physiologic plasticity suggested as a means by which individuals adapt to elevational heterogeneity [21]. Our findings suggest this interpretation is correct, but only across wider ranges of elevation than are typically encountered by PP trees growing on site-specific elevational transects.

Previous research has found that increased water stress and aridity are projected to lead to more mortality and senescence of PP growing at low elevations in the western United States [14,34,35]. The results of our study, however, suggest that this generalization may not be operational in the northern Rocky Mountains of western Montana. One plausible explanation for our differing results is the location of our study sites. Lascoux et al. [36] documented a divergence of PP approximately

250,000 years ago into two varieties: eastern and western. These varieties moved northward over time and reconvened in what Latta and Mitton [37] (p. 769) described as "west-central Montana". Latta and Mitton [37] (p. 769) further noted that at this transition zone, "gene flow between the two varieties will introduce genes to potentially different adaptive regimes. The varieties are interfertile . . . ". This finding suggests that PP growing in the transition zone may possess certain genetic adaptations that allow for greater resiliency in a changing climate, and this might explain why trees from our study sites reveal differing broader climate responses to changing climatic conditions.

Detecting and/or modeling PP health/mortality, specifically at lower elevations, in response to warmer and drier climate conditions in the western United States, has been successful [15,34]. For example, Van Mantgem et al. [34] modeled tree mortality rates across 76 forest plots in western North America. They found significant increases in mortality rates through time in all regions and species, across all diameter classes of trees, and at all elevations. For elevation, they found the greatest increases in mortality in the mid-elevation ranges (1000–2000 m). However, other investigations of PP health have been unable to either detect or confidently make linkages to large-scale climate change. For example, McCullough et al. [38] investigated the climate response of 161 PP chronologies by grouping sites with similar climate response. They generally found that trees growing in the more western populations were overall less sensitive to climate than eastern populations. More importantly, McCullough et al. [38] suggested that making generalizations about a species that occupies large elevation and spatial gradients comes at the risk of oversimplification. They concluded by saying they believe that most PP ecosystems will experience changing climate regimes, yet the responses of trees will be based on local conditions. Our findings, which show no within site difference in climate-growth or growth-elevation responses, but do show elevational differences when trees across a wider range of elevations are grouped, suggest that local (i.e., site-specific) environmental differences are less important to PP radial growth rates than those experienced over a wider range.

5. Conclusions

PP occupy large elevational gradients in the western United States, making it an ideal species to investigate the elevational effects of a changing climate. Previous research has indicated that PP exhibit physiologic plasticity, which enables the species to express elevational heterogeneity in climate response [21]. Thus, PP trees growing in lower elevations should be better adapted to withstand warm, dry periods while trees at high elevations are better suited for cool, wet conditions. Our primary conclusions are that: (1) when examining raw ring widths, no elevational bias in growth response occurs during either wet or dry years; (2) standardized radial growth of PP does exhibit significant climate-growth responses that are dependent on elevation, but only when considering broader elevation ranges of trees that are typical of the ranges found from valley bottom to ridge top in Western Montana with lower elevation trees outperforming higher elevation trees during dry periods; and, (3) the results using both raw and standardized data suggest that lower elevation trees may be able to successfully adapt to the projected 21st century summertime climatic changes (i.e., warmer and drier) within the region.

Author Contributions: E.E.M. assisted with methodology development, formal analysis, investigation, Writing—Original draft preparation and Writing—Review and editing, as well as visualization. P.T.S. was a part of conceptualization, methodology development, Writing—Original draft preparation, Writing—Review and editing, and supervision. P.A.K. assisted with conceptualization, methodology development, writing review and editing, and supervision. J.T.M. assisted with conceptualization, methodology development, and Writing—Review and editing.

Funding: Funding was provided by a University Research Council grant from Appalachian State University, and by a Proposal Preparation Program grant from the University of North Carolina, Greensboro.

Acknowledgments: We thank Steve Shelly of the United States Forest Service for site selection and fieldwork assistance and David Austin, Lindsay Cummings, Thomas Patterson, and Phil White for fieldwork and laboratory assistance.

Conflicts of Interest: The authors declare no conflict of interest.

References

1. Pinus ponderosa—The Gymonsperm Database. Available online: https://www.conifers.org/pi/Pinus_ponderosa.php (accessed on 2 September 2019).
2. Plant Guide. Available online: https://plants.usda.gov/plantguide/pdf/pg_pipo.pdf (accessed on 7 September 2019).
3. Joyce, L.A.; Talbert, M.; Sharp, D.; Stevenson, J. Historical and Projected Climate in the Northern Rockies Region. In *Climate Change and Rocky Mountain Ecosystems. Advances in Global Change Research*; Halofsky, J., Peterson, D., Eds.; Springer: New York, NY, USA, 2018; pp. 17–23.
4. Littell, J.S.; McKenzie, D.; Kerns, B.K.; Cushman, S.; Shaw, C.G. Managing uncertainty in climate-driven ecological models to inform adaptation to climate change. *Ecosphere* **2011**, *9*, 1–19. [CrossRef]
5. Riley, K.L.; Loehman, R.A. Mid-21st-century climate changes increase predicted fire occurrence and fire season length, Northern Rocky Mountains, United States. *Ecosphere* **2016**, *7*, 1–19. [CrossRef]
6. Keane, R.E.; Mahalovich, M.F. Effects of Climate Change on Forest Vegetation in the Northern Rockies Region. *Aspen Bibliogr.* **2018**, *374*, 128–273.
7. Whitlock, C.; Cross, W.; Maxwell, B.; Silverman, N.; Wade, A.A. *Montana Climate Assessment*; Montana Institute on Ecosystems, Montana State University and University of Montana, Bozeman and Missoula: Missoula, MT, USA, 2017.
8. Westerling, A.L.; Hidalgo, H.G.; Cayan, D.R.; Swetnam, T.W. Warming and earlier spring increase western US forest wildfire activity. *Science* **2006**, *5789*, 940–943. [CrossRef]
9. Hamlet, A.F.; Lettenmaier, D.P. Effects of 20th century warming and climate variability on flood risk in the western US. *Water Resour Res.* **2007**, *6*, 1–17.
10. Soulé, P.T.; Knapp, P.A. Radial Growth Rates of Two Co-occurring Coniferous Trees in the Northern Rockies during the Past Century. *J. Arid Environ.* **2013**, *94*, 87–95. [CrossRef]
11. Soulé, P.T.; Knapp, P.A. Analyses of intrinsic water-use efficiency indicate performance differences of ponderosa pine and Douglas-fir in response to CO_2 enrichment. *J. Biogeogr.* **2015**, *42*, 144–155. [CrossRef]
12. Petrie, M.D.; Wildeman, A.M.; Bradford, J.B.; Hubbard, R.M.; Lauenroth, W.K. A review of precipitation and temperature control on seedling emergence and establishment for ponderosa and lodgepole pine forest regeneration. *For. Ecol. Manag.* **2016**, *361*, 328–338. [CrossRef]
13. Marquardt, P.E. Investigation into Climatic Effects on the Growth and Genetic Structure of Sky Island Ponderosa Pine. Ph.D. Thesis, Michigan State University, East Lansing, MI, USA, 2018.
14. Allen, C.D.; Macalady, A.K.; Chenchouni, H.; Bachelet, D.; McDowell, N.; Vennetier, M.; Kitzberger, T.; Rigling, A.; Breshears, D.D.; Hogg, E.T.; et al. A global overview of drought and heat-induced tree mortality reveals emerging climate change risks for forests. *For. Ecol. Manag.* **2010**, *259*, 660–684. [CrossRef]
15. McDowell, N.G.; Allen, C.D.; Marshall, L. Growth, carbon-isotope discrimination, and drought-associated mortality across a Pinus ponderosa elevational transect. *Glob. Chang. Biol.* **2010**, *16*, 399–415. [CrossRef]
16. Ganey, J.L.; Vojta, S.C. Tree mortality in drought-stressed mixed-conifer and ponderosa pine forests, Arizona, USA. *For. Ecol. Manag.* **2011**, *261*, 162–168. [CrossRef]
17. Tague, C.L.; McDowell, N.G.; Allen, C.D. An integrated model of environmental effects on growth, carbohydrate balance, and mortality of Pinus ponderosa forests in the southern Rocky Mountains. *PLoS ONE* **2013**, *8*, e80286. [CrossRef] [PubMed]
18. Lloret, F.; Keeling, E.G.; Sala, A. Components of tree resilience: Effects of successive low-growth episodes in old ponderosa pine forests. *Oikos* **2011**, *120*, 1909–1920. [CrossRef]
19. Rehfeldt, G.E. Genetic differentiation among populations of Pinus ponderosa from the upper Colorado River Basin. *Bot. Gaz.* **1990**, *1*, 125–137. [CrossRef]
20. Zhang, J.; Cregg, B.M. Growth and physiological responses to varied environments among populations of Pinus ponderosa. *For. Ecol. Manag.* **2005**, *1*, 1–12. [CrossRef]
21. De Luis, M.; Čufar, K.; Di Filippo, A.; Novak, K.; Papadopoulos, A.; Piovesan, G.; Rathgeber, C.B.; Raventós, J.; Saz, M.A.; Smith, K.T. Plasticity in dendroclimatic response across the distribution range of Aleppo pine (*Pinus halepensis*). *PLoS ONE* **2013**, *12*, e83550. [CrossRef]
22. Evenden, A.G.; Moeur, M.; Shelly, J.S.; Kimball, S.F.; Wellner, C.A. *Research natural areas on National Forest System lands in Idaho, Montana, Nevada, Utah, and western Wyoming: A Guidebook for Scientists, Managers, and Educators*; General Technical Report RMRS-GTR-69; US Department of Agriculture, Forest Service, Rocky Mountain Research Station: Ogden, UT, USA, 2001; p. 69.

23. PRISM Climate Group, Oregon State University. Available online: http://prism.oregonstate.edu (accessed on 23 October 2019).

24. Stokes, M.A. *An Introduction to Tree-Ring Dating*; University of Arizona Press: Tucson, AZ, USA, 1996.

25. Yamaguchi, D.K. A simple method for cross-dating increment cores from living trees. *Can. J. For. Res.* **1991**, *3*, 414–416. [CrossRef]

26. Regent Instruments Canada Inc. *WINDENDRO for Tree-Ring Analysis*; Canada Inc.: Québec, QC, Canada, 2013.

27. Holmes, R.L. Computer-assisted quality control in tree-ring dating and measurement. *Tree Ring Bull.* **1983**, *43*, 51–67.

28. Dannenberg, M.P.; Wise, E.K. Seasonal climate signals from multiple tree ring metrics: A case study of Pinus ponderosa in the upper Columbia River Basin. *J Geophys. Res. Biogeosci.* **2016**, *121*, 1178–1189. [CrossRef]

29. Zhang, J.; Finley, K.A.; Johnson, N.G.; Ritchie, M.W. Lowering stand density enhances resiliency of ponderosa pine forests to disturbances and climate change. *For. Sci.* **2019**, *65*, 496–507. [CrossRef]

30. Cook, E.R.; Peters, K. The smoothing spline: A new approach to standardizing forest interior tree-ring width series for dendroclimatic studies. *Tree Ring Bull.* **1981**, *41*, 45–53.

31. Palmer, W.C. *Meteorological Drought*; Paper 45, 58 pp; US Department of Commerce, Weather Bureau: Washington, DC, USA, 1965.

32. Stout, D.L.; Sala, A. Xylem vulnerability to cavitation in *Pseudotsuga menziesii* and *Pinus ponderosa* from contrasting habitats. *Tree Physiol.* **2003**, *23*, 43–50. [CrossRef] [PubMed]

33. Affolter, P.; Büntgen, U.; Esper, J.; Rigling, A.; Weber, P.; Luterbacher, J.; Frank, D. Inner Alpine conifer response to 20th century drought swings. *Eur. J. For. Res.* **2010**, *129*, 289–298. [CrossRef]

34. Van Mantgem, P.J.; Stephenson, N.L.; Byrne, J.C.; Daniels, L.D.; Franklin, J.F.; Fulé, P.Z.; Harmon, M.E.; Larson, A.J.; Smith, J.M.; Taylor, A.H.; et al. Widespread Increase of Tree Mortality Rates in the Western United States. *Science* **2009**, *323*, 521–524. [CrossRef]

35. Knutson, K.C.; Pyke, D.A. Western juniper and ponderosa pine ecotonal climate–growth relationships across landscape gradients in southern Oregon. *Can. J. For. Res.* **2008**, *38*, 3021–3032. [CrossRef]

36. Lascoux, M.; Palmé, A.E.; Cheddadi, R.; Latta, R.G. Impact of Ice Ages on the genetic structure of trees and shrubs. *Philos. Trans. R. Soc. Lond. Ser. B* **2004**, *359*, 197–207. [CrossRef]

37. Latta, R.G.; Mitton, J.B. Historical separation and present gene flow through a zone of secondary contact in ponderosa pine. *Evolution* **1999**, *53*, 769–776. [CrossRef]

38. McCullough, I.M.; Davis, F.W.; Williams, A.P. A range of possibilities: Assessing geographic variation in climate sensitivity of ponderosa pine using tree rings. *For. Ecol. Manag.* **2017**, *402*, 223–233. [CrossRef]

Article

Microsites and Climate Zones: Seedling Regeneration in the Alpine Treeline Ecotone Worldwide

Adelaide C. Johnson [1,*] and J. Alan Yeakley [2]

[1] U.S.D.A., Forest Service, Pacific Northwest Research Station, Forestry Sciences Laboratory, 11305 Glacier Hwy, Juneau, AK 99801, USA

[2] Department of Geography and Environmental Systems, University of Maryland Baltimore County, Baltimore, MD 21250, USA; yeakley@umbc.edu

* Correspondence: ajohnson03@fs.fed.us

Received: 9 September 2019; Accepted: 27 September 2019; Published: 3 October 2019

Abstract: Microsites, local features having the potential to alter the environment for seedling regeneration, may help to define likely trends in high-elevation forest regeneration pattern. Although multiple microsites may exist in any alpine treeline ecotone (ATE) on any continent, some microsites appear to enhance density of seedling regeneration better than others. Known seedling regeneration stresses in the ATE include low temperature, low substrate moisture, high radiation, drought, wind, and both high and low snowfall amount. Relationships among various microsite types, annual temperature, annual precipitation, and tree genera groups were assessed by synthesizing 52 studies from 26 countries spanning six continents. By categorization of four main microsite types (convex, concave, object, and wood) by mean annual precipitation and temperature, four major climatic zone associations were distinguished: cold & dry, cold & wet, warm & dry, warm & wet. Successful tree recruitment varied among microsite types and by climatic zones. In general, elevated convex sites and/or decayed wood facilitated earlier snow melt for seedlings located in cold & wet climates with abundant snowfall, depressions or concave sites enhanced summer moisture and protected seedlings from wind chill exposure for seedlings growing in cold & dry locations, and objects protected seedlings from excessive radiation and wind in warm & dry high locations. Our study results suggest that climate change will most benefit seedling regeneration in cold & wet locations and will most limit seedling regeneration in warm & dry locations given likely increases in fire and drought. Study results suggest that high-elevation mountain forests with water-limited growing seasons are likely to experience recruitment declines or, at best, no new recruitment advantage as climate warms. Climate envelope models, generally focusing on adult trees rather than seedling requirements, often assume that a warming climate will move tree species upward. Study results suggest that climate models may benefit from more physically-based considerations of microsites, climate, and current seedling regeneration limitations.

Keywords: climate change; climate zone; environmental stress; forest edge; precipitation; tree regeneration; tree seedling recruitment; upward advance

---

1. Introduction

Alpine treeline ecotone (ATE) locations are conspicuous high-elevation forest edges located on all continents except Antarctica [1]. The ATE extends from subalpine forest with trees of large stature moderating climatic stresses [2], up to treeline, the upper limit for tree growth (Figure 1). Between the forest line and treeline, the transitional ecotone may vary in width from tens to hundreds of meters [3]. ATE form can be abrupt, with an immediate transition from subalpine to alpine tundra or diffuse, a gradual transition from subalpine to alpine meadows, and may contain krummholz (i.e., stunted struggling trees), and tree islands [4]. Growing season temperature is generally assumed to be the main limitation for seedling regeneration and forest growth at the treeline [5,6]. Other factors also determine the forest distribution locally, such as viability of seed supply, animal and insect herbivory, substrate moisture availability, solar radiation, wind, snow cover, snow glide, snow avalanching, rock fall, human influence, competition from neighbors, unfavorable soil properties, fire, and microsite availability [1,5,7–9].

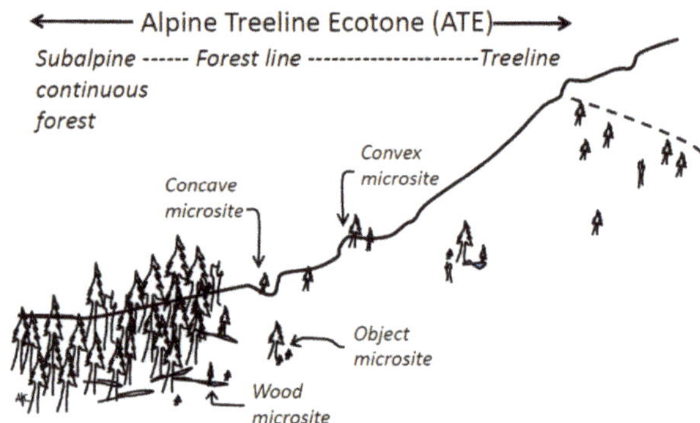

Figure 1. Timberline location and microsite type within the alpine treeline ecotone (ATE). Four microsite types summarized in this study are highlighted. Note ATE width, varying considerably worldwide, varies from abrupt to diffuse in structure. A diffuse ATE ecotone, typical of cool/wet climates is illustrated here.

Regeneration of seedlings represents a period of life where trees suffer the greatest mortality of all life stages [10]. Although climate projections often assume that the climate suitable for adult trees will be adequate for forest regeneration, seedling recruitment is often ignored, and given low seedling regeneration success at many locations, a demographic bottleneck for upward tree migration may exist [11]. Seedling regeneration and possible upward advance of forests associated with climate change is dependent on a few site-specific factors occurring within the broad ATE; these factors are highly heterogeneous spatially and temporally [12].

Microsites are local features/structures on the scale of centimeters to meters having the potential to alter the environment for seedling regeneration. Although multiple microsites may exist in any ATE on any continent, locally, some microsites appear to enhance density of seedling regeneration better than others. The type of microsite most associated with seedling regeneration may vary by local climate [5,7,8]. In general, enhancement of local moisture, increased growing season, and protection of seedlings from frost stress and wind are emphasized as key reasons for seedling survival on specific microsites. For example, in the U.S. Rocky Mountains and Ecuador, *Polylepis*, *Abies lasciocarpa* and *Picea engelmannii* were associated with facilitating effects of tree clumps [13–15]. In some areas, seedling regeneration occurs where objects such as rocks or canopy shade the ground, which reduce winter/spring sun radiation, night time heat loss, frost occurrence, photoinhibition, and subsequent tree death [16]. *Pinus albicaulis* survival in the Beartooth Plateau, Montana has been linked to caches of the Clark's nutcracker on the leeward side of trees [17]. In the Caucasus Mountains of Georgia [18], Finnish Lapland [19], and the Cordoba Mountains of Argentina [20], *Betula litwinowi*, *Picea abies* and *Polylepis australis* seedlings, respectively, have been associated with small valleys and furrow microsites. In the Pacific Northwest (PNW), *Tsuga mertensiana* and *Abies lasciocarpa* have been associated with ridges and convexities [21,22]. Others have noted significant regeneration of *Abies lasciocarpa*, *Abies densa*, *Picea abies*, and *Rhododendron calophytum* on microsites formed by decayed wood in PNW alpine meadows, European Alps, and mountainous southwest China [23–25].

Although enhanced tree regeneration in alpine treeline ecotones (ATE) is associated with favorable microsites [26–28], there is no known global summary linking microsite type with local climate.

Our study had three main objectives: (1) summarize main microsite types found worldwide, (2) determine any microsite and annual temperature and annual precipitation relationships that most often lead to successful seedling regeneration, and (3) evaluate evidence supporting the role of microsites in upward forest advance potential.

2. Materials and Methods

2.1. Global Dataset

The initial world-wide search of the scientific literature to discern key microsites in the ATE (objective 1) included search terms including: microsite, timberline, forest line, tree line, subalpine, alpine, mountain, seedlings, advance, precipitation, facilitation and regeneration. Alongside searches using general-purpose search engines, we used two main abstracting/indexing systems—ISI Web of Knowledge and Scopus—with a wide range of search strings. For each microsite type, site characteristics including location, elevation, dominant tree genus associated with regeneration, and type of regeneration were noted. For example, if an article highlighted significant regeneration in depressions (concavities), distance of seedling regeneration above the upper limit of the continuous forest was described as well as the type of forest edge (i.e., abrupt or dispersed), if possible.

When compiling the dataset, emphasis was placed on assembling a wide range of microsites worldwide. Hence, only one location/microsite/tree genus combination per region was selected. For example, only one of the multiple studies of object (or shade) microsites with *Picea* sp. among multiple studies on Niwot Ridge, Colorado, USA, was selected for the analysis. We drew a study from each combination at random. Sites with noted human disturbance and animal grazing were not selected for use in the dataset. Parameters compiled in the dataset included: microsite type, mean annual precipitation, mean annual temperature, elevation, and dominant tree genus (Supplemental Information). Other information, if listed, was compiled including: number of seedlings per microsite type, indications of positive tree regeneration by microsite. About 80% of the references had climate information. If precipitation or temperature was missing, either another local reference, DayMet (in western USA) [29], or WorldClim (global database) [30] was used.

Creation of the dataset involved some subjective interpretation. For example, if it was stated that most seedlings grew on convex microsites and regeneration appeared or was also implied to be influenced by overstory canopy, convex microsites were used in the analysis because this reference appeared most dominant. If a range of elevations were part of a study in the ATE, the highest microsite elevation was used in the analyses.

2.2. Data Analysis

Once the dataset was compiled, assessments of relationships among microsite type, annual temperature, annual precipitation, genus, and elevation were evaluated in three steps (objective 2) by: (1) testing for significant differences, (2) plotting 95% confidence intervals of annual temperature and precipitation for microsite types, and (3) visually splitting microsite populations to assess possible climate zone categorization. Upward forest advance potential (objective 3) was assessed by summarizing microsite—climate associations including: seedling density information; distance of regenerating seedlings from the forest; noted physiologic stresses; positive, negative, and negligible microsite attributes; and noted upward advance. Once summarized, implications of these associations were evaluated in the Discussion.

Significant statistical differences were evaluated among microsite type, tree genus, and mean annual precipitation and temperature by conducting separate one-way analysis of variance (ANOVA) tests among microsite type (four categories) with mean annual precipitation, mean annual temperature, and dominant genus groups (*Abies, Betula, Nothofagus, Picea, Pinus, Polylepis,* and *Tsuga*). Tests for meeting assumptions for normality and uniform variation were conducted and Post-hoc multiple comparison tests (Tukey's) used when significant differences were detected. Statistical significance was assessed with α = 0.05 and all statistical analyses were conducted with the R statistics package [31].

3. Results

Four microsite categories were summarized from the literature including: (1) convex (also elevations and mounds), (2) concave (also including furrows, depressions, and mires), (3) object (associated with vegetation, rocks or wood), and (4) wood substrates (including nurse logs) (Figure 1). The resulting dataset consisted of 52 ATE locations with climate data (Supplemental Information), spanning all continents (n = 6) except for Antarctica (Figure 2).

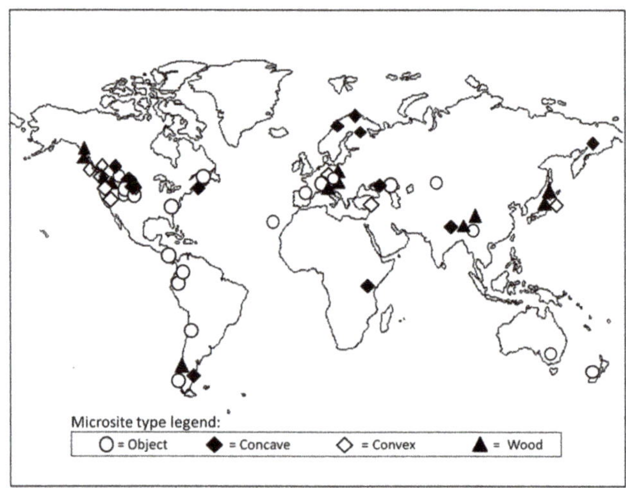

Figure 2. Dominant microsite types for seedling regeneration in the alpine treeline ecotone (ATE) around the world. Sites were selected by searching literature with initial search terms including: microsite, timberline, forest line, treeline, subalpine, alpine, mountain, seedlings, advance, precipitation, facilitation, and regeneration. Relevant literature was differentiated into four ATE microsite types (see Supplemental Information for associated references).

Mean annual temperature and precipitation at microsites ranged from −4.0 °C to 12 °C and 14 cm to 439 cm, respectively (Table 1). Mean annual temperature was significantly different for microsites (p = 0.01) with object microsites having significantly greater temperatures than concave microsites (p = 0.01). Mean annual precipitation was different for microsite types (p < 0.001). Convex and wood microsites had more precipitation than concave microsites (p < 0.01 and p < 0.001, respectively), convex and wood microsites had greater precipitation than object microsites (p = 0.03 and p < 0.01, respectively). Mean annual precipitation for the *Abies* genus group was greater than the *Pinus* genus group (p = 0.03) and mean annual temperature for the *Pinus* genus group was greater than that of the *Picea* and *Betula* groups (p < 0.01, p < 0.01, respectively). Elevation was not associated with climate zone, microsite type, or genus group (p > 0.07).

Microsite type, positively associated with tree recruitment in certain climatic zones, was categorized by visually splitting 95% confidence intervals of microsite annual temperature and precipitation distributions (Figure 3). Climate categorization was as follows: warm & wet (n = 8), cold & wet (n = 12), warm & dry (n = 14), and cold & dry (n = 18). Climate zones separated wood/convex microsites, concave microsites, and object microsites. An annual temperature at approximately 4 °C separated warm from cold climates and an annual precipitation of about 145 cm separated dry and wet climates. The warm & wet climate zone had least noted microsite dominance (n = 1 for wood, n = 4 for object, and n = 3 for convex microsites) (Supplemental Information).

Table 1. Alpine treeline ecotone (ATE) microsite types enhancing seedling regeneration with mean annual temperature, precipitation, and elevation.

Microsite Type	Number Observations	Temperature Range, °C (mean/st.dev.)	Precipitation Range, cm (mean/st.dev.)	Elevation Range, m (mean/st.dev.)	Microsite Characteristics
Wood	12	0.1–5.0 (3.06/1.78)	86–439 (227/93.3)	873–3300 (1814/862)	Seedlings growing in decayed wood lying on ground.
Convex	10	−2.4–5.0 (2.77/1.65)	66–350 (208/99.8)	1195–3100 (1964/529)	Seedlings growing on elevations or mounds on ground surface.
Concave	11	−3.6–8 (1.37/4.1)	14–167 (87.4/34.5)	460–4200 (2185/901)	Seedlings growing in depressions, valleys, and furrows in the ground surface.
Object	19	−4.0–12 (6.1/4.98)	22–225 (123/55.6)	700–4100 (2465/896)	Association of seedlings with trees, plants, wood, or rocks.

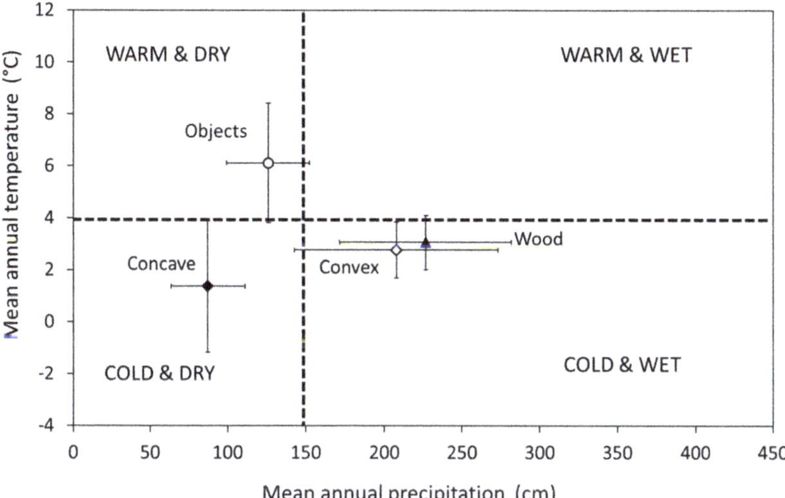

Figure 3. Alpine ecotone treeline microsite presence within four climate zones. Climate zones were differentiated (as indicated by dashed lines) by significant annual temperature differences between concavity and object microsites and significant annual precipitation differences for wood/convexity microsites and concavity/object microsites. Dashed lines indicate visual splits of precipitation and temperature based on microsite population groupings. Error bars indicate 95% confidence intervals.

For *Picea*, *Abies*, *Pinus*, and *Betula*, the most represented genus groups (4 out of 7), associations among continent, climate, and microsite were summarized. *Picea* ($n = 14$) was found mostly in cold & wet climates ($n = 7$) on wood microsites ($n = 4$) in the countries of Europe, North America, and Asia. *Abies* ($n = 10$), located in North America and Asia and in all climate zones and on all microsites, was located mostly in cold & wet elevations ($n = 5$). *Abies* was found on both convex microsites ($n = 3$) and wood microsite ($n = 2$) in North America and Asia. *Pinus* ($n = 7$), located in Europe, North America, and Asia, was found mostly in Europe ($n = 4$) at warm & dry locations in association with object microsites. *Betula* ($n = 5$), found in Asia, Europe, and North America, was located mostly in concave sites ($n = 4$) in cold & dry locations.

Microsite—climate associations along with noted seedling limiting factors and noted physiologic responses were summarized from the literature (Table 2). Warm & wet and warm & dry locations were generally associated with ample carbon assimilation [7,27,28], were limited by drought, had noted fire, were limited by humus substrates, and in some cases were limited by late snow melt. Seedlings in cold & wet and cold & dry locations had limits on carbon assimilation, infrequent seed crops, and suffered from stresses including photoinhibition, snow mold, and physical disturbance by snow.

Table 2. Summary of microsite type, associated climate zone, and noted site limitations.

Climate Category	Dominant Genus Noted, Typical Site Stress Ameliorated by Microsite	Noted upward Migration, Seedling Density, Seedling Survival Trends, Microsite Association	Noted Physiologic Response [2,27,28]
Warm & wet (wood/object/convex–no dominance)	Six species, no species dominance, moderate snow pack, moist	Seedling regeneration limited by humus late snow melt. Seedling regeneration greater on wood, object, and convex sites as compared to adjacent substrates.	Ample carbon assimilation.
Warm & dry (object)	Pinus Drought, fire	Greater seedling regeneration in proximity of trees or other objects; "low", "downward", and "unlikely".	Ample carbon assimilation, photoinhibition, drought stress.
Cold & wet (convexity/wood)	Picea, Abies, Tsuga, high spring snowpack, low summer soil moisture	Seedling regeneration occurs beyond forests; "notable", "possible", "continuous recruitment".	Limits on carbon assimilation, death by snow mold, death by snow damage, infrequent seed crops.
Cold & dry	Picea, Low growing season moisture and low temperature, high radiation.	Seedling regeneration occurs; "some in depressions", "some where crypogams are present".	Limits on carbon assimilation, poor, infrequent seed crops, photoinhibition, infrequent seed crops, drought stress.

4. Discussion

This global-scope microsite assessment serves to summarize essential seedling regeneration strategies that ameliorate site climate-related stresses in the alpine treeline ecotone (ATE). Although multiple microsites can be present in various climates, this study found tendencies of a specific microsite favoring regeneration to be in a particular climate zone. In general, elevated convex sites and/or decayed wood facilitate earlier snow melt for seedlings located in cold & wet climates with abundant snowfall, depressions or concave sites both enhance moisture and protect seedlings from wind chill exposure for seedlings growing in cold & dry locations, and objects protect seedlings from excessive radiation in warm & dry high locations. Microsite strategies enhancing successful regeneration: (1) help to categorize common climate-related physiologic stresses, (2) aid in assessments of possible upward movement of ATE, and (3) highlight possible climate adaptation strategies and restoration efforts.

Microsite—climate associations highlight typical physiologic stresses in the ATE (Table 2). For example, for sites in cold &wet locations, there is limited carbon uptake (carbon assimilation) of seedlings due to cold temperatures and time under snow [27], and these stresses appear to be ameliorated by convex sites [22]. Further, in cold & wet locations, snow mold infests seedlings particularly on concave sites, but seedlings are less prone to infestations on convex sites [32]. In particular, *Pinus cembra* can be attacked by *Phacidium infestans*, a fungus that infects needles and shoots of pines that are covered by snow for too long [33,34]. In warm & dry areas, objects such as trees and rocks serve to protect/facilitate seedling growth by protection from wind and high radiation pattern [1,35] and damage or death by photoinhibition. Photoinhibition, associated with periods of night frost followed by clear-sky days [36], is a notable cause of seedling death for seedlings in ATEs of Australia [36], Spain [37], and the North American Rocky Mountains [16]. Seedling association with objects that provide shade leads to less photoinhibition [16], substantial increases in photosynthetic carbon gain throughout the summer growth period, enhanced root growth, amelioration of drought stress, and increased seedling survival [3]. In the colder and wetter locations such as the US Pacific Northwest, photoinhibition does not appear to be as common [23]. Mechanical damage of seedlings and trees is associated with wind abrasion of needle cuticles, apical bud damage, snow loading and frost heaving causing tissue and whole-tree mortality [7], stresses found to be ameliorated by both object microsites and concave microsites [7]. In the Dawodang Mountains of China and the Cascade Mountains of the US Pacific Northwest, greater stomatal conductance, for *Rhododendron calophytum* and *Abies lasciocarpa*, was attributed to the higher moisture provided by wood microsites than by adjacent soil substrates [23,25]. Future research could be enhanced by evaluation of both dominant and sub-dominant microsites having some effect on the pattern of seedling regeneration. Animal grazing, a dominant factor controlling treelines in many regions worldwide, was not in the scope of our analysis, but warrants further investigation in future studies.

Once physiologic stresses are alleviated, seedlings successfully regenerate. If trees survive to adulthood, there is potential for advance of the ATE. Although successful seedling generation does not mean that seedlings will grow to maturity, upward migration of forests to higher altitude must initially depend on new seedling establishment above the existing forest line into the treeline ecotone [3]. Given that global predictions of climate change forecast an increase in the frequency of extreme events such as heatwaves and frost, there could be a decrease in seedling establishment at alpine treelines [38,39]. Consistent with other research, our results suggest that high-elevation mountain forests with water-limited growing seasons are likely to experience recruitment declines or, at best, no new recruitment advantage as climate warms [11,40].

Although only a few (<10%) of the studies in this analysis explicitly describe potential for ATE advance, locations likely having least, moderate, and most upward advance is possible given the categorization of microsite type, climate zones, and current stresses (Table 2). Seedlings growing in concave and object microsites in locations that are most warm and dry likely have least potential for seedling regeneration as supported by the associated statements: "unlikely forest expansion" [20], "bottlenecks for tree recruitment" [41] and even "downward treeline migration" [42], attributes associated with fire and drought. Microsites in wetter warm & dry climate zones may indicate moderate, but steady regeneration with "little regeneration beyond 5m of trees" [10], where "greater survivorship of young seedlings was observed in microsites closer to tree islands or having overhead structures such as fallen stems" [43]. Where fire continues in such locations as the páramo of the Andes Mountains, treelines will not shift upwards; where fires are suppressed but no radiation-tolerant tree species are present, treelines will also not shift upwards; and even where fires are suppressed, and some radiation-tolerant tree species are present, treelines will shift upwards only slowly [44]. Cold & dry locations are expected to have moderate potential for upward advance, largely depending on presence of locations where species such as *Betula* grows in "wind-sheltered and steep concave slopes." [45]. As shown by research in Nepal, little regeneration is occurring; what little regeneration does occur happens in association with periods of higher rainfall [46]. Cold & wet zones, particularly in ATE locations with > 150 cm/year precipitation likely have greatest potential for regeneration with climate warming given an increase in growing season with less snow. Seedling regeneration in cold & wet zones, where earlier snowmelt is facilitated, is described as "fairly continuous", with a "disproportionate number of trees on mounds" [22,47], particularly in areas characterized by diffuse ATE form where regeneration is, "particularly responsive to overall temperature increases" [4].

Climate models, ecological theories, and suggestions for future research are illuminated by this worldwide microsite summary. With a general focus on growing requirements of adult trees rather than requirements of seedlings, climate envelope models frequently and inappropriately assume that a warming climate will move tree species upward [11]. This study, by focusing on favorable conditions for seedling survival as indicated by microsite type, provides a means to describe possible upward advance of forests. Microsite presence could be considered an ecological limiting factor helping to define specific growth limitations at any given time. Considering that microsites ameliorate growth limitations in environments where seedling interactions are facultative and non-competitive in contrast to adult tree environments, more research describing both scale and optimal use of resources is warranted [48,49]. In addition, further research is needed to decipher the physiologic stresses occurring in the transition between seedlings to adult trees. As trees grow taller than ~3m, microsite effects may become negligible since taller crowns decouple trees from the near-ground surface climate and are more deeply immersed within the prevailing atmosphere [40,50], when the local climate (mesoclimate) associated with topography begins to determine tree growth [51]. At some transitional stage, trees may be influenced by landforms larger than microsites, but smaller than mountain slopes. Additional information is needed on seedling requirements during both the summer and winter. In some cases, seedling establishment is more sensitive to winter conditions, notably to the length of snow cover than summer temperatures, so regional-scale factors such as winter climate and biotic interactions should be included in modelling exercises to improve future treeline location forecasts [52,53]. Further, while uniformity of climatic data (i.e. use of the same observation periods), comparison of seedling density on different microsites, and examination of time and scale of microsite/climate associations was not in the scope of this study, these methods would help to make comparisons among microsites. Given climate warming and greater evaporative demand, associated microsite preferences may be altered. For example, current seedling regeneration on some wood microsites may transition to greater seedling regeneration on object microsites [23,54]. Study results highlight some practical applications of microsite use in climate adaptation strategies and restoration actions. Given a large variation of micro-climatic conditions in alpine landscapes, areas of refugia will likely be present, so rather than forcing all species upslope to track climatic warming [55], these refugia may offer climate adaptation options. For example, one measure for improving *Pinus albicaulis* seedling regeneration success includes planting seedlings near standing trees [56]. Further, wood microsites, providing moister and warmer substrates, can be used in restoration practices. For example, in boreal and alpine areas where histories of fire suppression, forest harvest, and mechanical degradation of alpine meadows have occurred, newer policies encouraging controlled burns and natural succession processes, restoration practices have been benefitted by utilizing downed wood and wood fabric [54,57,58]. At these locations, wood microsites have effectively enhanced the rate of plant colonization.

5. Conclusions

By summarizing annual temperature and precipitation for key microsite types (object, concave, convex, and decayed wood) in a global context in the alpine treeline ecotone (ATE), four climate zones were illuminated: cold & dry, warm & dry, cold & wet, and warm & wet. Although microsite type and genus groups can be present in various climates, our findings suggest general tendencies of a specific microsite favoring seedling regeneration to be found within specific climates zones, and aid in better understanding of the potential for upper forest advance of the ATE. Warm & wet and warm & dry locations with warmth enabling seedling carbon assimilation were limited by drought, fire, humus substrates, and late snow melt; these are sites where regeneration was enhanced by object microsites. Seedlings in cold & wet (dominated by wood and convex microsites) and cold & dry (dominated by concave microsites) locations had limits on carbon assimilation, infrequent seed crops, and suffered from stresses including photoinhibition, snow mold, and physical disturbance by snow. Although the level that a microsite can ameliorate local stresses will vary in a changing climate, cold & wet climate zones with *Abies*, *Tsuga*, and *Picea* seedlings in ATE locations with > 150 cm/year precipitation likely have greatest potential for regeneration with climate warming given an increase in growing season with less snow. Study results highlight some practical applications of microsite use in climate adaptation strategies and restoration actions including use of objects providing shade particularly for *Pinus* seedling regeneration and use of rotten wood for *Picea* and *Abies* seedling regeneration. Considering that seedling mortality is much greater than for that of adult trees, further investigation of multiple driving mechanisms for advance and retreat is needed for better prediction of Earth's future forests in alpine areas.

Supplementary Materials: The following are available online at http://www.mdpi.com/1999-4907/10/10/864/s1.

Author Contributions: Conceptualization, A.C.J. and J.A.Y.; Methodology, A.C.J.; Software, A.C.J.; Validation, A.C.J. and J.A.Y; Formal Analysis, A.C.J.; Investigation, A.C.J.; Resources, A.C.J.; Data Curation, A.C.J.; Writing—Original Draft Preparation, A.C.J.; Writing—Review & Editing, J.A.Y.; Visualization, A.C.J.; Supervision, J.A.Y.; Project Administration, A.C.J.; Funding Acquisition, A.C.J.

Acknowledgments: This manuscript was benefitted from review and discussions with Ryan Bellmore, Alina Canter, Sarah Eppley, Andrew Fountain, Joseph Maser, Yangdong Pan, and by comments received by two anonymous reviewers.

Conflicts of Interest: The authors declare no conflict of interest.

References

1. Holtmeier, F. Mountain timberlines: Ecology, patchiness, dynamics. In *Advances in Global Change Research*; Springer Science & Business Media: Berlin, Germany, 2009; Volume 36, pp. 5–10.
2. Tranquillini, W. *Physiological Ecology of the Alpine Timberline: Tree Existence at High Altitudes with Special Reference to the European Alps*; Springer Science & Business Media: Berlin, Germany, 2012; Volume 31.
3. Smith, W.K.; Germino, M.J.; Hancock, T.E.; Johnson, D.M. Another perspective on altitudinal limits of alpine timberlines. *Tree Physiol.* **2003**, *23*, 1101–1112. [CrossRef] [PubMed]
4. Harsch, M.A.; Bader, M.Y. Treeline form—A potential key to understanding treeline dynamics. *Global Ecol. Biogeogr.* **2011**, *20*, 582–596. [CrossRef]
5. Holtmeier, F.; Broll, G. The influence of tree islands and microtopography on pedoecological conditions in the forest-alpine tundra ecotone on Niwot Ridge, Colorado Front Range, USA. *Arctic Alpine Res.* **1992**, *24*, 216–228. [CrossRef]
6. Körner, C.; Paulsen, J. A world-wide study of high altitude treeline temperatures. *J. Biogeogr.* **2004**, *31*, 713–732. [CrossRef]
7. Malanson, G.P.; Butler, D.R.; Fagre, D.B.; Walsh, S.J.; Tomback, D.F.; Daniels, L.D.; Bunn, A.G. Alpine treeline of western North America: Linking organism-to-landscape dynamics. *Phys. Geogr.* **2007**, *28*, 378–396. [CrossRef]
8. Johnson, A.; Yeakley, A. Wood microsites at timberline-alpine meadow borders: Implications for conifer seedling regeneration and alpine meadow conifer invasion. *Northwest Sci.* **2013**, *87*, 140–160. [CrossRef]
9. Andrus, R.A.; Harvey, B.J.; Rodman, K.C.; Hart, S.J.; Veblen, T.T. Moisture availability limits subalpine tree establishment. *Ecology* **2018**, *99*, 567–575. [CrossRef] [PubMed]
10. Germino, M.J.; Smith, M.K.; Resor, C.A. Conifer seedling distribution and survival in an alpine-treeline ecotone. *Plant Ecol.* **2002**, *162*, 157–168. [CrossRef]
11. Kueppers, L.M.; Conlisk, E.; Castanha, C.; Moyes, A.B.; Germino, M.J.; De Valpine, P.; Torn, M.S.; Mitton, J.B. Warming and provenance limit tree recruitment across and beyond the elevation range of subalpine forest. *Glob. Chang. Biol.* **2017**, *23*, 2383–2395. [CrossRef]
12. Walther, G.R.; Post, E.; Convey, P.; Menzel, A.; Parmesan, C.; Beebee, T.J.; Fromentin, J.-M.; Hoegh-Guldberg, O.; Bairlein, F. Ecological responses to recent climate change. *Nature* **2002**, *416*, 389–395. [CrossRef] [PubMed]
13. Cierjacks, A.; Iglesias, J.E.; Wesche, K.; Hensen, I. Impact of sowing, canopy cover and litter on seedling dynamics of two Polylepis species at upper tree lines in central Ecuador. *J. Trop. Ecol.* **2007**, *23*, 309. [CrossRef]
14. Daly, C.; Shankman, D. Seedling establishment by conifers above tree limit on Niwot Ridge, Front Range, Colorado, USA. *Arctic Alpine Res.* **1985**, *17*, 389–400. [CrossRef]
15. Maher, E.L.; Germino, M.J.; Hasselquist, N.J. Interactive effects of tree and herb cover on survivorship, physiology, and microclimate of conifer seedlings at the alpine tree-line ecotone. *Can. J. Forest Res.* **2005**, *35*, 567–574. [CrossRef]
16. Germino, M.J.; Smith, M.K. Sky exposure, crown architecture, and low-temperature photoinhibition in conifer seedlings at alpine treeline. *Plant Cell Environ.* **1999**, *22*, 407–415. [CrossRef]
17. Mellmann-Brown, S. Regeneration of whitebark pine in the timberline ecotone of the Beartooth Plateau, USA: Spatial distribution and responsible agents. In *Mountain Ecosystems: Studies in Treeline Ecology*; Broll, G., Keplin, B., Eds.; Springer: New York, NY, USA, 2005.
18. Hughes, N.M.; Johnson, D.M.; Akhalkatsl, M.; Abdaladze, O. Characterizing *Betula litwinowii* seedling microsites at the alpine-treeline ecotone, central Greater Caucasus Mountains, Georgia. *Arct. Antarct. Alpine Res.* **2009**, *41*, 112–118. [CrossRef]
19. Autio, J.; Colpaert, A. The impact of elevation, topography and snow load damage of trees on the position of the actual timberline on the fells in central Finnish Lapland. *Fennia* **2005**, *183*, 15–36.
20. Enrico, L.; Funes, G.; Cabido, M. Regeneration of *Polylepis australis* Bitt. in the mountains of central Argentina. *For. Ecol. Manag.* **2004**, *190*, 301–309. [CrossRef]
21. Lowery, R.F. Ecology of subalpine zone tree clumps in the north Cascade Mountains of Washington. Ph.D Thesis, University of Washington, Seattle, WA, USA, 1972.
22. Rochefort, R.M.; Peterson, D.L. Temporal and spatial distribution of trees in subalpine meadows of Mount Rainier National Park, Washington, USA. *Arctic Alpine Res.* **1996**, *28*, 52–59. [CrossRef]

23. Johnson, A.C.; Yeakley, J.A. Seedling regeneration in the alpine treeline ecotone: Comparison of wood microsites and adjacent soil substrates. *J. Mt. Res. Development* **2016**, *36*, 443–452. [CrossRef]

24. Motta, R.; Berretti, R.; Lingua, E.; Piussi, P. Coarse woody debris, forest structure and regeneration in the Valbona Forest Reserve, Paneveggio, Italian Alps. *Forest Ecol. Manag.* **2006**, *235*, 155–163. [CrossRef]

25. Ran, F.; Wu, C.; Peng, G.; Korpelainen, H.; Li, C. Physiological differences in Rhododendron calophytum seedlings regenerated in mineral soil or on fallen dead wood of different decaying stages. *Plant Soil* **2010**, *337*, 205–215. [CrossRef]

26. Rochefort, R.M.; Little, R.T.; Woodward, A.; Peterson, D.L. Changes in subalpine tree distribution in western North America: A review of climatic and other causal factors. *Holocene* **2004**, *4*, 89–100. [CrossRef]

27. Franklin, J.F.; Moir, W.H.; Douglas, G.W.; Wiberg, C. Invasion of subalpine meadows by trees in the Cascade Range, Washington and Oregon. *Arct. Alpine Res.* **1971**, *3*, 215–224. [CrossRef]

28. Moir, W.H.; Rochelle, S.G.; Schoettle, A.W. Microscale patterns of tree establishment near upper treeline, Snowy Range, Wyoming, USA. *Arct. Antarct. Alpine Res.* **1999**, *31*, 379–388. [CrossRef]

29. DAYMET. Available online: www.daymet.org (accessed on 2 February 2012).

30. WorldClim. Available online: www.worldclim.org (accessed on 1 July 2018).

31. R Core Development Team. *R: A Language and Environment for Statistical Computing*; R Foundation for Statistical Computing: Vienna, Austria, 2005; Available online: http://www.R-project.org (accessed on 10 August 2018).

32. Hiller, B.; Müterthies, A. Humus forms and reforestation of an abandoned pasture at the alpine timberline (Upper Engadine, Central Alps Switzerland). In *Mountain Ecosystems: Studies in Treeline Ecology*; Springer: Berlin/Heidelberg, Germany, 2005; Volume 2, pp. 203–218.

33. Roll-Hansen, F. *Phacidium infestans*—A literature review. *Eur. J. For. Pathol.* **1980**, *19*, 237–250. [CrossRef]

34. Barbeito, I.; Brücker, R.L.; Rixen, C.; Bebi, P. Snow fungi—Induced mortality of *Pinus cembra* at the alpine treeline: Evidence from plantations. *Arct. Antarct. Alpine Res.* **2013**, *45*, 455–470. [CrossRef]

35. Hunziker, U.; Brang, P. Microsite patterns of conifer seedling establishment and growth in a mixed stand in the southern Alps. *For. Ecol. Manag.* **2005**, *210*, 67–79. [CrossRef]

36. Ball, M.C.; Hodges, V.S.; Laughlin, G.P. Cold induced photoinhibition limits regeneration of snow gum at tree-line. *Funct. Ecol.* **1991**, *5*, 665–668. [CrossRef]

37. Batllori, E.; Camarero, J.J.; Ninot, J.M.; Gutiérrez, E. Seedling recruitment, survival and facilitation in alpine Pinus uncinata tree line ecotones. Implications and potential responses to climate warming. *Glob. Ecol. Biogeogr.* **2009**, *18*, 460–472. [CrossRef]

38. Inouye, D.W. The ecological and evolutionary significance of frost in the context of climate change. *Ecol. Lett.* **2000**, *3*, 457–463. [CrossRef]

39. IPCC. Climate Change 2013: The physical science basis. In *Contribution of Working Group I to the Fifth Assessment Report of the Intergovernmental Panel on Climate Change*; Stocker, T.F., Qin, D., Plattner, G.K., Tignor, M., Allen, S.K., Boschung, J., Naueks, A., Xia, Y., Bex, V., Midgley, P.M., Eds.; Cambridge University Press: Cambridge, UK, 2013; p. 1535.

40. Wieser, G.; Holtmeier, F.-K.; Smith, W.K. Treelines in a changing global environment. In *Trees in a Changing Environment*; Springer: Dordrecht, The Netherlands, 2014; pp. 221–263.

41. Cuevas, J.G. Tree recruitment at the Nothofagus pumilio alpine timberline in Tierra del Fuego, Chile. *J. Ecol.* **2000**, *88*, 840–855. [CrossRef]

42. Hemp, A. Climate change-driven forest fires marginalize the impact of ice cap wasting on Kilimanjaro. *Glob. Change Biol.* **2005**, *11*, 1013–1023. [CrossRef]

43. Wearne, L.J.; Morgan, J.W. Recent forest encroachment into subalpine grasslands near Mount Hotham, Victoria, Australia. *Arct. Antarct. Alpine Res.* **2001**, *33*, 369–377. [CrossRef]

44. Bader, M.Y.; Ruijten, J.J. A topography-based model of forest cover at the alpine tree line in the tropical Andes. *J. Biogeogr.* **2008**, *35*, 711–723. [CrossRef]

45. Kullman, L.; Öberg, L. Post-Little Ice Age tree line rise and climate warming in the Swedish Scandes: A landscape ecological perspective. *J. Ecol.* **2009**, *97*, 415–429. [CrossRef]

46. Liang, E.; Dawadi, B.; Pederson, N.; Eckstein, D. Is the growth of birch at the upper timberline in the Himalayas limited by moisture or by temperature? *Ecology* **2014**, *95*, 2453–2465. [CrossRef]

47. Brett, R.B.; Klinka, K. A transition from gap to tree-island regeneration patterns in the subalpine forest of south-coastal British Columbia. *Can. J. Forest Res.* **1998**, *28*, 1825–1831. [CrossRef]

48. Danger, M.; Daufresne, T.; Lucas, F.; Pissard, S.; Lacroix, G. Does Liebig's law of the minimum scale up from species to communities? *Oikos* **2008**, *117*, 1741–1751. [CrossRef]

49. Choler, P.; Michalet, R.; Callaway, R.M. Facilitation and competition on gradients in alpine plant communities. *Ecology* **2001**, *82*, 3295–3308. [CrossRef]

50. Yu, D.; Wang, Q.; Wang, X.; Dai, L.; Li, M. Microsite Effects on Physiological Performance of Betula ermanii at and Beyond an Alpine Treeline Site on Changbai Mountain in Northeast China. *Forests* **2019**, *10*, 400. [CrossRef]

51. Li, X.; Liang, E.; Gricar, J.; Rossi, S.; Cufar, K.; Ellison, A. Critical Minimum Temperature Limits Xylogenesis and Maintains Treelines on the Tibetan Plateau. Available online: https://www.biorxiv.org/content/early/2016/12/13/093781.full.pdf (accessed on 7 April 2019).

52. Hijmans, R.J.; Graham, C.H. The ability of climate envelope models to predict the effect of climate change on species distributions. *Glob. Chang. Biol.* **2006**, *12*, 2272–2281. [CrossRef]

53. Renard, S.M.; McIntire, E.J.; Fajardo, A. Winter conditions–not summer temperature–influence establishment of seedlings at white spruce alpine treeline in Eastern Quebec. *J. Veg. Sci.* **2016**, *27*, 29–39. [CrossRef]

54. Marzano, R.; Garbarino, M.; Marcolin, E.; Pividori, M.; Lingua, E. Deadwood anisotropic facilitation on seedling establishment after a stand-replacing wildfire in Aosta Valley (NW Italy). *Ecol. Eng.* **2013**, *51*, 117–122. [CrossRef]

55. Scherrer, D.; Körner, C. Topographically controlled thermal-habitat differentiation buffers alpine plant diversity against climate warming. *J. Biogeogr.* **2011**, *38*, 406–416. [CrossRef]

56. Keane, R.E.; Tomback, D.F.; Aubry, C.A.; Bower, A.D.; Campbell, E.M.; Cripps, C.L.; Jenkins, M.B.; Mahalovich, M.F.; Manning, M.; McKinney, S.T.; et al. *A Range-Wide Restoration Strategy for Whitebark Pine (Pinus albicaulis)*; General Technical Report RMRS-GTR-279; US Department of Agriculture, Forest Service, Rocky Mountain Research Station: Fort Collins, CO, USA, 2012; 108p.

57. Fattorini, M. Establishment of transplants on machine-graded ski runs above timberline in the Swiss Alps. *Restor. Ecol.* **2001**, *9*, 119–126. [CrossRef]

58. Vanha-Majamaa, I.; Lilja, S.; Ryöma, R.; Kotiaho, J.S.; Laaka-Lindberg, S. Rehabilitating boreal forest structure and species composition in Finland through logging, dead wood creation and fire: The EVO experiment. *For. Ecol. Manag.* **2007**, *250*, 77–88. [CrossRef]

 *forests*

Review

Treeline Research—From the Roots of the Past to Present Time. A Review

Friedrich-Karl Holtmeier [1] and Gabriele Broll [2],*

[1]   Institute of Landscape Ecology, Westfälische Wilhelms-University Münster, 48149 Münster, Germany; fkholtmeier@arcor.de
[2]   Institute of Geography, University of Osnabrück, 49074 Osnabrueck, Germany
*   Correspondence: Gabriele.Broll@uni-osnabrueck.de

Received: 30 November 2019; Accepted: 17 December 2019; Published: 26 December 2019

Abstract: Elevational and polar treelines have been studied for more than two centuries. The aim of the present article is to highlight in retrospect the scope of treeline research, scientific approaches and hypotheses on treeline causation, its spatial structures and temporal change. Systematic treeline research dates back to the end of the 19th century. The abundance of global, regional, and local studies has provided a complex picture of the great variety and heterogeneity of both altitudinal and polar treelines. Modern treeline research started in the 1930s, with experimental field and laboratory studies on the trees' physiological response to the treeline environment. During the following decades, researchers' interest increasingly focused on the altitudinal and polar treeline dynamics to climate warming since the Little Ice Age. Since the 1970s interest in treeline dynamics again increased and has considerably intensified from the 1990s to today. At the same time, remote sensing techniques and GIS application have essentially supported previous analyses of treeline spatial patterns and temporal variation. Simultaneously, the modelling of treeline has been rapidly increasing, often related to the current treeline shift and and its implications for biodiversity, and the ecosystem function and services of high-elevation forests. It appears, that many seemingly 'new ideas' already originated many decades ago and just confirm what has been known for a long time. Suggestions for further research are outlined.

Keywords: history of treeline research; elevational treeline; polar treeline; treeline dynamics

---

1. Introduction

Elevational and polar treelines, the most conspicuous vegetation boundaries in high mountains and in the Subarctic, have attracted researchers from numerous disciplines. Thus, scientific approaches to treeline have become increasingly complex. Treeline research dates back to more than two centuries. Since the end of the 20th century, treeline publications have been rapidly increasing. Nearly 60% of the articles and books from the 1930s onward were published during the two last decades. Older contributions have gradually become disregarded in recent articles, or have been cited from secondary sources. The objective of the present article is to highlight in retrospect the scope of treeline research, scientific approaches and hypotheses on treeline causation, and its spatial and temporal structures.

The cited references can only represent a very small selection of an abundance of relevant publications. This also means, that we could not cover all topics in treeline research. Anthropogenic pressure (pastoralism, use of fire, mining, etc.), for example, is only randomly considered, although it has controlled treeline position and spatial pattern in the inhabited world for thousands of years, often overruling the influence of climate. Moreover, the multiple influences of wild animals, pathogens, and diseases have not been concentrated on, as the species' distribution, populations and kinds of impact vary locally and regionally. Extensive reviews on this particular topic and the anthropogenic impact were presented by [1–4].

Early publications were usually descriptive, and the results were not tested statistically. Anyway, going 'back to the roots', the reader will soon realize that many 'new ideas' seemingly representing the actual 'frontline' of treeline research originated many decades ago and just confirm what has been known for a long time such as the fundamental functional role of heat deficiency in treeline control [5,6], or the advantage of short plant tree stature in high mountain and subarctic/arctic environments (e.g., [7,8]).

2. Scope and Topics of Treeline Research

2.1. Early Treeline Research

Early treeline research started around the end of the 18th century and the beginning of the 20th century. In the beginning, treeline research was closely linked with the exploration of high-mountains and the northern forest-tundra. There were occasional and more general observations, including the physiognomy and distribution of vegetation and hypotheses on the influence of climate on treeline (e.g., [9–16]) (Figure 1).

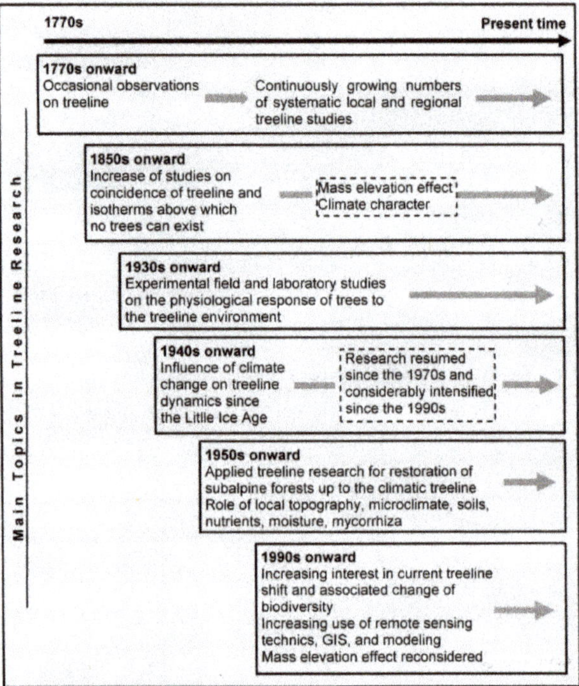

Figure 1. Scope and main topics of treeline research in the course of time.

Systematic research dates back to the end of the 19th century/beginning of the 20th century [5,7,17–37] (Figure 1). Coincidence of treeline position and certain isotherms of mean growing season temperatures became apparent which clearly reflected the influence of heat deficiency. Gannet's article on timberline [38] is regarded as the beginning of systematic treeline research in North America [39]. Later, Griggs [40] published an overview on North American treeline which was followed by a more popularized introduction to timberlines by Arno [41]. This book is focused on North America but also gives a short worldwide view.

Early researchers ([2] for review, [22,27,33–35,42,43]) already pointed to the treeline raising mass-elevation effect (MEE). Brockmann-Jerosch [33] found MEE often overlapping with the positive influence of the relatively continental climate character in the central parts of big mountain masses (see also [34]). There, the combined effects of greater daily thermal amplitude associated with high elevation, high solar radiation loads, reduced cloudiness, and lower precipitation, resulting in warmer topsoils and microclimate near the ground allow treeline trees to exist at a lower mean growing season air temperature (or annual temperature in the tropics) than in the outer low ranges exposed to moisture-carrying air masses and on isolated mountains ('summit phenomenon', sensu [44]). The mechanism of MEE has been approached again in numerous studies largely confirming the previous hypotheses ([2] for literature until 2005). During the last 14 years the discussion on the MEE came up again (e.g., [45–52]).

Growth forms of trees were found to be indicators of the harsh treeline environment. Dwarfed growth of trees, for example, and wide-spacing of trees in treeline ecotones had already been described in detail by early researchers as a characteristic of the climatic treeline (e.g., [7,21]). They considered 'mats' and similar low growth forms as 'trees' suppressed by recurrent winter injury (desiccation, frost, abrasion) above the protective winter snowpack. The beneficial effect of the relatively warmer conditions (growing season) near the ground surface was also well known (see also e.g., [5,7,8]) and studied later in detail [53].

2.2. Modern Treeline Research

Modern treeline research began in the 1930s, with emphasis on the trees' physiological response to the harsh treeline environment (Figure 1). A biological study by Däniker [54] on treeline causation in the Alps, with special regard to the influence of climate and tree anatomy, had already opened perspectives for future research. Experimental studies in the field and laboratory became of major importance (e.g., [55–65]). During World War II (1939–1945), the number of publications on treeline at the regional scale considerably decreased, especially in Eurasia, where the political situation made field studies very difficult or even impossible. Nevertheless, a few articles also date from this period [66–71].

Disastrous avalanche catastrophes in the European Alps during the extremely snow-rich winters 1951/1952 and 1953/1954 gave a fresh impetus to basic and applied treeline research. In Austria and Switzerland, basic research on the physiological reponse of treeline trees to their environment combined with the analyses of the functional role of the most relevant site factors and processes created scientific basics for high-elevation forest management and maintenance, including afforestation up to the potential tree limit (e.g., [72–74]) (Figure 2). For practical use in high-altitude afforestation in the central Alps ecograms were developed [75,76] showing a schematic transect from convex to concave microtopography with the distribution of the characteristic plant communities as indicators of the site conditions varying along the transect see also [1,77].

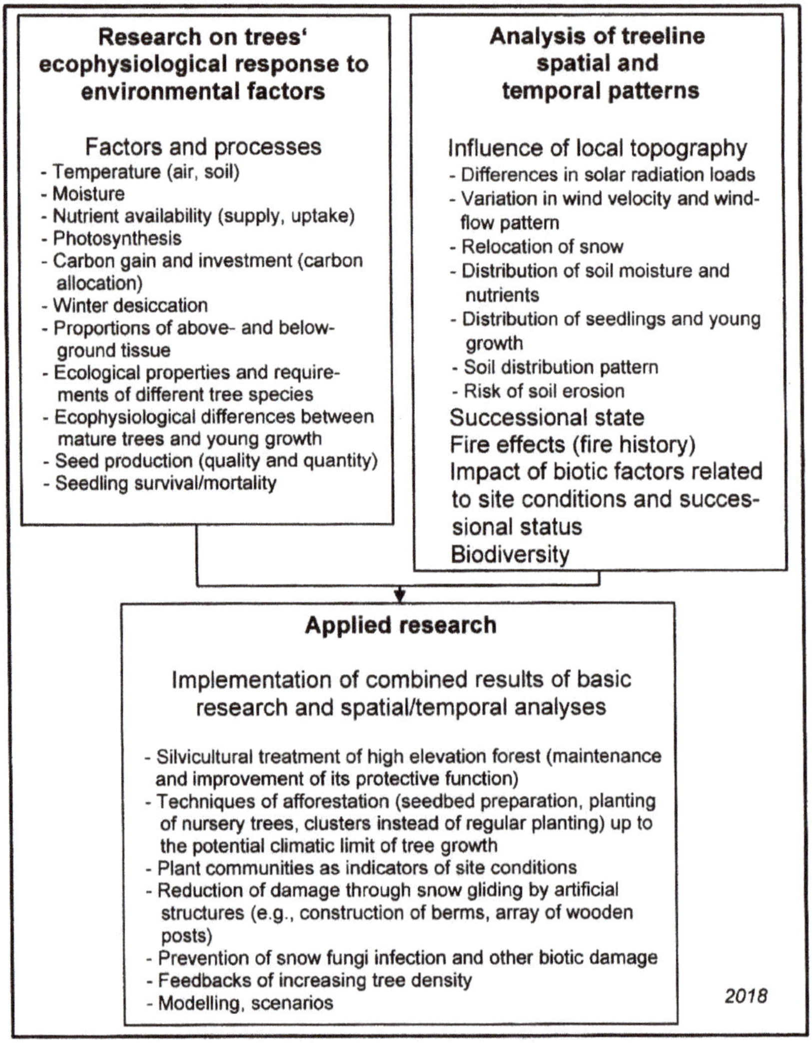

Figure 2. Fundamentals of treeline research and practical implementation.

2.2.1. Carbon

Limited carbon gain due to low temperature and a short growing season, has long been considered the main ecophysiological cause of climatic treeline (e.g., [78,79]), the inhibition of carbon investment gained importance (e.g., [80–82]). Increasing CO_2 in the atmosphere and associated ecophysiological effects on the treeline environments and trees has again stimulated research on carbon dynamics during the last three decades (e.g., [83–98]).

2.2.2. Winter Desiccation

Winter desiccation (WD) is an additional recurrent topic. It was already considered by early researchers as the bottleneck in the performance of young trees at the alpine and polar treelines (e.g., [7,99–102]), where WD had been observed mainly on strongly wind-swept local topography (see also [103]). WD was blamed for the high cuticular transpiration loss through immature needle foliage with both frozen soil and the conductive tissue preventing water uptake (e.g., [56,104–120]). This plausible 'mechanism' has been readily accepted as a key factor in treeline causation in biology, forestry, and geography textbooks (e.g., [121–123]). Although reduced cuticle thickness and cuticular water loss are common at the treeline in winter-cold mountains, they are not necessarily associated with lethal damage (e.g., [124–126]). If needles and shoots are not killed, they may rapidly rehydrate in the following spring [127,128]. It appears that on extremely windy terrain mechanical injuries (e.g., abrasion by blowing snow/ice and sand, breakage of frozen needles, and shoots) can increase the susceptibility of needle foliage and shoots to WD (e.g., [129–133] for review). Frequent freeze-thaw events may also be involved, as well as early or late frost (e.g., [63,73,134,135]). On the whole it appears that WD cannot be attributed to insufficient cuticle development alone, and its role in treeline causation needs to be relativized.

In 1979 Tranquillini [103] compiled the results of his own and that of others in an often cited book, which represented the state of knowledge about the physiological ecology of treeline at that time. Wieser and Tausz [136] edited a volume with contributions of 11 experts on the eco-physiology of trees at their upper limits. The main focus of this book is on the European Alps. Five years later, Körner [93] gave a concise overview of the present state of knowledge about the ecophysiological response of trees to the environmental constraints at treeline.

2.2.3. Treeline and Temperature

The relationship between treeline and temperature, well known already to early researchers, has also been in the focus of modern research. As root zone temperatures during the growing season correlate better than more widely fluctuating air temperatures with worldwide treeline position (e.g., [86,137]), they were suggested as the universal factor in treeline causation ([86] onward). The relative effects of soil temperature on trees, and especially on tree seedlings, may considerably vary depending on local site conditions (e.g., available moisture, organic matter, decomposition, nutrients, and the species' specific requirements). In a worldwide view, the location of the potential climatic treeline has been associated with the length of a growing season of at least 94 d, with a daily minimum temperature of just above freezing (0.9 °C) and a mean of 6.4 °C during this period [138]. This largely corresponds to the previous statement by Ellenberg [121] that the altitudinal position of the climatic treeline is associated with an air temperature exceeding 5 °C for at least 100 days. A critical mean root zone temperature of about 6 °C during the growing season comes close to the critical thermal threshold of 5 °C, when biochemical processes are generally impeded [139,140] and tree growth is interrupted [90].

Although the mean growing season temperature is not identical with a real physical threshold temperature below which no trees can exist (e.g., [141,142]), its relation to treeline position once again supports the well substantiated early finding that heat deficiency is the globally dominating constraint on tree existence at the elevational and polar treelines [143]. Therefore, linking treeline with growing season mean soil or air temperature allows an approximate projection of the future climatic treeline position at broad scales (global, zonal) (e.g., [138,144]).

Regional and local variations, however, are more difficult to foresee, as thermal differences may occur on small scales, that are as great in magnitude as those that occur over thousands of kilometers in the lowland [115,145]. Exposure of mountain slopes and microtopography to incident solar radiation and prevailing winds is of major importance in this respect (e.g., [2,115,146–159]). These factors and many additional physical and biotic disturbances (e.g., [134,160]) may prevent tree growth from reaching the temperature-controlled climatic treeline projected by models, as for example the model of [138]. In addition, historical displacement of treeline by humans and a delay of treeline response to climate change also play a major role in the respect. Hence, at finer scales, treeline position is often out-of-phase with climate (e.g., [1,161–176]). Global overviews, necessarily disregarding the local differences, may easily overemphasize coarse drivers such as temperature [167,177,178]

Overall, the assessment of the underlying possible causes at finer scales will be a great challenge to treeline researchers also in the future (e.g., [179]). Modelling the influence of abiotic factors on the New Zealand treeline [180], for example, showed 82% of treeline variation at regional scale being associated with thermal conditions, whereas only about 50% could be attributed to temperature at finer scales.

2.2.4. Treeline Fluctuations

Treeline fluctuations due to warming and cooling periods after the end of the Little Ice Age and during the 20th century were already considered in numerous studies ([2], for literature). However, it appears that the climatically driven current treeline advance to greater elevation and to northern latitude has even attracted greater attention (Figure 1). During the last three decades, especially since the 1990s, publications on this topic have rapidly increased (by about 90%), partly in a broader context with the change of vegetation and biodiversity, and the expected implications for the ecosystem functions and services of high-elevation forests (e.g., [88,134,167,181–205]) (Figure 3). Dendrochronology, pollen analysis, sediment analysis, and radiocarbon-dating of fossil wood remains (mega fossils) have provided a profound insight into Holocene treeline fluctuations (e.g., [203,206–224]). In not a few cases their after-effects have lastingly influenced the current treeline position and spatial patterns (e.g., [1,2] and references therein).

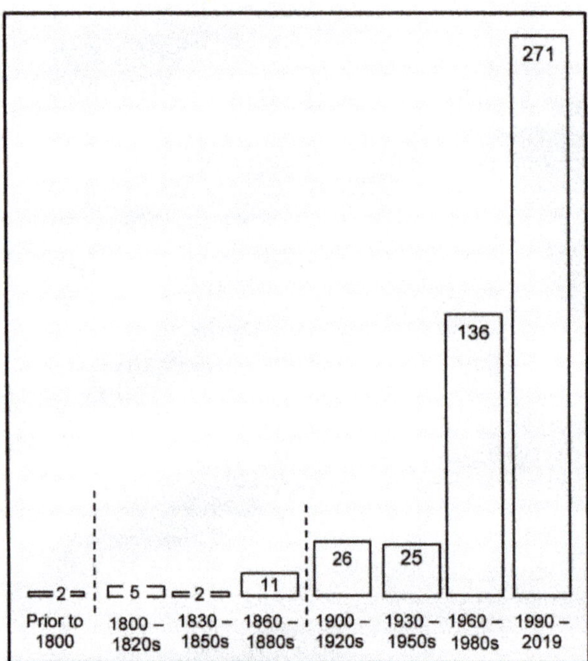

Figure 3. Number of treeline-related publications during consecutive 30-years periods, compiled from the reference list of the current article. Note the exorbitant increase since the 1960s and especially since the 1990s, when the number of publications doubled compared to the previous 30 years (1960–1980s).

Treeline studies by geographers, landscape ecologists, foresters, and geobotanists have referred in particular to treeline in the landscape context aiming to explain spatial treeline structures and their development under the influence of multiple, partly interacting abiotic and biotic factors (Figure 2) [225–229]; for more publications prior to 2008 see [2,77,199,230–233]. In this context, natural and anthropogenic disturbances have been increasingly studied (e.g., [2,3,160,166,177,215,232,234–245]).

These studies have contributed to the assessment of treeline fluctuations in a more holistic view. As both natural and anthropogenic disturbances are closely linked with a multitude of locally and regionally varying preconditions, a generalization, however, is problematic.

Traditional ground-based repeat photography (e.g., [168,171,203,246–251]), remote sensing techniques (oblique air photos, satellite images) and GIS data (e.g., [181,242,252–270]) have effectively supported the analysis of current (and also of historical) treeline spatial patterns and temporal variation such as treeline fluctuations, especially in remote areas and areas difficult to access, such as steep and rugged mountain terrain [270,271]. These studies have also contributed to a more complex view of the driving factors and also underlined that factors and processes vary by scale of consideration (e.g., [159,180,184,205,272,273]) (Figure 4). Thus, in addition to the numerous studies on the physiological response of treeline trees to heat deficiency at the broader scales (global/zonal/regional), the influences of local topography (landforms) on treeline spatial patterns and associated ecological processes have been increasingly studied, particularly in Austria and Switzerland (e.g., [75,274–277]) (Figure 4). During the last decades, this issue has moved again in the focus of treeline research worldwide (e.g., [147–149,156,227,229,278–293]).

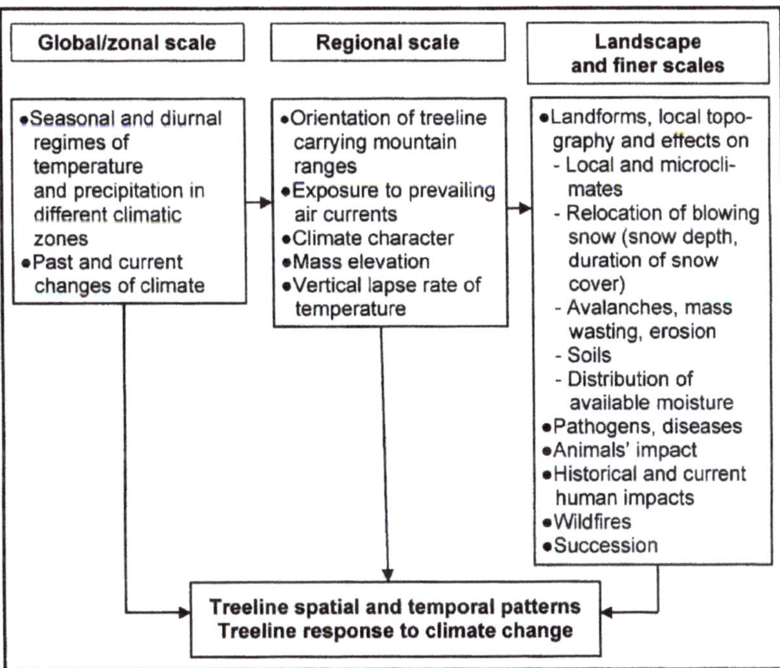

Figure 4. Combined results of research at different scales explain treeline spatio-temporal response to climate change.

2.2.5. Soils in the Treeline Ecotone and in the Alpine Zone

The need for maintenance and restoration of high elevation protective forests and afforestation above the present subalpine forest has also fuelled research on soils in the treeline ecotone and in the alpine zone (e.g., [139,279,280,294–312]). High-elevation soils have been considered in a worldwide view by a FAO report [313]. In addition, Egli and Poulenard [314] published a review on soils of mountainous landscapes. However, in most treeline studies soils are considered only briefly, with just a few exceptions (e.g., [256,309,310,315]). Based on many treeline studies in the Alps, northern Europe and North America, the present authors have come to the conclusion that no real treeline-specific soil types exist [2,273,316]. Instead, the treeline ecotone is usually characterized by a mosaic of soil types, closely related to the locally varying conditions [256].

2.2.6. Precipitation and Soil Moisture

While heat deficiency has generally been accepted as the main factor in global treeline control, the role of precipitation and related soil moisture supply at treeline has remained less clear. This may be due to the great regional and local variation. During the last two decades, however, the role of precipitation has been increasingly dealt with, mainly in the context with climatically-driven treeline shifts and the growing number of studies on treeline in mountains influenced by seasonal drought. The impacts of heat deficiency and drought often overlap, and drought periods may overrule the positive effects of climate warming [178,257,263,317–327]. A detailed review on the physiological response of trees to drought and new insights in understanding of trees' hydraulic function in general was recently presented by Choat et al. [328].

While warming climate is often associated with increasing summer drought, moisture carrying warm air currents may bring about increased winter snowfall at high elevation (e.g., Alps, Himalaya, Alaska). There are great variations, however, due to climate regimes, elevation and distance from the oceans. Anyway, when big snow masses accumulate during a few days of extreme synoptic conditions more destructive (wet snow) avalanches are likely [273,329–332].

Moisture availability is only partly associated with the total amount of precipitation and may considerably vary depending on microtopography, soil physical conditions, and soil organic matter. Moreover, the tolerance of moisture deficiency of tree species differs (e.g., [333]). Seedlings are more sensitive to lack of moisture than deep-rooted mature trees (e.g., [334–337]) which may be attributed to slow initial growth and great reliance on seed reserves [338].

On several tropical/subtropical islands, drought stress affects trees and seedlings above the trade-wind inversion. There may however be many additional factors involved (e.g., geological and vegetation history, climate variability, human impact, etc.) (e.g., [46,339–343]). Thus, the low position of the treeline on some remote ocean islands has also been ascribed to the absence of hardy tree species which could not reach these islands [110,343].

Growing concern about climatically-driven impact by drought, fire, mass-outbreaks of bark beetle, pathogens, and increasing outdoor activities on high mountain resources has fuelled treeline research, often with emphasis on high-elevation forest management ([344] and literature within). High-elevation forests including treeline ecotones are particularly valuable as wildlife habitats and because of their protective function such as avalanche and erosion control (e.g., [160,345]). Not least they serve as an environmental indicator.

2.2.7. Natural Regeneration

While, in general, response of mature trees to climate warming is in the focus, climate change has also stimulated new research on natural regeneration and its particular role as a driving force in worldwide treeline upward and poleward shifts (see [2] for literature prior to 2008 (e.g., [201,346–366]). Publications on regeneration at treeline have increased by more than 60% since the last two decades.

Normally, both seed quantity and quality (viability) decrease when approaching the treeline. Occasionally, however, trees growing even far above the mountain forest produce viable seeds, from which scattered seedlings may emerge at safe sites. As the regeneration process usually extends over several years, it is highly prone to disturbances and may therefore fail at any stage [2,336,367].

It has been and still is being debated, whether the scarcity of viable seeds or paucity of safe sites are more restrictive to tree establishment within and above the current treeline ecotone (e.g., [93,109,168,171,182,186,361,368–378]). Whatsoever, increased production and availability of viable seeds will not necessarily initiate successful seedling recruitment (e.g., [171,182,186,261,367,379–381]) Altogether, seed-based regeneration in and above the treeline ecotone depends on the availability of viable seeds, varying locally, and on the distance from the seed sources, suitable seed beds, and multiple disturbances (e.g., [2,3,160,191,261,353,355,361,366,367]). Last but not least, mycorrhization is essential for a successful seedling establishment in and especially above the treeline ecotone, where it is closely related to locally varying site conditions (e.g., [382–386]). It appears, that only trees with ectomycorrhiza are able to exist up to the climatic limit of tree growth [382,387]. In the end, however, effective natural regeneration depends on the hardiness of tree seedlings and whether these will attain full tree size or at least survive as suppressed growth form. Overall, operating with long-term growing season temperature alone does not allow a fairly well assessment of whether seed-based regeneration within the treeline ecotone and in the lower alpine zone will be successful or fail.

As to current treeline dynamics, the role of layering (the formation of adventitious roots) and also of stump sprouts and root suckers, well known already to early researchers (e.g., [7,21,388]), deserves closer attention [389]. Layering still is active at low temperatures which would prevent seed-based regeneration (e.g., [390–392]). Clonal groups which developed at high elevation under temporary favourable conditions and survived subsequent cooling (after the Little Ice Age) may now serve as an effective seed source far above the current subalpine forest (e.g., [157]).

2.2.8. Feedbacks of Increasing Tree Population

The feedbacks of increasing tree population in the treeline ecotone on their close environment (Figure 5) are being increasingly considered (e.g., [147–149,155,157,258,278,393,394]). Increasing stem density may be associated with both positive and negative effects. Thus, trees and clonal groups growing to greater height reduce wind velocity and increase the deposition of blowing snow, which may provide shelter to young growth from climatic injury in winter-cold climates (e.g., [368,395]), while infection of evergreen conifers by parasitic snow fungi increases and can be fatal for seedlings and saplings. The increasing crown cover will reduce growing season soil temperature in the rooting zone and may thus impede root growth. Whatsoever, field studies on the response of Swiss stone pines to low soil temperatures as a result of self-shading have provided evidence that fine roots abundance and dynamics at the treeline is not affected by self-shading [396].

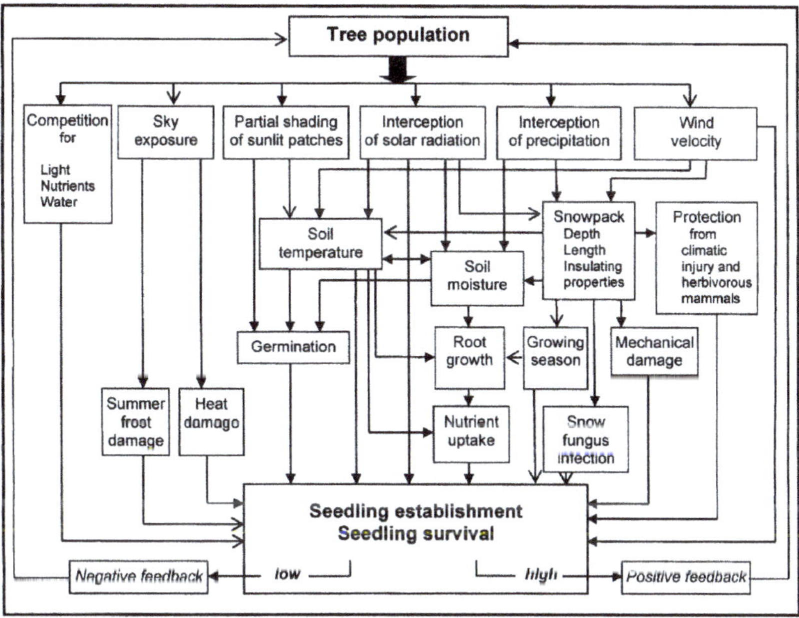

Figure 5. Feedbacks of increasing tree population on site conditions, seedling erstablishment, and survival (modified from [2]).

It has even been argued that densely grouped trees themselves would reduce their possible lifetime through shading the ground [86], whereas, sunlit patches between wide-spaced trees and tree groups usually exhibit higher daytime soil temperatures during the growing season (e.g., [20,22,23,27,131,147,397–401]), that might facilitate tree establishment. Nevertheless, adverse climatic factors [372,402–404] (e.g., recurrent frost damage, precocious dehardening in winter, summer frosts, strong prevailing winds, ice-particle abrasion, and photooxidative stress) can outweigh the advantage of reduced shading between wide-spaced trees (e.g., [2,103,178,359,372,403,405–410]).

Moreover, increasing tree density in the treeline ecotone may aggravate the competition for moisture, nutrients, and light between adult trees and juvenile trees (e.g., [157,293,411,412]). Not least, the competition between young trees, dwarf shrub, and ground vegetation may impede the establishment of trees [293,365,401,413–419]. Competition may even be more important than direct climatic effects, such as the length of the snow-free period, for example [359], or soil temperature. More systematic studies on competition as an ecological factor in the treeline ecotone are needed for a better understanding of treeline spatial and temporal dynamics.

2.2.9. Modelling Treeline and its Environmental Constraints

Since the 1990s, the modelling of treeline and its environmental constraints has rapidly increased. Modelling, in particular when combined with field- and laboratory experiments, allows alternative scenarios of possible treeline response to environmental pressure (Figure 6) (e.g., [87,137,138,145,172,173,258,289,350,365,373,409,420–431]). Due to rapidly growing numbers and the great diversity of modelling techniques it may be difficult, however, to find the most appropriate method (e.g., [432–434]). To apply the results to other environments proves to be problematic or even impossible as models and scenarios are usually based on local data sets collected within a limited period of time (see also [2,205,433]). Thus, model-based projections of future treeline have often failed. At the treeline of Scots pine in northernmost Europe, for example, climate warming did not seem to be sufficient to compensate abiotic and biotic pressures on treeline trees (e.g., [321]). In addition, in north-central Canada (west of Hudson Bay) the climatically-driven northward advance of treeline as predicted by various models has not yet occurred [326].

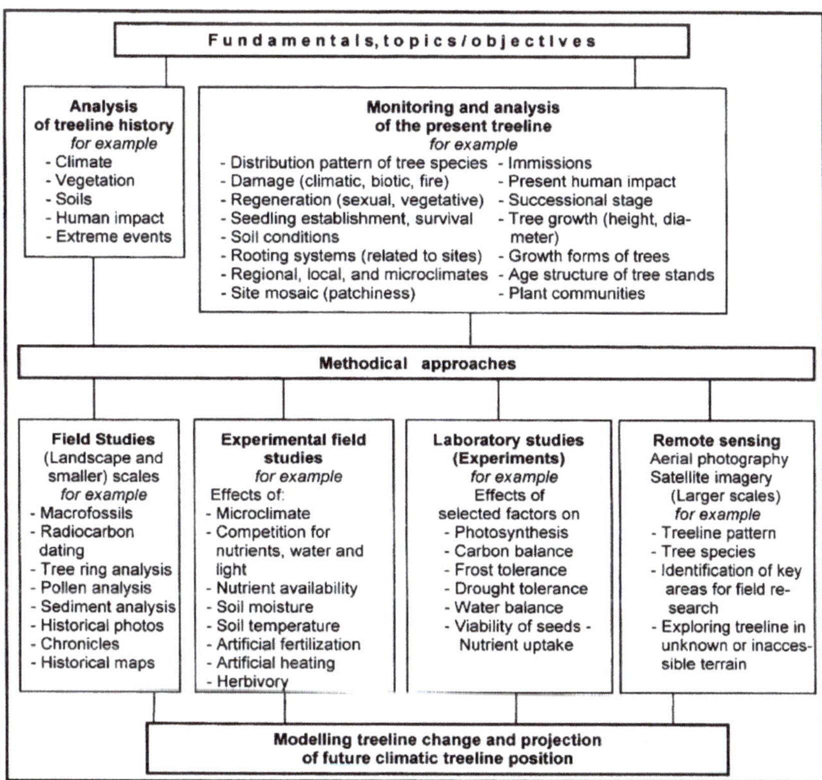

Figure 6. Fundamentals, topics, and objectives of modelling treeline (modified from [2]).

3. Conclusions and Perspectives

This research review, stemming from the roots of early research in the 1700s to modern research beginning in the 1930s, contributes to a multi-facetted view of the treeline (Figures 1, 2 and 4). Although the number of studies per year and the level of complexity has increased over time, many 'new ideas' seemingly representing the actual 'frontline' of treeline research, originated decades earlier and just confirm what has been known a long time (Figures 1–3).

Some established hypotheses need to be reassessed, such as global and regional overviews that have often overemphasized coarse drivers such as temperature. A focal point of early treeline research was linking the worldwide treeline position with an empirically found thermal average (mean growing season temperature) seemingly limiting tree growth. Not a few recent treeline studies have still been following the same idea and come to not really 'new' insights into trees' functional response to the treeline environment. Especially during the last two decades, the mass-elevation effect (MEE) was reconsidered in many studies largely confirming the previous hypotheses dating from the early 1900s.

Additional work is needed to decipher the fundamental functional role of heat deficiency in treeline control. The same applies to the role of precipitation (rain, snow) and available soil moisture (dependent on soil texture, humus content, and microtopography) during the growing season in different climatic regions. More research at local and regional scales is required to broaden and corroborate knowledge in this field.

Disastrous avalanche catastrophes during two successive extremely snow-rich winters in the 1950s and current climate warming have become important drivers of modern treeline research. The extremely snow-rich winters in particular stimulated basic and applied treeline research (Figure 2).

As afforestation and maintenance of high-elevation protective forests was the main objective, the nature of the entire elevational belt (treeline ecotone) itself was increasingly considered. Finding out which factors prevent trees in many places from reaching its potential climatic limit has been a strong challenge for researchers since then.

Scales of consideration play an important role in treeline research, as treeline heterogeneity, spatial mosaic, and ecological variety increase from coarse (global, zonal) to finer (regional, landscape, micro) scales (Figure 4). Finer scale assessments of the relative importance and function of the treeline-relevant factors and processes (human impact included) are fundamental for better understanding of treeline causation and its possible response to climate (environmental) change.

Research on episodic reproduction and treeline fluctuations in response to climate warming since the end of the Little Ice Age has steadily increased from the 1940s onward (Figure 1). Since the 1990s, the effects of current climate warming on treeline have considerably fuelled treeline research comparable to the effect of the avalanche catastrophes in the Europen Alps during the snow-rich winters in the 1950s.

Changes of biodiversity and structural diversity due to feedbacks of increasing tree population (Figure 5) within the current treeline ecotone and in the adjacent lower alpine zone, as well as the resulting implications for the ecological conditions, also require local studies.

As the impact of climate change and treeline response often overlaps with anthropogenic influences and natural factors, it is suggested that future treeline-related research could benefit from means to successfully disentangle the effects of these factors on treeline position, spatial structures and dynamics.

Remote sensing techniques and GIS have become excellent tools for monitoring treeline spatial patterns and structures at coarse and finer scales, particularly in areas difficult to access. Such innovative research techniques will be needed to come to a better assessment of the multiple and often mutually influencing factors and processes controlling treeline causation, spatial patterns, and dynamics. Thus, they will also promote projection of future treeline position (Figure 6).

In addition to specialized research on the physiological functions and response of trees to the harsh treeline environment, studies based on a more holistic approach may considerably contribute to complete our knowledge on nature of treeline. The results from the different disciplines and experts involved in treeline research must be combined and integrated into the varying spatio-temporal patterns of treelines. In this context, research on the effects of recurrent disturbances, disregarded whether natural or anthropogenic, appear as important as tree physiological research.

Author Contributions: Both authors have contributed to this article at equal parts. All authors have read and agreed to the published version of the manuscript.

Funding: This research received no external funding.

Acknowledgments: We are grateful to Ingrid Lüllau (Ismaing) for revising the English text.

Conflicts of Interest: There are no conflicts of interest.

References

1. Holtmeier, F.-K. Geoökologische Beobachtungen und Studien an der subarktischen und alpinen Waldgrenze in vergleichender Sicht (nördliches Fennoskandien/Zentralalpen). In *Erdwissenschaftliche Forschung*; Steiner Verlag: Wiesbaden, Germany, 1974; Volume 8.
2. Holtmeier, F.-K. Mountain Timberlines. Ecology, Patchiness, and Dynamics. In *Advances in Global Change Research*; Springer: Dordrecht, The Netherlands, 2009; Volume 36.
3. Holtmeier, F.-K. Impact of wild herbivorous mammals and birds on the altitudinal and northern treeline ecotones. *Landsc. Online* **2012**, *30*, 1–28. [CrossRef]
4. Crawford, R.M.M. Tundra-Taiga Biology. In *Human, Plant, and Animal Survival in the Arctic*; Oxford University Press: Oxford, UK; New York, NY, USA, 2014.
5. Wegener, A. Das Wesen der Baumgrenze. *Meteorol. Z.* **1923**, *40*, 371–372.

6.	Daubenmire, R. Alpine timberlines in the Americas and their interpretation. *Butl. Univ. Bot. Stud.* **1954**, *11*, 119–136.

7.	Kihlman, A.O. *Pflanzenbiologische Studien aus Russisch-Lappland*; Weilin & Göös: Helsinki, Finland, 1890.

8.	Raunkiaer, C. Types biologiques pour la géographie botanique. *Bull. Acad. R. Sci.* **1905**, *5*, 347–437.

9.	Von Haller, A. *Historia Stirpium Indigenarum Helvetiae Inchoate*; Sumptibus Societas Typographica: Bernae, Switzerland, 1768.

10.	Hacquet, B. Mineral—Bot. Lustreise von dem Berge Terglu Krain zu dem Berge Glockner Tirol. 1779.

11.	Zschokke, H. Beobachtungen im Hochgebirge auf einer Alpenreise im Sommer 1803. In *Isis, Wochenschrift von deutschen und schweizerischen Gelehrten*; Band I/II: Zürich, Switzerland, 1805.

12.	Von Humboldt, A.; Bonpland, A. *Essai sur la géographie des Planters*; Chez Levraud, Schoell et Compagnie: Paris, France, 1805.

13.	Wahlenberg, G. *Flora lapponica*; Berolini: Berlin, Germany, 1812.

14.	Kasthofer, K. *Bemerkungen auf einer Alpenreise über den Susten, Gotthard, Bernhardin, und über die Oberalp, Furka und Grimsel*; Aarau, Switzerland, 1822.

15.	Kasthofer, K. *Bemerkungen auf einer Alpen-Reise über den Brünig, Bragel, Kirenzenberg, die Flüela, den Maloja und Splügen*; Bern, Suisse, 1925.

16.	Von Middendorf, A.T. *Sibirische Reise. Band IV. Teil 1. Übersicht der Natur Nord- und Ost-Sibiriens*; Die Gewächse Sibiriens; Vierte Lieferung: St. Petersburg, Russia, 1864.

17.	Sendtner, O. *Die Vegetationsverhälnisse Südbayerns*; Literarisch-Artistische Anstalt: München, Germany, 1854.

18.	Landolt, E. *Bericht an den hohen schweizerischen Bundesrat über die Untersuchung der schweizerischen Hochgebirgswaldungen, vorgenommen in den Jahren 1858, 1959 und 1860*; Bern, Switzerland, 1862.

19.	Kerner, A. Studien über die obere Waldgrenze in den österreichischen Alpen. *Österreichische Revue*, 1894/1895. 2/3.

20.	Fritzsch, M. Über Höhengrenzen in den Ortler Alpen. In *Wissenschaftliche Veröffentlichungen für Erdkunde 2*; Wien, Austria, 1894; pp. 105–292.

21.	Roder, K. Die polare Waldgrenze. Ph.D. Thesis, Philosophische Fakultät der Universität Leipzig, Leipzig, Germany, 1895.

22.	Imhof, E. Die Waldgrenze in der Schweiz. *Gerlands Beiträge zur Geophys.* **1900**, *2*, 241–330.

23.	Fankhauser, F. Der oberste Baumwuchs. *Schweiz. Z. -fFür Forstwes.* **1901**, *2*, 1–5.

24.	Rikli, M. Versuch einer pflanzengeographischen Gliederung der arktischen Wald-und Baumgrenze. *Vierteljahresschr. der Nat. Ges.* **1904**, *49*, 128–142.

25.	Pohle, R. Vegetationsbilder aus Nord-Rußland. In *Vegetationsbilder*; Karsten, G., Schenk, H., Eds.; Gustav Fischer: Jena, Germany, 1907; Volume 5.

26.	Koch, M. Beiträge zur Kenntnis der Höhengrenzen der Vegetation im Mittelmeergebiet. Ph.D. Thesis, Universität Halle-Wittenberg, Halle, Germany, 1909.

27.	Marek, R. Waldgrenzstudien in den österreichischen Alpen. *Petermanns Geogr. Mitt.* **1910**, *48*, 403–425.

28.	Cleve-Euler, A. Skogträdens höjdgränser I trakten af Stora Sjöfallet. *Sven. Bot. Tisk.* **1912**, *6*, 496–509.

29.	Holmgren, A. *Studier öfore Nordligaste Skandinaviens Björkskogar*; Norstedts Förlag: Stockholm, Sweden, 1912.

30.	Renvall, A. Die periodischen Erscheinungen der Reproduktion der Kiefer an der polaren Waldgrenze. *Acta For. Fenn.* **1912**, *29*, 154. [CrossRef]

31.	Regel, K. Zur Kenntnis des Baumwuchses und der polaren Waldgrenze. *Sitz. der Nat. Ges. Dorpat* **1915**, *24*, 3–31.

32.	Pohle, R. Wald-und Baumgrenze in Nord-Russsland. *Z. der Ges. für Erdkd. Berl.* **1917**, *4*, 1–25.

33.	Brockmann-Jerosch, H. *Baumgrenze und Klimacharakter*; Beiträge zur geobotanischen Landesaufnahme, Raecher: Zürich, Switzerland, 1919

34.	Köppen, W. Baumgrenze und Lufttemperatur. *Petermanns Geogr. Mitt.* **1919**, *6*, 201–203.

35.	Köppen, W. Verhältnis der Baumgrenze zur Lufttemperatur. *Meteorol. Z.* **1920**, *37*, 39–42.

36.	Regel, K. Die Lebensformen der Holzgewächse an der polaren Wald- und Baumgrenze. *Sitzungsber.Naturfr. Ges. Dorpat* **1921**, *28*, 1–15.

37.	Hannerz, A.G. Die Waldgrenzen in den östlichen Teilen von Schwedisch Lappland. *Sven. Bot. Tidskr.* **1923**, *17*, 1–29.

38.	Gannet, H. The timber-line. *J. Am. Geogr. Soc. N. Y.* **1899**, *31*, 118–122. [CrossRef]

39.	Whiteside, C.J. The unknown father of American timberline research. *Prog. Phys. Geogr.* **2018**, *42*, 406–412. [CrossRef]

40. Griggs, R.F. The timberlines of northern America and their interpretation. *Ecology* **1946**, *27*, 275–289. [CrossRef]
41. Arno, S.F. *Timberline, Mountain and Arctic Frontiers*; The Mountaineers: Seattle, WA, USA, 1984.
42. Schlagintweit, A.; Schlagintweit, H. *Neue Untersuchungen über die physikalische Geographie und die Geologie der Alpen*; Weigel: Leipzig, Germany, 1854.
43. De Quervain, A. Die Hebung der atmosphärischen Isothermen in den Schweizer Alpen und ihre Beziehung zu den Höhengrenzen. *Gerlands Beiträge zur Geophys.* **1904**, *6*, 481–533.
44. Scharfetter, R. *Das Pflanzenleben der Ostalpen*; Deuticke: Wien, Austria, 1938.
45. Han, F.; Yao, Y.; Dai, S.; Wang, C.; Sub, R.; Xu, J.; Zhang, B. Mass elevation effect and its forcing in timberline altitude. *J. Geogr. Sci.* **2012**, *22*, 609–616. [CrossRef]
46. Irl, S.D.H.; Anthelme, F.; Harter, D.E.V.; Jentsch, A.; Lotter, E.; Steinbauer, M.J.; Beierkuhnlein, C. Patterns of island treeline elevation—A global perspective. *Ecography* **2016**, *39*, 427–436. [CrossRef]
47. Odland, A. Effect of latitude and mountain height on the timberline (Betula pubescens ssp. czerepanovii) elevation along the central Scandinavian mountain range. *Fennia* **2015**, *193*, 260–270.
48. Yao, Y.; Zhang, B. The mass elevation effect of the Tibetan Plateau and its implications for alpine treelines. *Int. J. Climatol.* **2015**, *35*, 1833–1846. [CrossRef]
49. He, W.; Zhang, B.; Zhao, F.; Zhang, S.; Qi, W.; Whang, J.; Zhang, W. The mass-elevation effect of the central Andes and its implications for the southern hemisphere's highest treeline. *Mt. Res. Dev.* **2016**, *36*, 213–221. [CrossRef]
50. Kašpar, J.; Treml, V. Thermal characteristics of alpine treelines in Central Europe north of the Alps. *Clim. Res.* **2016**, *68*, 1–12. [CrossRef]
51. Zhao, F.; Zhang, B.; Zhang, S.; Qi, W.; He, W.; Wang, J.; Yao, Y. Contribution of mass elevation effect to the altitudinal distribution of global treelines. *J. Mt. Sci.* **2015**, *12*, 289–297. [CrossRef]
52. Han, F.; Zhang, B.; Zhao, F.; Wan, L.; Tan, J.; Liang, T. Characterizing the mass-elevation effect across the Tibetan Plateau. *J. Mt. Sci.* **2018**, *15*, 2651–2665. [CrossRef]
53. Wilson, C.; Grace, J.; Allen, S.; Slack, F. Temperature and stature: A study of temperature in montane vegetation. *Funct. Ecol.* **1987**, *1*, 405–413. [CrossRef]
54. Däniker, A. Biologische Studien über Wald- und Baumgrenze, insbesondere über die klimatischen Ursachen und deren Zusammenhänge. *Vierteljahresschr. Naturfr. Ges. Zürich* **1923**, *63*, 1–102.
55. Stocker, O. Transpiration und Wasserhaushalt in verschiedenen Klimazonen. I. Untersuchungen an der arktischen Baumgrenze in Schwedisch Lappland. *Jahrb. für Wiss. Bot.* **1931**, *75*, 494–549.
56. Michaelis, P. Ökologische Studien an der alpinen Baumgrenze. II. Die Schichtung der Windgeschwindigkeit, Luftemperatur und Evaporation über einer Schneefläche. *Beih. Zum Bot. Zent.* **1934**, *52*, 310–332.
57. Michaelis, P. Ökologische Studien and der alpinen Baumgrenze. III. Über die winterlichen Temperaturen der pflanzlichen Organe, insbesondere der Fichte. *Beih. Bot. Zent.* **1934**, *52*, 333–377.
58. Michaelis, P. Ökologische Studien an der alpinen Baumgrenze. IV. Zur Kenntnis des winterlichen Wasserhaushaltes. *Jahrb. Wiss. Bot.* **1934**, *80*, 169–247.
59. Michaelis, P. Ökologische Studien an der alpinen Waldgrenze. V. Osmotischer Wert und Wassergehalt während des Winters in verschiedenen Höhenlagen. *Jahrb. Für Wiss. Bot.* **1934**, *80*, 337–362.
60. Steiner, M. Winterliches Bioklima und Wasserhaushalt an der alpinen Waldgrenze. *Bioklim. Beiblätter* **1935**, *2*, 57–65.
61. Schmidt, E. Baumgrenzstudien am Feldberg im Schwarzwald. *Tharandter Forstl. Jahrb.* **1936**, *87*, 143.
62. Pisek, A.; Cartellieri, E. Zur Kenntnis der Wasserhaushaltes der Pflanzen. IV. Bäume und Sträucher. *Jahrb. der Wiss. Bot.* **1939**, *88*, 222–268.
63. Pisek, A.; Schiessl, R. Die Temperaturbeeinflußbarkeit der Frosthärte von Nadelhölzern und Zwergsträuchern an der alpinen Waldgrenze. *Berd. es Nat. Med. Ver. Innsbr.* **1946**, *47*, 33–52.
64. Winkler, E. Klimaelemente für Innsbruck (582 m) und Patscherkofel (1909 m) im Zusammenhang mit der Assimilation von Fichten in verschiedenen Höhenlagen. *Veröffentlichungen des Mus. Ferdinandeum Innsbr.* **1957**, *37*, 19–48.
65. Pisek, A.; Winkler, E. Assimilationsvermögen und Respiration der Fichte (Picea excelsa Link) in verschiedener Seehöhe und der Zirbe (*Pinus cembra* L.) an der alpinen Waldgrenze. *Planta* **1958**, *51*, 518–543. [CrossRef]
66. Blüthgen, J. Baumgrenzen in Lappland. *Geogr. Anz.* **1937**, *1*, 532–533.
67. Hustich, I. Pflanzengeographische Studien im Gebiet der Niederen Fjelde im westlichen Finnischen Lappland. Teil I. Ph.D. Thesis, Societas pro fauna et flora Fennica, Helsingfors, Finland, 1937.

68. Aario, L. Waldgrenzen und subrezente Pollenspektren in Petsamo-Lappland. *Ann. Acad. Sci. Fenn.* **1940**, *54*, 1–120.

69. Blüthgen, J. Die polare Baumgrenze. Veröffentlichungen des Dtsch. *Wiss. Inst. zu Kph.* **1942**, *10*, 1–80.

70. Blüthgen, J. Dynamik der polaren Baumgrenze in Lappland. *Forsch. und Fortschr.* **1943**, *19*, 158–160.

71. Hustich, I. The scotch pine in northernmost Finland and ist dependence on the climate in the last decades. *Acta Bot. Fenn.* **1948**, *42*, 75.

72. Tranquillini, W. Standortsklima, Wasserbilanz und CO_2-Gaswechsel junger Zirben (Pinus cembra L.) an der alpinen Waldgrenze. *Planta* **1957**, *49*, 612–661. [CrossRef]

73. Tranquillini, W. Die Frosthärte der Zirbe unter besonderer Berücksichtigung autochthoner und aus Forstgärten stammender Jungpflanzen. *Forstwiss. Centralbl.* **1958**, *77*, 89–105. [CrossRef]

74. Ott, E.; Frehner, M.; Frey, H.-U.; Lüscher, P. *Gebirgsnadelwäder. Ein praxisorientierter Leitfaden für eine standortsgerechte Waldbehandlung*; Haupt: Bern, Switzerland, 1997.

75. Aulitzky, H. Grundlagen und Anwendung des vorläufigen Wind-Schnee-Ökogramms. *Mitt. der Forstl. Bundesvers. Mariabrunn* **1963**, *60*, 763–834.

76. Turner, H.; Rochat, P.; Streule, A. Thermische Charakteristik von Hangstandortstypen im Bereich der oberen Waldgrenze (Stillberg, Dischmatal bei Davos). *Mitt.der Eidg. Anst. für das Forstl. Vers.* **1975**, *51*, 95–111.

77. Holtmeier, F.K.; Broll, G. Altitudinal and polar treelines in the northern hemisphere—Causes and response to climate change. *Polarforschung* **2009**, *79*, 139–153.

78. Boysen-Jensen, P. *Die Stoffproduktion der Pflanze*; Gustav Fischer Verlag: Jena, Germany, 1932.

79. Ungerson, J.; Scherdin, G. Jahresgang von Photosynthese und Atmung unter natürlichen Bedigungen von Pinus sylvestris L. an ihrer Nordgrenze in der Subarktis. *Flora* **1968**, *157*, 391–434.

80. Dahl, E.; Mork, E. Om sambandet mellom temperatur, anding of vekst hos Gran (Picea abies (L.) Karst.). *Medd. Fran Det Nor. Skogförsöksvesen* **1959**, *53*, 82–93.

81. Kozlowski, T.T. *Growth and development of trees. Vol. II*; Academic Press: New York, NY, USA, 1971.

82. Skre, O. High temperature demands for growth and development in Norway Spruce (Picea abies (L.) Karst.) in Scandinavia. *Meldinder Fra Nor. Landbr.* **1972**, *51*, 1–29.

83. Stevens, G.C.; Fox, J.F. The causes of treeline. *Annu. Rev. Ecol. Syst.* **1991**, *22*, 177–191. [CrossRef]

84. Sveinbjörnsson, B.; Kauhanen, H.; Nordell, O. Treeline ecology of mountain birch in the Torneträsk area. *Ecol. Bull.* **1996**, *45*, 65–70.

85. Wieser, G. Carbon dioxide gas exchange of cembran pine (Pinus cembra) at the alpine timberline during winter. *Tree Physiol.* **1997**, *17*, 473–477. [CrossRef] [PubMed]

86. Körner, C. A re-assessment of high elevation treeline positions and their explanation. *Oecologia* **1998**, *115*, 445–459. [CrossRef]

87. Cairns, D.M.; Malanson, G.P. Environmental variables influencing the carbon balance at the alpine treeline: A modeling approach. *J. Veg. Sci.* **1998**, *9*, 679–692. [CrossRef]

88. Grace, J.; Berninger, F.; Nagy, L. Impacts of climate change on the treeline. *Ann. Bot.* **2002**, *90*, 537–544. [CrossRef]

89. Hoch, G.; Körner, C. The carbon charging of pines at the climatic treeline: A global comparison. *Oecologia* **2003**, *135*, 10–21. [CrossRef]

90. Körner, C.; Paulsen, J. A world-wide study of high altitude treeline temperatures. *J. Biogeogr.* **2004**, *31*, 713–732. [CrossRef]

91. Bugmann, H.; Zierl, B.; Schumacher, S. Projecting the impacts of climate change on mountain forests and landscapes. In Global change and mountain regions. An overview of current knowledge. In *Advances in Global Change Research*; Huber, U.M., Bugmann, H.K.M., Reasoner, M.A., Eds.; Springer: New York, NY, USA, 2005; pp. 477–487.

92. Shi, P.; Körner, C.; Hoch, G. End of season carbon supply status of wooden species near the treeline in western China. *Basic Appl. Ecol.* **2006**, *7*, 370–377. [CrossRef]

93. Körner, C. Alpine treeline. In *Functional Ecology of the Global High Elevation Tree Limits*; Springer: Basel, Suisse, 2012.

94. Handa, I.T.; Körner, C.; Hättenschwiler, S. A test of the treeline carbon limitation hypothesis by in situ CO_2 enrichment and defoliation. *Ecology* **2005**, *86*, 1288–1300. [CrossRef]

95. Kammer, A.; Hagedorn, F.; Shevchenko, I.; Leifeld, J.; Guggenberger, G.; Gorcheva, T.; Rigling, A.; Moiseev, P. Treeline shifts in the Ural mountains affect soil organic matter dynamics. *Glob. Chang. Biol.* **2009**, *15*, 1570–1583. [CrossRef]

96. Hagedorn, F.; Martin, M.; Rixen, C.; Rusch, S.; Bebi, P.; Zürcher, A.; Hättenschwiler, S. Short-term responses to ecosystem carbon fluxes to experimental soil warming at the Swiss alpine treeline. *Biogeochemistry* **2010**, *97*, 7–19. [CrossRef]

97. Wang, G.; Ran, F.; Chang, R.; Yang, Y.; Luo, J.; Fan, J. Variations in the biomass and carbon pools of Abies georgei along an elevation gradient on the Tibetan Plateau, China. *For. Ecol. Manag.* **2014**, *329*, 255–263. [CrossRef]

98. Grafius, D.R.; Malanson, G.P. Biomass distribution in dwarf tree, krummholz, and tundra vegetation in the alpine treeline ecotone. *Phys. Geogr.* **2015**, *36*, 337–352. [CrossRef]

99. Ebermayer, E. *Die physikalischen Einwirkungen des Waldes auf Luft und Boden. Band 1. Anhang: Die Ursachen der Schüttekrankheit junger Kiefernpflanzen*; C. Krebs: Aschaffenburg, Germany, 1873; pp. 251–261.

100. Shaw, C.H. The causes of timberline on mountains: The role of snow. *Plant World* **1909**, *12*, 169–181.

101. Shaw, C.H. Present problems in plant ecology: III Vegetation and altitude. *Am. Nat.* **1909**, *43*, 420–431. [CrossRef]

102. Pisek, A. Zur Kenntnis der Frosthärte alpiner Pflanzen. *Die Naturwissenschaften.* **1952**, *39*, 73–78. [CrossRef]

103. Tranquillini, W. *Physiological Ecology of the Alpine Timberline—Tree Existence at High Altitudes*; Springer: Berlin/Heidelberg, Germany, 1979.

104. Tranquillini, W. Über die physiologischen Ursachen der Wald- und Baumgrenze. *Mitt. der Forstl. Bundesvers. Mariabrunn* **1967**, *75*, 457–487.

105. Müller-Stoll, W.R. Beiträge zur Ökologie der Waldgrenze am Feldberg im Schwarzwald. *Angew. Pflanzensoziol.* **1954**, *2*, 824–847.

106. Larcher, W. Frosttrocknis an der Waldgrenze und in der alpinen Zwergstrauchheide. *Veröffentlichungen Mus. Ferdinandeum* **1957**, *37*, 49–81.

107. Holzer, K. Winterliche Schäden an Zirben nahe der alpinen Waldgrenze. *Centralbl. für das gesamte Forstwes.* **1959**, *76*, 232–244.

108. Wardle, P. Engelmannn spruce (Picea engelmannii) at its upper limits on the Front Range, Colorado. *Ecology* **1968**, *49*, 483–495. [CrossRef]

109. Holtmeier, F.-K. Waldgrenzstudien im nördlichen Finnish-Lappland und angrenzenden Nordnorwegen. *Rep. Kevo Subarct. Res. Stn.* **1971**, *8*, 53–62.

110. Wardle, P. An explanation for alpine timberlines. *N. Z. J. Bot.* **1971**, *9*, 371–402. [CrossRef]

111. Baig, M.N.; Tranquillini, W.; Havranek, W. Cuticuläre Transpiration von Picea abies und Pinus cembra-Zweigen aus verschiedener Seehöhe und ihre Bedeutung für die winterliche Austrocknung der Bäume an der alpinen Waldgrenze. *Centralbl. für das gesamte Forstwes.* **1974**, *91*, 195–211.

112. Tranquillini, W. Der Einfluß von Seehöhe und Länge der Vegetationszeit auf das cutikuläre Transpirationsvermögen von Fichtensämlingen im Winter. *Ber.Dtsch. Bot. Ges.* **1974**, *87*, 175–184.

113. Wardle, P. Alpine Timberlines. In *Arctic and Alpine Environment*; Ives, J.D., Barry, R.G., Eds.; Methuen young books: London, UK, 1974; pp. 371–402.

114. Platter, W. Wasserhaushalt, cuticuläres Transpirationsvermögen und Dicke der Cutinschichten einiger Nadelholzarten in verschiedendenen Höhenlagen und experimentelle Verkürzung der Vegetationszeit. Ph.D. Thesis, Innsbruck University, Innsbruck, Austria, 1976.

115. Turner, H. Types of microclimate at high elevations. In *Mountain Environments and Subalpine Tree Growth*; Benecke, U., Davis, M., Eds.; New Zealand Forest Service Technical Paper; New Zealand Forest Service: Welligton, New Zealand, 1980; Volume 70, pp. 21–26.

116. Sowell, J.P.; Koutnik, K.; Lansing, A.J. Cuticular transpiration of whitebark pine (Pinus albicaulis) within an Sierra Nevadan timberline ecotone, U.S.A. *Arct. Alp. Res.* **1982**, *14*, 97–103. [CrossRef]

117. Tranquillini, W. Frost-drought an its ecological significance. In *Physiological Plant Ecology II—Encyclopedia of Plant Physiology, New Series*; Lange, O.L., Nobel, P.S., Osmond, C.B., Eds.; Springer: Berlin/Heidelberg, Germany, 1982; pp. 379–400.

118. DeLucia, E.V.; Berlyn, G.P. The effect of mountain elevation on leaf cuticle thickness and cuticular transpiration in balsam fir. *Can. J. Bot.* **1984**, *62*, 2423–2431. [CrossRef]

119. Sowell, J.B. Winter relations of trees at alpine timberline. In *Establishment and Tending of Subalpine Forersts: Research and Management, Proceeding 3rd IUFRO Workshop*; Turner, H., Tranquillini, W., Eds.; Eidgenössische Anstalt für das forstliche Versuchswesen: Birmensdorf, Germany, 1985; pp. 71–77.

120. Cairns, D.M. Patterns of winter desiccation in krummholz forms of Abies lasiocarpa at treeline sites in Glacier National Park, Montana, USA. *Geogr. Ann.* **2001**, *83*, 157–168. [CrossRef]

121. Ellenberg, H. *Vegetation Mitteleuropas mit den Alpen in ökologischer Sicht*, 2nd ed.; Verlag Eugen Ulmer: Stuttgart, Deutschland, 1963; p. 981.

122. Walter, H.; Breckle, S. *Ökologie der Erde, Band 3, Spezielle Ökologie der Gemäßigten und arktischen Zonen Euro-Nordasiens*; Gustav Fischer: Stuttgart, Germany, 1986.

123. Richter, M. *Vegetationszonen der Erde*; Klett-Perthes: Gotha/Stuttgart, Germany, 2001.

124. Barclay, A.M.; Crawford, R.M.M. Winter desiccation stress and resting bud viability in relation to high altitude survival in *Sorbus aucuparia* L. *Flora* **1982**, *172*, 21–34. [CrossRef]

125. Cochrane, P.M.; Slatyer, R.O. Water relations of Eucaplytus pauciflora near the alpine tree line in winter. *Tree Physiol.* **1988**, *4*, 45–52. [CrossRef] [PubMed]

126. Slatyer, R.O.; Noble, I.R. Dynamics of Montane Treelines. In *Landscape Boundaries: Consequences for Biotic Diversity and Ecological Flows, Ecological Studies*; Hansen, A., Di Castri, F., Eds.; Springer: New York, NY, USA, 1992; pp. 346–359.

127. Marchand, P.J.; Chabot, B.F. Winter water relations of the tree-line plant species, New Hampshire. *Arct. Alp. Res.* **1978**, *10*, 105–116. [CrossRef]

128. Holtmeier, F.-K. Influence of wind on tree physiognomy at the upper timberline in the Colorado Front Range. In *Mountain Environments and Subalpine Tree Growth, Proceedings of the IUFRO Workshop, Christchurch, New Zealand, 1979*; Benecke, D., Ed.; New Zealand Forest Service Technical Paper: Wellington, New Zealand, 1980; pp. 247–262.

129. Marchand, P.J. Causes of coniferous timberline in the northern Appalachian mountains. In *Mountain Environments and Subalpine Tree Growth*; Benecke, U., Davis, M.R., Eds.; New Zealand Forest Service Technical Paper: Wellington, New Zealand, 1980; pp. 231–246.

130. Hadley, J.L.; Smith, W.K. Influence of wind exposure on needle deciccation and mortality for timberline conifers in Wyoming, U.S.A. *Arct. Alp. Res.* **1983**, *15*, 127–135. [CrossRef]

131. Hadley, J.L.; Smith, W.K. Wind effects on needles of timberline conifers: Seasonal influences on mortality. *Ecology* **1986**, *67*, 12–18. [CrossRef]

132. Grace, J. Cuticular water loss unlikely to explain treeline in Scotland. *Oecologia* **1990**, *84*, 64–68. [CrossRef]

133. Holtmeier, F.-K. Blowing snow and sand blast shaping tree physiognomy at certain treeline microsites—A review. *Geoöko* **2016**, *37*, 201–233.

134. Holtmeier, F.-K.; Broll, G. Treeline advance—Driving processes and adverse factors. *Landsc. Online* **2007**, *1*, 1–33. [CrossRef]

135. Neuner, G. Frost resistance at the upper timberline. In *Trees at Their Upper Limit. Treelife Limitation at the Alpine Timberline—Plant Ecophysiology*; Wieser, G., Tausz, M., Eds.; Springer: Dordrecht, The Netherlands, 2007; pp. 171–180.

136. Wieser, G.; Tausz, M. Current concepts for treelife limitation at the upper timberline. In *Trees at Their Upper Limit. Treelife Limitation at the Alpine Timberline—Plant Ecophysiology*; Wieser, G., Tausz, M., Eds.; Springer: Dordrecht, The Netherlands, 2007; pp. 1–18.

137. Gehrig-Fasel, J.; Guisan, A.; Zimmermann, N.E. Evaluating thermal treeline indicators based on air and soil temperature using air-to-soil temperature transfer model. *Ecol. Model.* **2008**, *213*, 345–355. [CrossRef]

138. Paulsen, J.; Körner, C. A climate-based model to predict treeline position around the globe. *Alp. Bot.* **2014**, *124*, 1–2. [CrossRef]

139. Retzer, J.L. Alpine soils. In *Arctic and Alpine Environment*; Ives, J.D., Barry, R.G., Eds.; Cambridge University Press: Cambridge, UK, 1974; pp. 771–802.

140. Karlsson, P.S.; Nordell, O. Effects of soil temperature on nitrogen economy and growth of mountain birch seedlings near its presumed low temperature distribution limit. *Ecoscience* **1996**, *3*, 181–189. [CrossRef]

141. Hermes, K. Die Lage der oberen Waldgrenze in den Gebirgen der Erde und ihr Abstand zur Schneegrenze. *Kölner Geogr. Arb.* **1955**, *5*, 277.

142. Holtmeier, F.-K. Ecological aspects of climatically-caused timberline fluctuations. In *Mountain Environments in Changing Climates*; Beniston, M., Ed.; Routledge: London, UK, 1994; pp. 220–233.

143. Büntgen, U.; Frank, D.C.; Kaczka, R.; Verstege, A.; Zwijacz-Kozica, T.; Esper, J. Growth response to climate in a multi-species tree-ring network in the western Carpathian Tatra Mountains, Poland, Slovakia. *Tree Physiol.* **2007**, *27*, 687–702. [CrossRef] [PubMed]

144. Ohsawa, M. An interpretation of latitudinal patterns of forest limits in south and east Asian mountains. *J. Ecol.* **1990**, *78*, 326–329. [CrossRef]

145. Bueno de Mesquita, C.P.; Tillmann, L.S.; Bernard, C.D.; Rosemond, K.C.; Molotsch, N.P.; Suding, K.N. Topographic heterogeneity explains patterns of vegetation response to climate change (1972-2008) across a mountain landscape, Niwot Ridge, Colorado. *Arct. Antarct. Alp. Res.* **2018**, *50*. [CrossRef]

146. Turner, H. Die globale Hangbestrahlung als Standortfaktor bei Aufforstungen in der subalpinen Stufe. *Mitt. der Schweiz. Anst. für das Forstl. Vers.* **1966**, *42*, 1110–1686.

147. Holtmeier, F.-K.; Broll, G. The influence of tree islands and microtopography on pedoecological conditions in the forest-alpine tundra ecotone on Niwot Ridge, Colorado Front Range, U.S.A. *Arct. Alp. Res.* **1992**, *24*, 216–228. [CrossRef]

148. Broll, G.; Holtmeier, F.-K. Die Entwicklung von Kleinreliefstrukturen im Waldgrenzökoton der Front Range (Colorado, USA) unter dem Einfuss leewärts wandernder Ablegergruppen (Picea engelmannii und Abies lasiocarpa). *Erdkunde* **1994**, *48*, 48–59. [CrossRef]

149. Hiemstra, C.A.; Liston, G.E.; Reiners, W.A. Snow redistribution by wind and interacting with vegetation at upper treeline in the Medicine Bow Mountains, Wyoming, U.S.A. *Arct. Antarct. Alp. Res.* **2002**, *34*, 262–273. [CrossRef]

150. Bekker, M. Positive feedback between tree establishment and patterns of subalpine forests advancement, Glacier National Park, Montana, U.S.A. *Arct. Antarct. Alp. Res.* **2005**, *37*, 97–107. [CrossRef]

151. Alftine, K.J.; Malanson, G.P. Directional positive feedback and pattern at an alpine treeline. *J. Veg. Sci.* **2004**, *15*, 3–12. [CrossRef]

152. Holtmeier, F.-K. Relocation of snow and its effects in the treeline ecotone with special regard to the Rocky Mountains, the Alps and Northern Europe. *Die Erde* **2005**, *136*, 343–373.

153. Wiegand, T.; Camarero, J.J.; Rüger, N.; Gutiérrez, E. Abrupt population changes at treeline ecotones along smooth gradients. *J. Ecol.* **2006**, *94*, 880–892. [CrossRef]

154. Humphries, H.C.; Bourgeron, P.S.; Mujica-Crapanzano, L.R. Tree spatial patterns and environmental relationships in the forest-alpine tundra ecotone at Niwot Ridge, Colorado, USA. *Ecol. Res.* **2008**, *23*, 589–605. [CrossRef]

155. Holtmeier, F.-K.; Broll, G. Wind as an ecological agent at treelines in North America, the Alps, and the European Subarctic. *Phys. Geogr.* **2010**, *31*, 203–233. [CrossRef]

156. Elliott, G.P.; Kipfmueller, K.F. Multi-scale influences of slope aspect and spatial pattern on ecotonal dynamics at upper treeline in the Southern Rocky Mountains, U.S.A. *Arct. Antarct. Alp. Res.* **2010**, *42*, 45–56. [CrossRef]

157. Holtmeier, F.-K.; Broll, G. Feedbacks of clonal groups and tree clusters on site conditions at the treeline: Implications for treeline dynamics. *Clim. Res.* **2017**, *73*, 85–96. [CrossRef]

158. Montpellier, E.E.; Soulé, P.T.; Knapp, P.A.; Shelley, J.S. Divergent growth rates of Larix lyallii Parl) in response to microenvironmental variability. *Arct. Antarct. Alp. Res.* **2018**, *50*. [CrossRef]

159. Maguire, A.; Eitel, J.U.H.; Vierling, L.A.; Johnson, D.M.; Griffin, K.L.; Bodman, N.; Jensen, J.E.; Greaves, H.E.; Meddens, A.J.H. Terrestrial lidarscanning reveals fine-scale linkages between microstructure and photosynthetic functioning of small-stature spruce trees at the forest-tundra ecotone. *Agric. For. Meteorol.* **2019**, *269*, 157–168. [CrossRef]

160. Holtmeier, F.-K.; Broll, G. Subalpine forest and treeline ecotone under the influence of disturbances: A review. *J. Environ. Prot.* **2018**, *9*, 815–845. [CrossRef]

161. Tolmachev, A.I. Die Erforschung einer entfernten Waldinsel in der Großlandtundra. *Colloq. Geogr.* **1970**, *12*, 98–103.

162. Payette, S.; Gagnon, R. Tree-line dynamics in Ungava Peninsula, northern Québec. *Holarct. Ecol.* **1979**, *2*, 239–248. [CrossRef]

163. Ives, J.D.; Hansen-Bristow, K.J. Stability and instability of natural and modified upper timberline landscapes in the Colorado Rocky Mountains, U.S.A. *Mt. Res. Dev.* **1983**, *3*, 149–155. [CrossRef]

164. Holtmeier, F.-K. Human impacts on high altitude forests and upper timberline with special reference to middle latitudes. In *Human Impacts and Management of Mountain Forests*; Fujimori, T., Kimura, M., Eds.; Forestry and Forest Products Research Institute: Ibaraki, Japan, 1987; pp. 9–20.

165. Löffler, J.; Lundberg, A.; Rössler, O.; Bräuning, A.; Jung, G.; Pape, R.; Wundram, D. The alpine treeline under changing land use and changing climate: Approach and preliminary results from continental Norway. *Nor. Geogr. Tidskr.* **2004**, *58*, 183–193. [CrossRef]

166. Leonelli, G.; Pelfini, M.; Battipaglia, G.; Cherubini, P. Site aspect influence on climatic sensitivity over time of a high-altitude Pinus cembra tree-ring network. *Clim. Chang.* **2009**, *96*, 185–201. [CrossRef]

167. Harsh, M.A.; Hulme, P.E.; McGlone, M.S.; Duncan, R.P. Are treelines advancing? A global meta-analysis of treeline response to climate warming. *Ecol. Lett.* **2009**, *12*, 1040–1045. [CrossRef]

168. Kullman, L. Norway spruce (Picea abies (L.) Karst.) treeline ecotone performance since the mid-1970s in the Swedish Scandes—Evidence of stability and minor change from repeat surveys and photography. *Geoöko* **2007**, *35*, 23–53.

169. Kullman, L. One century of treeline change and stability—An illustrated view from the Swedish Scandes. *Landsc. Online* **2010**, *17*, 1–31. [CrossRef]

170. Kullman, L. The alpine treeline ecotone in the southwestern Swedish Scandes: Dynamics on different scales. In *Ecotones between Forest and Grassland*; Myster, R.W., Ed.; Springer: New York, NY, USA, 2012.

171. Kullman, L. Higher-than-present Medieval pine (Pinus sylvestris) treeline along the Swedish Scandes. *Landscape online* **2015**, *42*, 1–14. [CrossRef]

172. Bruening, J.M.; Tran, T.J.; Bunn, A.G.; Weiss, S.B.; Salzer, M.W. Fine scale modeling a bristlecone treeline position in the Great Basin, USA. *Environ. Res. Lett.* **2017**, *12*. [CrossRef]

173. Wieczorek, M.; Kruse, S.; Eppl, L.; Kolmogorov, A.; Nikolaev, A.N.; Heinrich, E.; Jeltsch, F.; Pestryakova, L.A.; Zibulski, R.; Herzschuh, U. Dissimilar response of larch stands in northern Siberia to increasing temperatures—A field and simulation based study. *Ecology* **2017**, *98*, 2343–2355. [CrossRef] [PubMed]

174. Bonanomi, G.; Rita, A.; Allevato, E.; Cesarano, G.; Saulino, L.; Di Pasquale, G.; Alegrezza, M.; Pesaresi, S.; Borghetii, M.; Rossi, S. Anthropogenic and environmental factors affect the treeline position of Fagus sylvatica along the Apennines (Italy). *J. Biogeogr.* **2018**, *45*, 2595–2608. [CrossRef]

175. Singh, S.P.; Sharma, S.; Dghyani, P.P. Himalayan arc and treeline: Distribution, climate change responses and ecosystem properties. *Biodivers. Conserv.* **2019**, *28*, 1997–2016. [CrossRef]

176. Wieser, G.; Oberhuber, W.; Gruber, A. Effects of climate change at treeline: Lessons from space-for-time studies, manipulative experiments, and long-term observational records in the Central Austrian Alps. *Forests* **2019**, *10*, 508. [CrossRef]

177. Rössler, O.; Löffler, J. Uncertainties of treeline alterations due to climatic change during the past century in the Norwegian Scandes. *Geöko* **2007**, *28*, 104–114.

178. Moyes, A.B.; Germino, M.J.; Kueppers, L.M. Moisture rivals temperature in limiting phyotosynthesis by trees establishing beyond their cold-edge range limit under ambient and warm conditions. *New Phytol.* **2015**, *4*, 1005–1014. [CrossRef]

179. Hellmann, L.; Agafonov, L.; Ljungqvist, F.C.; Churakova, O.; Düthorn, E.; Esper, J.; Hülsmann, L.; Kirdyanov, A.V.; Moiseev, P.; Myglan, V.S.; et al. Diverse growth trends and climate responses across Eurasian's boreal forest. *Environ. Res. Lett.* **2016**, *11*. [CrossRef]

180. Case, B.; Duncan, R. A novel frame for disentangling the scale-dependent influence of abiotic factors on alpine treeline position. *Ecography* **2014**, *37*, 838–851. [CrossRef]

181. Callaghan, T.V.; Werkman, B.R.; Crawford, R.M.M. The tundra-taiga interface and its dynamics, concepts and applications. *Ambio Spec. Rep.* **2002**, *12*, 6–14.

182. Juntunen, V.; Neuvonen, S.; Norokorpi, Y.; Tasanen, T. Potential for timberline advance in northern Finland as revealed by monitoring during 1983–1999. *Arctic* **2002**, *55*, 348–361. [CrossRef]

183. Lloyd, A.H.; Fastie, C. Spatial and temporal variability in the growth and climate response of treeline trees in Alaska. *Clim. Chang.* **2002**, *52*, 481–509. [CrossRef]

184. Holtmeier, F.-K.; Broll, G. Sensitivity and response of northern hemisphere altitudinal and polar treelines to environmental change at landscape and local scales. *Glob. Ecol. Biogeogr.* **2005**, *14*, 395–410. [CrossRef]

185. Walther, G.-R.; Beißner, S.; Pott, R. Climate change and high mountain vegetation shifts. In *Mountain Ecosystems. Studies in Treeline Ecology*; Broll, G., Keplin, B., Eds.; Springer: Berlin/Heidelberg, Germany, 2005; pp. 77–95.

186. Juntunen, V.; Neuvonen, S. Natural regeneration of Scots pine and Norway spruce close to the timberline in Northern Finland. *Silva Fenn.* **2006**, *40*, 443–458. [CrossRef]

187. Wohlgemuth, T.; Bugmann, H.; Lische, H.; Tinner, W. Wie rasch ändert sich die Waldvegetation als Folge von raschen Klimaveränderungen? *Forum für Wissen* **2006**, *1*, 7–16.

188. Danby, R.K.; Hik, D.S. Variability, contingency and rapid change in recent subarctic alpine tree line dynamics. *J. Ecol.* **2007**, *95*, 352–363. [CrossRef]

189. Devi, N.; Hagedorn, F.; Moiseev, P.; Bugmann, H.; Shiyatov, S.; Mazepa, V.; Rigling, A. Expanding forest and changing growth forms in Siberian larch at the polar Urals treeline during the 20th century. *Glob. Chang. Biol.* **2008**, *14*, 1581–1591. [CrossRef]

190. MacDonald, G.; Kremenetski, K.V.; Beilman, D.W. Climate change and the northern Russian treeline zone. *Philos. Trans. R. Soc.* **2008**, *363*, 2285–2299. [CrossRef]

191. Smith, W.K.; Germino, M.J.; Johnson, D.M.; Reinhardt, K. The altitude of alpine treeline: A bellwether of climate change effects. *Bot. Rev.* **2009**, *75*, 163–190. [CrossRef]

192. Van Bogaert, R.; Haneca, K.; Hoogester, J.; Jonasson, C.; De Papper, M.; Callaghan, T.V. A century of tree line changes in subarctic Sweden shows local and regional variability and only minor influence of 20th century climate warming. *J. Biogeogr.* **2011**, *38*, 907–921. [CrossRef]

193. Batllori, E.; Camarero, J.J.; Gutiérrez, E. Climatic drivers of tree growth and recent recruitment at the Pyrenean alpine tree line ecotone. In *Ecotones*, 1st ed.; Myster, R.W., Ed.; Springer: New York, NY, USA, 2012; pp. 247–289.

194. Aakala, T.; Hari, P.; Dengel, S.; Newberry, S.L.; Mizunuma, T.; Grace, J. A prominent stepwise advance of the treeline in north-east Finland. *J. Ecol.* **2014**, *102*, 1582–1591. [CrossRef]

195. Greenwood, S.; Jump, A.S. Consequences if treeline shifts for the diversity and function of high altitude ecosystems. *Arct. Antarct. Alp. Res.* **2014**, *46*, 829–840. [CrossRef]

196. Broll, G.; Jokinen, M.; Aradottir, A.L.; Cudlin, P.; Dinca, L.; Gömöryová, E.; Grego, S.; Holtmeier, F.-K.; Karlinski, L.; Klopcic, M.; et al. Working Group 2: Indicators of changes in the treeline ecotone. In *SENSFOR Deliverable 5. COST Action ES 1203: Enhancing the Resilience Capacity of Sensitive Mountain Forest Ecosystems and Environmental Change*; EU: Brussels, Belgium, 2016.

197. Schickhoff, U.; Bobrowski, M.; Böhner, J.; Bürzle, B.; Chaudhary, R.P.; Gerlitzky, L.; Lange, J.; Müller, M.; Scholten, T.; Schwab, N. Climate change and treeline dynamics in the Himalaya. In *Climate Change, Glacier Response, and Vegetation Dynamics in the Himalaya*; Singh, R.B., Schickhoff, U., Mal, S., Eds.; Springer: Cham, Switzerland, 2016; pp. 271–306.

198. Cudlin, P.; Klopčič, M.; Tognetti, R.; Mális, F.; Alados, C.L.; Bebi, P.; Grunewald, K.; Zhiyanski, M.; Andonowski, V.; La Porta, N.; et al. Drivers o treeline shift in different European Mountains. Contribution to CR Special 34 SENSFOR: Resilience in SENSItive mountain FORest ecosystems under environmental change. *Clim. Res.* **2017**, *73*, 135–150.

199. Miller, A.E.; Wilson, T.L.; Sherriff, R.L.; Walton, J. Warming drives a front of white spruce establishment near western treeline, Alaska. *Glob. Chang. Biol.* **2017**, *23*, 5509–5522. [CrossRef]

200. Wielgolaski, F.E.; Hofgaard, A.; Holtmeier, F.-K. Sensitivity to environmental change of the treeline ecotone and its associated biodiversity in European mountains. *Clim. Res.* **2017**, *73*, 151–166. [CrossRef]

201. Brown, C.D.; Dufour Tremblay, G.; Jameson, R.G.; Mamet, S.; Trant, A.J.; Walker, X.J.; Boudreau, S.; Harper, K.; Henry, G.H.R.; Hermanutz, L.; et al. Reproduction as a bottleneck to treeline advance across the circumarctic forest tundra ecotone. *Ecography* **2018**. [CrossRef]

202. Jochner, M.; Bugmann, H.; Nötzli, M.; Bigler, C. Among-tree variability and feedback effects result in different growth responses to climate change at the upper treeline in the Swiss Alps. *Ecol. Evol.* **2017**, *7*, 7937–7953. [CrossRef] [PubMed]

203. Kullman, L.; Öberg, L. A one hundred-year study of the upper limit of tree growth (Terminus arboreus) in the Swedish Scandes—Updated and illustrated change in an historical perspective. *Int. J. Res. Geogr.* **2018**, *4*, 10–25.

204. Cazolla Gatti, R.; Velichevskaja, A.; Dudko, A.; Fabbio, L.; Battipaglia, G.; Liang, J. Accelerating upward treeline shift in the Altai Mountains under last-century climate change. *Sci. Rep.* **2019**, *9*, 7678. [CrossRef]

205. Hofgaard, A.; Ols, C.; Drobyshev, I.; Kirchhefer, A.; Sandberg, S.; Söderström, L. Non-stationary response of tree growth to climate trends along the Arctic margin. *Ecosystems* **2019**, *22*, 434–451. [CrossRef]

206. Nichols, H. *Palynological and Paleoclimatic Study of the Late Quaternary Displacement of the Boreal Forest-Tundra Ecotone in Keewatin and Mackenzie N.W.T., Canada*; Institute of Arctic and Alpine Research, University of Colorado: Boulder, CO, USA, 1975.

207. Vorren, K.-D.; Mørkved, B.; Bortenschlager, S. Human impact on the Holocene forest line in the Central Alps. *Veg. Hist. Archaeobotany* **1993**, *2*, 145–156. [CrossRef]

208. Aas, B.; Faarlund, T. The present and the Holocene birch belt in Norway. In *Paleoclimate Research*; Frenzel, B., Alm, V., Eds.; Fischer: Innsbruck, Austria, 1996; pp. 19–42.

209. Kullman, L.; Kjällgren, L. A coherent postglacial tree-limit chronology (Pinus sylverstris L.) for the Swedsish Scandes: Aspects of paleoclimatic and "recent warming", based on megafossil evidence. *Arct. Antarct. Alp. Res.* **2000**, *32*, 419–428.

210. Gervais, B.R.; Macdonald, G.M.; Snyder, J.A.; Kremenetski, C.V. Pinus sylvestris treeline development and movement on the Kola Peninsula of Russia: Pollen and stomate evidence. *J. Ecol.* **2002**, *90*, 627–638. [CrossRef]

211. Tinner, W.; Theurillat, J.-P. Uppermost limit, extent, and fluctuations of the timberline and treeline ecocline in the Swiss Alps during the past 11.500 years. *Arct. Antarct. Alp. Res.* **2003**, *35*, 158–169. [CrossRef]

212. Esper, J.; Shiyatov, S.G.; Mazepa, V.S.; Wilson, R.J.S.; Graybill, D.A.; Funkhouser, G. Temperature-sensitive Tien Shan tree ring chronologies show multi-centennial growth trends. *Clim. Dyn.* **2003**, *21*, 699–706. [CrossRef]

213. Tinner, W.; Kaltenrieder, P. Rapid responses of high-mountain vegetation to early Holocene environmental changes in the Swiss Alps. *J. Ecol.* **2005**, *93*, 936–947. [CrossRef]

214. Helama, S.; Timonen, M.; Holopainen, J.; Ogurtsov, M.; Mielikäinen; Eronen, M.; Lindholm, M.; Meriläinen, J. Summer temperature variations in Lapland during the Medieval Warm Period and the Little Ice Age relative to natural instability of multi-decadal and multi-centennial scales. *J. Quat. Sci.* **2009**, *24*, 450–456. [CrossRef]

215. Brown, A.G.; Hatton, J.; Selby, K.A.; Leng, N.J.; Christie, N. Multi-proxy study of Holocene environmental change and human activity in the Central Apennine Mountains, Italy. *J. Quat. Sci.* **2013**, *28*, 71–82. [CrossRef]

216. Kononov, Y.M.; Friedrich, M.; Boettger, T. Regional summer temperature reconstruction in the Khibiny low mountains (Kola Pensinsula, NW Russia) by means of tree-ring width during the last four centuries. *Arct. Antarct. Alp. Res.* **2009**, *41*, 460–468. [CrossRef]

217. Mazepa, V.S. Climate dependent dynamics of the upper treeline ecotone in the Polar Urals for the last millennium. In Proceedings of the XIII World Forestry Congress, Buenos Aires, Argentina, 18–23 October 2009.

218. Väliranta, M.; Kaakinen, A.; Kuhry, P.; Kultti, S.; Salonen, J.S.; Seppä, H. Scattered late-glacial and early Holocene tree populations as dispersal nuclei for forest development in north-eastern European Russia. *J. Biogeogr.* **2010**, *38*, 922–932. [CrossRef]

219. Staffler, H.; Nicolussi, K.; Patzelt, G. Postglaziale Waldgenzentwicklung in den Westtiroler Zentralalpen. *Gredleriana* **2011**, *11*, 93–114.

220. Magyari, E.K.; Jakab, G.; Bálint, M.; Kern, Z.; Buczko, K.; Braun, M. Rapid vegetation response to Lateglacial and early Holocene climatic fluctuation in the South Carpathian Mountains (Romania). *Quat. Sci. Rev.* **2012**, *32*, 116–130. [CrossRef]

221. Hofgaard, A.; Tømmervik, H.; Rees, G.; Hansen, F. Latitudinal forest advance in northernmost Norway since the early 20th century. *J. Biogeogr.* **2013**, *40*, 936–949. [CrossRef]

222. Badino, F.; Ravazzi, C.; Vallè, F.; Pini, R.; Aceiti, A.; Brunetti, M.; Champvillair, E.; Maggi, V.; Maspero, F.; Prego, E.; et al. 8800 years of high-altitude vegetation and climate history at the Rutor Glacier forefield, Italian Alps. Evidence of middle Holocene timberline rise and glacier contraction. *Quat. Sci. Rev.* **2018**, *195*, 41–68. [CrossRef]

223. Leúnda, M.; González-Sampéris, P.; Gil-Romera, G.; Bartolomé, M.; Belmonte-Ribas, A.; Goméz-García, D.; Kaltenrieder, P.; Rubiales, J.M.; Schwörer, C.; Tinner, W.; et al. Ice cave reveals environmental forcing of long-term Pyrenean tree line dynamics. *J. Ecol.* **2018**. [CrossRef]

224. Li, K.; Liao, M.; Ni, J.; Liu, X.; Wang, Y. Treeline composition and biodiversity change on the southeastern Tibetan Plateau during the past millennium inferred from high-resolution alpine pollen record. *Quat. Sci. Rev.* **2019**, *206*, 11–55. [CrossRef]

225. Autio, J.; Colpaert, A. The impact of elevation, topography and snow load damage of trees on the position of the actual timberline on the fells in central Finnish Lapland. *Fennia* **2005**, *183*, 11–52.

226. Schickhoff, U. The upper timberline in the Himalayas, Hindu Kush and Karakorum: A review of geographical and ecological aspects. In *Mountain Ecosystems. Studies in Treeline Ecology*; Broll, G., Keplin, B., Eds.; Springer: Berlin/Heidelberg, Germany; New York, NY, USA, 2005; pp. 275–354.

227. Resler, L.M. Geomorphic controls of spatial pattern and process at alpine treeline. *Prof. Geogr.* **2006**, *58*, 124–138. [CrossRef]

228. Brandes, R. *Waldgrenzen griechischer Hochgebirge unter besonderer Berücksichtigung des Taygetos, Südpeleponnes (Waldgrenzdynamik, dendrochronologische Untersuchungen)*; Erlanger Geographische Arbeiten: Erlangen, Deutschland, 2007.

229. Butler, D.R.; Malanson, G.P.; Walsh, S.J.; Fagre, D.B. Influence of geormorphology and geology on alpine treeeline in the American West—More important than climatic influences? *Phys. Geogr.* **2007**, *28*, 434–450. [CrossRef]

230. Malanson, G.P.; Brown, D.G.; Butler, D.R.; Cairns, D.M.; Fagre, D.B.; Walsh, S.J. Ecotone dynamics: Invasibilty of alpine tundra by tree species from the subalpine forest. In *The Changing Alpine Treeline: The Example of Glacier National Park, MT, USA: Developments in the Earth Surface Processes*; Butler, D.R., Malanson, G.P., Walsh, S.J., Fagre, D.B., Eds.; Elsevier: Boston, Heidelberg, 2009; pp. 35–61.

231. Holtmeier, F.-K.; Broll, G. Landform influences on treeline patchiness and dynamics in a changing climate. *Phys. Geogr.* **2012**, *33*, 403–437. [CrossRef]

232. Kaczka, R.J.; Czajka, B.; Łajczak, A.; Swagrzyk, J.; Nicia, P. The timberline as a result of the interactions among forest, abiotic environment and human activity in Baba Góra Mt., Western Carpathians. *Geogr. Pol.* **2015**, *88*, 177–191. [CrossRef]

233. Drollinger, S.; Müller, M.; Kobl, T.; Schwab, N.; Böhner, J.; Schickhoff, U.; Scholten, T. Decreasing nutrient concentration in soils and trees with increasing elevation across a treeline ecotone in Rolwaling Himal, Nepal. *J. Mt. Sci.* **2017**, *14*, 843–858. [CrossRef]

234. Sibold, J.S.; Veblen, T.T.; Gonzalez, M.E. Spatial and temporal variation on fire regimes in subalpine forests across the Colorado Front Range in Rocky Mountain National Park, Colorado, USA. *J. Biogeogr.* **2006**, *33*, 631–647. [CrossRef]

235. Potthoff, K. Grazing history affects the treeline-ecotone: A case study from Hardanger, western Norway. *Fennia* **2009**, *18*, 31–98.

236. Staland, H.; Salmonsson, J.; Hörnberg, G. A thousand years of human impact in the northern Scandinavian mountain range: Long-lasting effects on forest lines and vegetation. *Holocene* **2010**, *21*, 379–391. [CrossRef]

237. Cunill, R.; Soriana, J.M.; Bal, M.C.; Pèlachs, A.; Rodrígez, J.M.; Pérez-Obiol, T. Holocene high-altitude vegetation dynamics in the Pyrenees: A pedoanthroecology contribution to an interdisciplinary approach. *Quat. Int.* **2013**, *284*, 60–79. [CrossRef]

238. Garbarino, M.; Lingua, L.; Weisberg, P.J.; Boltero, A.; Meloni, F.; Motta, R. Land-use history and topographic gradients as driving factors of subalpine Larix decidua forests. *Landsc. Ecol.* **2013**, *28*, 305–817. [CrossRef]

239. Palombo, C.; Chirici, G.; Marchetti, M.; Tognetti, R. Is land abandonment affecting forest dynamics at high elevation in Mediterranean mountains more than climate change? *Plant Biosyst.* **2013**, *147*, 1–11. [CrossRef]

240. Higuera, P.E.; Briles, C.E.; Witlock, C. Fire-regime complacency amd sensitivity to centennial-through millienial-scale climate change in the Rocky Mountain subalpine forest, Colorado, USA. *J. Ecol.* **2014**, *102*, 1429–1441. [CrossRef]

241. Ameztegui, A.; Coll, L.; Brotons, L.; Ninot, J.M. Land-use legacies rather than climate change are driving the recent upward shift of the mountain treeline in the Pyrenees. *Glob. Ecol. Biogeogr.* **2015**, *25*, 263–273.

242. Masseroli, A.; Leonelli, G.; Bollati, I.; Trombino, L.; Pelfini, M. The influence of geomorphological processes on the treeline position in Upper Valtellina (Central Italian Alps). *Geogr. Fis. E Din. Quat.* **2016**, *39*, 171–182.

243. Cansler, C.A.; McKenzie, D.; Halpern, C.B. Fire enhances the complexity of forest structures in alpine treeline ecotones. *Ecosphere* **2018**, *9*, e02091. [CrossRef]

244. Singh, N.; Tewari, A.; Sah, S. Tree regeneration pattern and size class distribution in anthropogenically disturbed subalpine treeline areas of Indian Western Himalaya. *Int. J. Sci. Technol. Res.* **2019**, *8*, 537–546.

245. Vitali, A.; Garbarino, M.; Camarero, J.J.; Malandra, F.; Toromani, E.; Spalevic, V.; Čurović, M.; Urbinati, C. Pine recolonization dynamics in Mediterranean human-disturbed treeline ecotones. *For. Ecol. Manag.* **2019**, *435*, 28–37. [CrossRef]

246. Butler, D.R.; Malanson, G.P.; Cairns, D.M. Stability of alpine treeline in Montana, USA. *Phytocoenologia* **1994**, *22*, 485–500. [CrossRef]

247. Holtmeier, F.-K. Change in the timberline ecotone in northern Finnish Lapland during the last thirty years. *Rep. Kevo Subarct. Res. Stn.* **2005**, *23*, 97–113.

248. Roush, W.; Munroe, J.S.; Fagre, D.B. Development of a spatial analysis method using ground-based repeat photography to detect changes in the alpine treeline ecotone, Glacier National Park, Montana, U.S.A. *Arct. Antarct. Alp. Res.* **2007**, *39*, 297–308. [CrossRef]

249. Jacob, M.; Frankl, A.; Beeckman, H.; Mesfin, G.; Hendrickx, M.; Guyassa, E.; Nyssen, J. North Ethiopian afro-alpine treeline dynamics and forest-cover change since the early 20th century. *Land Degrad. Dev.* **2014**. [CrossRef]

250. Kullman, L. Recent treeline shift on Kebnekaise Mountains, northern Sweden – A climate change case. *Int. J. Curr. Res.* **2018**, *10*, 63736–63792.

251. Kullman, L. A review and analysis of factual change on the max rise of the Swedish Scandes treeline in relation to climate. *J. Ecol. Nat. Resour.* **2018**, *2*. [CrossRef]

252. Brown, D.G.; Cairns, D.M.; Malanson, G.P.; Walsh, S.J.; Butler, D.R. Remote sensing and GIS techniques for spatial and biophysical analyses of alpine treeline through process and empirical models. In *Environmental Information Management and Analysis*; Michener, W.K., Stafford, J., Bruns, J., Eds.; Taylor&Francis: London, UK, 1994; pp. 453–481.

253. Allen, T.R.; Walsh, S.J. Spatial and compositional pattern of alpine treeline, Glacier National Park, Montana. *Photogram. Eng. Remote Sens.* **1996**, *62*, 1261–1268.

254. Crawford, R.M.M.; Jeffree, C.E.; Rees, W.G. Paludification and forest retreat in northern oceanic environments. *Ann. Bot.* **2003**, *91*, 213–226. [CrossRef] [PubMed]

255. Rees, G.; Brown, I.; Mikkola, K.; Virtanen, T.; Werkman, B. How can the dynamics of the tundra-taiga boundary be remotely monitored? *Ambio Spec. Rep.* **2002**, *12*, 56–62.

256. Broll, G.; Holtmeier, F.-K.; Anschlag, K.; Brauckmann, H.-J.; Wald, S.; Drees, B. Landscape mosaic in the treeline ecotone on Mt. Rodjanoaivi, Subarctic Finland. *Fennia* **2007**, *185*, 89–105.

257. Dial, R.J.; Berg, E.E.; Timm, K.; McMahon, A.; Geck, J. Changes in the alpine forest-tundra ecotone commensurate with recent warming in southcentral Alaska: Evidence from orthophotos and field plots. *J. Geophys. Res.* **2007**, *112*. [CrossRef]

258. Bekker, M.F.; Malanson, G.P. Modeling feedback effects on linear patterns of subalpine forest advancement. In *The Changing Alpine Treeline—Development in Earth Surface Processes*; Butler, D.R., Malanson, G.P., Walsh, S.J., Fagre, D.B., Eds.; Elsevier: Heidelberg, Germany, 2009; pp. 167–190.

259. Walsh, S.J.; Brown, D.G.; Geddes, C.A.; Weiss, D.J.; McKnight, S.; Hammer, E.S.; Tuttle, J.P. Pattern-process relations in the alpine and subalpine environmets: A remote sensing and GISscience perspective. In *The Changing Alpine Treeline: The Example of Glacier National Park, MT, USA—Developments in Earth Surface Processes*; Butler, D.R., Malanson, G.P., Walsh, S.J., Fagre, D.B., Eds.; Elsevier: Heidelberg, Germany, 2009; pp. 11–34.

260. Olthof, I.; Pouliot, D. Treeline vegetation composition and damage in Canada's western Subarctic AVHRR and canopy reflectance modeling. *Remote Sens. Environ.* **2010**, *114*, 805–815. [CrossRef]

261. Stueve, K.; Isaacs, R.E.; Tyrrell, L.; Densmore, R.V. Spatial variability of biotic and abiotic tree establishment constraints across a treeline ecotone in the Alaska Range. *Ecology* **2011**, *92*, 496–506. [CrossRef]

262. Simms, É.L.; Ward, H. Multisensor NDVI-based monitoring of the tundra-taiga interface (Mealy Mountains, Labrador, Canada). *Remote Sens.* **2013**, *5*, 1066–1090. [CrossRef]

263. Salzer, M.W.; Larson, E.R.; Bunn, A.G.; Hughes, M.K. Changing climate response near-treeline bristlecone pine with elevation and aspect. *Environ. Res. Lett.* **2014**, *9*. [CrossRef]

264. Gaire, N.P.; Koirala, M.; Bhuju, D.R.; Carrer, M. Site- and site-specific treeline responses to climate variability in eastern Nepal Himalaya. *Dendrochronologia* **2016**, *41*, 44–56. [CrossRef]

265. Chhetri, P.K.; Shresta, K.B.; Cairns, D.M. Topography and human disturbances are major controlling factors in treeline pattern at Barun and Manang area in the Nepal Himalaya. *J. Mt. Sci.* **2017**, *14*, 119–127. [CrossRef]

266. Dinca, L.; Nita, M.D.; Hofgaard, A.; Alados, C.L.; Broll, G.; Borz, S.A.; Wertz, B.; Monteiro, A.T. Forest dynamics in the montane-alpine boundary: A comparative study using satellite imagery and climate data. *Clim. Res.* **2017**, *73*, 97–110. [CrossRef]

267. Potthoff, K. Spatiotemporal patterns of birch regrowth in a Western Norwegian treeline ecotone. *Landsc. Res.* **2017**, *42*, 63–77. [CrossRef]

268. Latwal, A.; Sah, P.; Sharma, S. A cartographic representation of a timberline, treeline and woody vegetation around a central himalayan summit using remote sensing method. *Trop. Ecol.* **2018**, *59*, 177–186

269. Chhetri, P.K.; Thai, E. Remorte sensing and geographic information systems techniques in studies on treeline ecotone dynamics. *J. For. Res.* **2019**. [CrossRef]

270. Morley, P.J.; Donoghue, D.; Chan, J.-C.; Jump, A. Quantifying structural diversity to better estimate change at mountain forest margins. *Remote Sens. Environ.* **2019**, *223*, 291–306. [CrossRef]

271. Rawat, D.S. Monitoring ecosystem boundaries in the Himalayas through an 'eye in the sky'. *Curr. Sci.* **2012**, *1012*, 1352–1354.

272. Meentemeyer, V.; Box, E.O. Scale effects in landscape studies. In *Landscape Heterogeneity and Disturbance. Ecological Studies*; Turner, M.G., Ed.; Springer: New York, NY, USA, 1987; pp. 15–34.

273. Holtmeier, F.-K.; Broll, G. Treelines—Approaches at different scales. *Sustainability* **2017**, *9*, 808. [CrossRef]
274. Turner, H. Maximaltemperaturen oberflächennaher Bodenschichten an der alpinen Waldgrenze. *Wetter und Leben* **1958**, *10*, 1–12.
275. Friedel, H. Schneedecken-Andauer und Vegetations-Verteilung im Gelände. *Mitt. der Forstl. Bundesvers. Mariabrunn* **1961**, *59*, 317–368.
276. Aulitzky, H. Bioklima und Hochlagenaufforstung in der subalpinen Stufe der Inneralpen. *Schweiz. Z. Forstwes.* **1963**, *114*, 125.
277. Friedel, H. Verlauf der alpinen Waldgrenze im Rahmen anliegender Gebirgsgelände. *Mitt. der Forstl. Bundesvsers. Wien* **1967**, *75*, 79–172.
278. Holtmeier, F.-K. Die bodennahen Winde in den Hochlagen der Indian Peaks Section (Colorado Front Range). *Münstersche Geogr. Arb.* **1978**, *3*, 4–47.
279. Burns, S.F. Alpine Soil distribution and Development. Ph.D. Thesis, University of Colorado, Boulder, CO, USA, 1980.
280. Burns, S.F.; Tonkin, P.J. Soil-geomorphic models and the spatial distribution and development of alpine soils. In *Space and Time in Geomorphology*; Thorn, C.E., Ed.; Allen and Unwin: London, UK, 1982; pp. 25–43.
281. Walsh, S.J.; Butler, D.R.; Allen, T.R.; Malanson, G.P. Influence of snow pattern and snow avalanches on the alpine treeline ecotone. *J. Veg. Sci.* **1994**, *5*, 657–672. [CrossRef]
282. Walder, U. Ausaperung und Vegetationsverteilung im Dischmatal. *Mitt. der Eidg. Anst. für fas Forstl. Vers.* **1983**, *59*, 80–2012.
283. Holtmeier, F.-K. Die Wirkungen des Windes in der subalpinen und alpinen Stufe der Front Range, Colorado, U.S.A. In *Beiträge aus den Arbeitsgebieten am Institut für Landschaftsökologie Münster*; Holtmeier, F.-K., Ed.; Institut für Landschaftsökologie: Münster, Germany, 1996; pp. 19–45.
284. Butler, D.R.; Malanson, G.P.; Bekker, M.F.; Resler, L.M. Lithologic structural and geomorphic controls on ribbon forest patterns in a glaciated mountain environment. *Geomorphology* **2003**, *55*, 203–217. [CrossRef]
285. Millar, C.I.; Westfall, R.D.; Delay, D.L.; King, J.C.; Graumlich, L.-J. Response of subalpine conifers in the Sierra Nevada, California, U.S.A., to 20th century warming and decadal climate variability. *Arct. Antarct. Alp. Res.* **2004**, *36*, 181–200. [CrossRef]
286. Kullman, L. Trädgränsen i Dalafjällen. Del I. Gamla og nya träd på hög nivå. *Svensk Bot. Tidskrift* **2005**, *99*, 315–330.
287. Resler, L.M.; Butler, D.R.; Malanson, G.P. Topographic shelter and conifer establishment and mortality in an alpine environment, Glacier National Parkk, Montana. *Phys. Geogr.* **2005**, *26*, 112–125. [CrossRef]
288. Malanson, G.P.; Butler, D.R.; Fagre, D.B.; Walsh, S.J.; Tomback, D.F.; Daniels, L.D.; Resler, L.M.; Smith, W.K.; Weiss, D.J.; Peterson, D.L.; et al. Linking organism-to-landscape dynamics. *Phys. Geogr.* **2007**, *28*, 378–396. [CrossRef]
289. Bekker, M.F.; Clark, J.; Jackson, M.W. Landscape metrics indicate differences in patterns and dominant control of ribbon forests in the Rocky Mountains, USA. *Appl. Veg. Sci.* **2009**, *12*, 237–249. [CrossRef]
290. Butler, D.R.; Malanson, G.P.; Resler, L.M.; Walsh, S.J.; Wilkerson, F.D.; Schmid, G.L.; Sawyer, C.F. Geomorphic patterns and processes at alpine treeline. In *The Changing Alpine Treeline: The Example of Glacier National Park, MT, USA—Developments in Earth Surface Processes*; Butler, D.R., Malanson, G.P., Walsh, S.J., Fagre, D.B., Eds.; Elsevier: Heidelberg, Germany, 2009; pp. 63–84.
291. Nicklen, E.F.; Roland, C.A.; Ruess, R.W.; Schmidt, J.H.; Lloyd, A.H. Local site conditions drive climate-growth responses of Picea mariana and Pincea glauca in interior Alaska. *Ecosphere* **2016**, *7*, e01507. [CrossRef]
292. Yu, D.; Wang, W.Q.; Liu, J.; Zhou, W.; Qi, L.; Wang, X.; Zhou, L.; Dai, L. Formation mechanisms at the alpine Erman's birch (Betula ermanii treeline on Changbai Mountain in Northeast China. *Trees* **2014**, *28*, 935–947. [CrossRef]
293. Dearborn, K.D.; Danby, R.K. Spatial analysis of forest-tundra ecotones reveals the influence of topography and vegetation in alpine timberline patterns in the Subarctic. *Ann. Am. Assoc. Geogr.* **2019**.
294. Neuwinger, I. Böden der subalpinen und alpinen Stufe in den Tiroler Alpen. *Mitt. Ostalpin-Dinarische Ges. für Veg.* **1970**, *11*, 135–150.
295. Blaser, P. Der Boden als Standortsfaktor bei Aufforstungen in der subalpinen Stufe (Stillberg, Davos). *Schweiz. Anst. für fas Forstl. Vers.* **1980**, *56*, 529–611.
296. Aulitzky, H.; Turner, H.; Mayer, H. Bioklimatische Grundlagen einer standortgemäßen Bewirtschaftung des subalpinen Lärchen-Arvenwaldes. *Eidgenössische Anst. fürdDas Forstl. Vers.* **1982**, *58*, 327–577.

297. Schönenberger, W. Performance of high altitude afforestation under various site conditions. In *Proceedings 3rd IUFRO Workshop*; Turner, H., Tranquillini, W., Eds.; Eidgenössische Anstalt für das forstliche Versuchswesen: Birmensdorf, Switzerland, 1985; Berichte 270; pp. 233–240.

298. Neuwinger, I.; Wieser, G.; Winklehner, W. Bodenwasseruntersuchungen in einer Hochlagenaufforstung bei Haggen im Sellraintal, Tirol. *Oesterr. Wasser Abfallwirtsch.* **1988**, *40*, 57–61.

299. Nimlos, T.J.; McConnell, R.C. Alpine soils in Montana. *Soil Sci.* **1965**, *99*, 310–321. [CrossRef]

300. Czell, A. Beitrag zum Wasserhaushalt subalpiner Böden. *Mitt. der Forstl. Bundesversuchsanstalt Wien* **1967**, *1*, 305–332.

301. Johnson, P.D.; Cline, A.J. Colorado mountain soils. *Adv. Agron.* **1965**, *17*, 233–287.

302. Blaser, P.; Reiser, M. Eine topographische Bodensequenz in subalpinen Lawinengassen auf Silikatgestein. *Mitt. der Eidg. Anst. für das Forstl. Vers.* **1975**, *51*, 1199–1214.

303. Neuwinger, I. Bodenökologische Untersuchungen im Gebiet Obergurgler Zirbenwald Hohe Mut. In *MaB-Projekt Obergurgl—Veröffentlichungen des Österreichischen MaB-Programm*; Patzelt, G., Ed.; Universitätsverlag Wagner: Innsbruck, Austria, 1987; pp. 173–190.

304. Litaor, M.I. The influence of eolian dust on the genesis of alpine soils in the Front Range, Colorado. *Soil Sci. Soc. Am. J.* **1987**, *51*, 142–147. [CrossRef]

305. Bowman, W.D.; Theodose, T.A.; Schardt, J.C.; Conant, R.T. Constraints on nutrient availability on primary production in two alpine tundra communities. *Ecology* **1993**, *74*, 2085–2097. [CrossRef]

306. Weiss, L.; Shiels, A.B.; Walker, L.T. Soil impacts of bristlecone pine (Pinus longaeva) tree islands on alpine tundra, Charleston Peak, Nevada. *West. North Am. Nat.* **2001**, *65*, 536–540.

307. McNown, R.W.; Sullivan, P.F.; Turnball, M. Low photosynthesis of treeline white spruce is associated with limited soil nitrogen availability in the western Brooks Range, Alaska. *Funct. Ecol.* **2013**, *27*, 672–683. [CrossRef]

308. Bockheim, J.G.; Munroe, J.S. Organic carbon and genesis of alpine soils with permafrost: A review. *Arct. Antarct. Alp. Res.* **2014**, *46*, 987–1006. [CrossRef]

309. Müller, M.; Schickhoff, U.; Scholten, T.; Drollinger, S.; Böhner, J.; Chaudhary, R.P. How do soil properties affect alpine treelines? General principles in a global perspective and novel findings from Rolwaling Himal, Nepal. *Prog. Phys. Geogr.* **2016**, *40*, 135–160. [CrossRef]

310. Moscatelli, M.C.; Bonifacio, E.; Chiti, T.; Cudlin, P.; Dinca, L.; Grego, S.; LaPorta, N.; Karlinski, L.; Pellis, G.; Rudawska, M.; et al. Soil properties as indicators of treeline dynamics in relation to anthropogenic pressure and climate change. *Clim. Res.* **2017**, *73*, 73–84. [CrossRef]

311. Davis, E.L.; Hager, H.A.; Gedalof, Z. Soil properties as constraints to seedling regeneration beyond alpine timberlines in the Canadian Rocky Mountains. *Arct. Antarct. Alp. Res.* **2018**, *50*. [CrossRef]

312. Liu, Y.; Wang, L.; He, R.; Chan, Y.; Xu, Z.; Tan, B.; Zhang, L.; Xiao, J.; Zhu, P.; Chen, L.; et al. Higher soil fauna abundance accelerates litter carbon release across an alpine forest-tundra ecotone. *Sci. Rep.* **2019**, *9*, 10561. [CrossRef] [PubMed]

313. FAO. *FAO-report. Understanding Mountain Soils: A Contribution from Mountain Areas to the International Year of Soils 2015*; Food and Agriculture Organization of the United Nations: Rome, Italy, 2015.

314. Egli, M.; Poulenard, J. Soils of mountainous landscapes. *Encycl. Geogr.* **2016**. [CrossRef]

315. Stöhr, D. Heterogeneity at a microcale. In *Trees at Their Upper Limit. Treelife Limitation at the Alpine Timberline—Plant Ecophysiology*; Wieser, G., Tausz, M., Eds.; Springer: Dordrecht, The Netherlands, 2007; pp. 37–56.

316. Holtmeier, F.-K. *Die Höhengrenze der Gebirgswälder*; Arb. aus dem Inst. für Landschaftsökologie der Westfälischen Wilhelms-Univ.: Münster, Germany, 2000; Volume 8, p. 337.

317. Pearson, R.G.; Phillips, S.J.; Loranty, M.M.; Neck, P.S.A.; Damoulas, T.; Knight, S.J.; Goetz, S.J. Shifts in Arctic vegetation and associated feedbacks under climate change. *Nat. Clim. Chang.* **2013**, *3*, 673–677. [CrossRef]

318. Liang, E.; Dawadi, B.; Pederson, N.; Eckstein, D. Is the growth of birch at the upper timberline in the Himalayas limited by moisture or by temperature? *Ecology* **2014**, *95*, 2453–2465. [CrossRef]

319. Roland, C.A.; Schmidt, J.H.; Johnstone, F. Climate sensitivity of reproduction in a mast-seeding boreal conifer across its distribution range from lowland to treeline forests. *Oecologia* **2014**, *174*, 665–677. [CrossRef]

320. Winkler, D.E.; Chapin, K.J.; Kueppers, L. Soil moisture mediates alpine life form and community productivity responses to warming. *Ecology* **2016**, *97*, 1533–1563. [CrossRef]

321. Franke, A.K.; Bräuning, A.; Timonen, M.; Rautio, P. Growth response of Scots pine in polar-alpine tree-line to a warming climate. *For. Ecol. Manag.* **2017**, *399*, 94–107. [CrossRef]

322. Kueppers, L.M.; Conlisk, E.; Castanha, C.; Moyes, A.; Germino, M.J.; de Valpine, P.; Torn, M.S.; Mitton, J. Warming and provenance limit tree recruitment across and beyond the elevation range of subalpine forests. *Glob. Chang. Biol.* **2017**, *23*, 2383–2395. [CrossRef]

323. Millar, C.I.; Westfall, R.D.; Delany, R.L.; Flint, A.L.; Flint, L.E. (2015): Recruitment patterns and growth of high-elevation pines in response to climatic variability (1883-2013) in the western Great Basin, USA. *Can. J. For. Res.* **2015**, *45*, 1299–1312. [CrossRef]

324. Andrus, R.A.; Harvey, B.J.; Rodman, K.C.; Hart, S.J.; Veblen, T.T. Moisture availability limit subalpine tree establishment. *Ecology* **2018**, *99*, 567–575. [CrossRef] [PubMed]

325. Berner, L.T.; Beck, P.S.; Bunn, A.G.; Goetz, S.J. Plant response to climate change along the forest-tundra ecotone in northeastern Siberia. *Glob. Chang. Biol.* **2018**, *19*, 3449–3462. [CrossRef] [PubMed]

326. Timoney, K.P.; Mamet, S.D.; Cheng, R.; Lee, P.; Robinson, A.L.; Downing, D.; Wein, R. Tree cover response to climate change in the forest-tundra ecotone of north-central Canada: Fire-driven decline, not northward advance. *Ecoscience* **2018**, *26*, 133–148. [CrossRef]

327. De Boer, H.J.; Roberston, I.; Clisby, R.; Loader, N.J.; Gagen, M.M.; Giles, H.F.; Young, G.H.F.; Wagner-Cremer, F.; Hipkin, C.R.; McCarroll, D. Tree-ring isotopes suggest atmospheric drying limits temperature-growth responses of treeline bristlecone pine. *Tree Physiol.* **2019**, *39*, 983–999. [CrossRef] [PubMed]

328. Choat, B.; Brodripp, T.; Brodersen, C.R.; Duursma, R.A.; López, R.; Medin, B.E. Triggers of tree mortality under drought. *Nature* **2018**, *558*, 531–539. [CrossRef]

329. Beniston, M. The effect of global warming on mountain regions: A summary of the 1995 Report of the Intergovernmental Panel on Climate Change. In *Global Change and Protected Areas—Advances in Global Change Research*; Viskonti, G., Beniston, M., Ianovelli, E.D., Varba, D., Eds.; Springer: New York, NY, USA, 2001; pp. 155–185.

330. Reardon, B.A.; Pederson, G.T.; Caruso, C.J.; Fagre, D.B. Spatial reconstructions and comparisons of historic snow avalanche frequency and extent using tree rings in Glacier National Park, Montana, U.S.A. *Arct. Antarct. Alp. Res.* **2008**, *40*, 148–160. [CrossRef]

331. IPCC. Climate Change 2014: Synthesis Report. In *Contribution of Working Groups I, II and III to the Fifth Assessment Report of the Intergovernmental Panel on Climate Change*; Core Writing Team, Pachauri, R.K., Meyer, L.A., Eds.; IPCC: Geneva, Switzerland, 2014; 151p.

332. Ballestero-Canovas, J.A.; Trappmann, D.; Madrigal-Gonzáles, J.; Eckert, N.; Stoffel, M. Climate warming enhances snow avalanche risk in the Western Himalayas. *Proc. Natl. Acad. Sci. USA* **2018**, *115*, 3410–3415. [CrossRef]

333. Treml, V.; Veblen, T.T. Does tree growth sensitivity to warming trends vary according to treeline form? *J. Biogeogr.* **2017**. [CrossRef]

334. Black, R.A.; Bliss, L.C. Reproductiion ecology of Picea mariana (Mill,) BSP. at treeline near Inuvik, Northwest Territories, Canada. *Ecol. Monogr.* **1980**, *50*, 331–354. [CrossRef]

335. Brito, P.; Lorenzo, J.R.; González-Rodriguez, A.M.; Morales, P.; Wieser, G.; Jiménez, M.S. Canopy transpiration of a semi arid Pinus canariensis forest at a treeline ecotone in two hydrologically contrasting years. *Agric. For. Meteorol.* **2015**, *201*, 120–127. [CrossRef]

336. Wieser, G.; Holtmeier, F.-K.; Smith, W.K. Treelines in a changing global environment: Ecophysiology, adaptation and future survival. In *Plant Ecophysiology*; Tausz, N., Grulke, N., Eds.; Springer: Dordrecht, The Netherlands, 2014; pp. 221–263.

337. Wieser, G.; Brito, P.; Lorenzo, J.R.; González-Rodriguez, A.M.; Morales, D.; Jiménez, M.S. Canary island pine (Pinus canariensis) and evergreen species in a semiarid treeline. In *Progress in Botany*; Canovas, F., Luettge, R., Matyssek, R., Eds.; Springer International Publishing: Cham, Switzerland, 2016; pp. 416–435.

338. Lazarus, B.E.; Castanha, C.; Germino, M.J.; Kueppers, L.M.; Moyes, A.B. Growth strategies and threshold responses to water deficit modulate effects of warming on tree seedlings from forest to alpine. *J. Ecol.* **2018**, *106*, 571–585. [CrossRef]

339. Höllermann, P.W. Geoecological aspects of the upper timberline in Tenerife, Canary Islands. *Arct. Alp. Res.* **1978**, *19*, 365–382. [CrossRef]

340. Leuschner, C.; Schulte, M. Microclimatical investigations in the tropical alpine scrub of Maui, Hawaii: Evidence for a drought-induced alpine timberline. *Pac. Sci.* **1991**, *45*, 152–168.

341. Gieger, T.; Leuschner, C. Altitudinal change in needle water relations of Pinus canariensis and possible evidence of a drought-induced alpine timberline on Mt. Teide, Tenerife. *Flora Morphol. Distrib. Funct. Ecol. Plants* **2004**, *199*, 100–109. [CrossRef]

342. Bello-Rodríguez, V.; Cubas, J.; Del Arco, M.J.; Martín, J.L.; González-Macebo, J.M. Elevational and structural shifts in the treeline of an oceanic island (Tenerife, Canary islands) in the context of global warming. *Int. J. Appl. Earth Obs. Geoinf.* **2019**, *82*. [CrossRef]

343. Karger, D.N.; Kessler, M.; Conrad, D.; Weigelt, P.; Kreft, H.; König, C.; Zimmermann, N.E. Why treelines are lower on islands—Climatic and biographic effects hold the answer. *Glob. Ecol. Biogeogr.* **2019**. [CrossRef]

344. Rehfeldt, G.E.; Worrall, J.J.; Marchettti, S.B.; Crookston, N.L. Adaptive forest management to climate change using bioclimate models with topographic drivers. *Forestry* **2015**, *88*, 528–539. [CrossRef]

345. McClung, D.; Schaerer, P. *The Avalanche Handbook*, 3rd ed.; The Mointaineers Books: Seattle, DC, USA, 2006.

346. Anschlag, K.; Broll, G.; Holtmeier, F.-K. Mountain birch seedlings in the treeline ecotone, subarctic Finland: Variation in above- and below-ground growth depending on microtopography. *Arct. Antarct. Alp. Res.* **2008**, *40*, 609–616. [CrossRef]

347. Aune, S.; Hofgaard, A.; Söderström, L. Contrasting climate and land-use driven tree encroachment pattern of subarctic tundra in northern Norway and the Kola Peninsula. *Can. J. For. Res.* **2011**, *41*, 437–449. [CrossRef]

348. Leonelli, G.; Pelfini, M.; Morra di Cella, U.; Caravaglia, V. Climate warming and the recent treeline shift in the European Alps: The role of geomorphological factors a high altitude. *Ambio* **2011**, *40*, 264–273. [CrossRef]

349. Carrer, M.; Soraruf, L.; Lingua, E. Convergent space-time tree regeneration patterns along an elevational gradient at high altitude in the Alps. *For. Ecol. Manag.* **2013**, *304*, 1–9. [CrossRef]

350. Grau, O.; Ninot, J.M.; Cornelissen, J.H.C.; Callaghan, T.V. Similar tree seedling responses to shrubs and to simulated environmental changes of Pyrenean and subarctic treelines. *Plant Ecol. Divers.* **2013**, *6*, 329–342. [CrossRef]

351. Camarero, J.J.; García-Ruiz, J.; Sangüesa-Barrado, G.; Galván, J.D.; Alla, A.Q.; Sanjuán, Y.; Beguería, S.; Gutiérrez, E. Recent and intense dynamics in a formerly static Pyrean treeline. *Arct. Antarct. Alp. Res.* **2015**, *47*, 773–783. [CrossRef]

352. Shresta, K.B.; Hofgaard, A.; Vandvik, V. Tree-growth responses to climatic variability in two climatically contrasting treeline ecotone areas, central Himalaya, Nepal. *Can. J. For. Res.* **2015**, *45*, 1634–1653. [CrossRef]

353. Johnson, A.C.; Yeakley, J.A. Microsites and climatic zones: Seedling regeneration in the alpine treeline ecotone worldwide. *Forests* **2019**, *10*, 864. [CrossRef]

354. Renard, S.M.; McIntire, E.J.B.; Fajardo, A. Winter conditions—Not summer temperature—Influence establishment of seedlings of white spruce alpine treeline in Eastern Quebec. *J. Veg. Sci.* **2016**, *27*, 29–39. [CrossRef]

355. Wang, J.S.; Feng, J.G.; Chen, B.X.; Shi, P.-L.; Zhang, J.-L.; Fang, J.-P.; Wang, Z.-K.; Yao, S.-C.; Ding, L.-B. Controls of seed quantity and quality on seedling recruitment of Smith fir along altitudinal gradient in southwestern Tibetan Plateau. *J. Mt. Sci.* **2016**, *13*. [CrossRef]

356. Loranger, H.; Zotz, G.; Bader, M.Y. Competition or facilitation? The ambiguous role of alpine grassland for the early establishment of tree seedlings at treeline. *Oikos* **2017**, *126*, 1625–1636. [CrossRef]

357. Neuschulz, E.L.; Merges, D.; Bollmann, K.; Gugerli, F.; Böhning-Gaese, K. Biotic interactions and seed deposition rather than abiotic factors determine recruitment at elevational range limits of an alpine treeline. *J. Ecol.* **2017**, *106*, 948–959. [CrossRef]

358. Walker, X.; Henry, G.H.R.; McLeod, K.; Hofgaard, A. Reproduction and seedling establishment of Pice glauca across the northernmost forest-tundra region in Canada. *Glob. Chang. Biol.* **2017**, *18*, 322–3211.

359. Bader, M.Y.; Loranger, H.; Zotz, G.; Mendieta-Leiva, G. Responses of tree seedlings near the alpine treeline to delayed snowmelt and reduced sky exposure. *forests* **2018**, *9*, 12. [CrossRef]

360. Burzle, B.; Schlckhoff, U.; Schwab, N.; Vernicke, L.M.; Müller, Y.K.; Böhner, J.; Chaudhary, R.P.; Scholten, T.; Oldeland, J. Seedling recruitment and facilitation dependence on safe sites characteristics in a Himalayan treeline ecotone. *Plant Ecol.* **2018**, *219*, 115–132. [CrossRef]

361. Davis, E.L.; Gedalof, Z. Limited prospects fo future alpine treeline advance in the Canadian Rocky Mountains. *Glob. Chang. Biol.* **2018**, *24*, 4489–4504. [CrossRef] [PubMed]

362. Elliott, G.P.; Petrucelli, C.A. Tree recruitment at the treeline across the continental divide in the northern Rocky Mountains, USA: The role of spring cover and autumn climate. *Plant Ecol. Divers.* **2018**. [CrossRef]

363. Kambo, D.; Danby, R.K. Factors influencing the establishment and growth of tree seedlings at Subarctic alpine treeline. *Ecosphere* **2018**, *9*. [CrossRef]

364. Lett, S.; Dorrepaal, E. Global drivers of tree seedling establishment at alpine treelines in a changing climate. *Funct. Ecol.* **2018**, *32*, 1666–1680. [CrossRef]
365. Angulo, M.A.; Ninot, J.M.; Peñuelas, J.; Cornelissen, J.H.C.; Grau, O. Tree sapling response to 10 years of experimental manipulation of temperature, nutrient availability, and shrub cover at the Pyreneen treeline. *Front. Plant Sci.* **2019**, *9*. [CrossRef]
366. Brodersen, C.R.; Germino, M.J.; Johnson, D.M.; Reinhardt, K.; Smith, W.K.; Resler, L.M.; Bader, M.Y.; Sala, A.; Kueppers, L.M.; Broll, G.; et al. Seedling survival at timberline critical to conifer mountain forest elevation and extent. *Front. For. Glob. Chang.* **2019**, *2*. [CrossRef]
367. Holtmeier, F.-K. Der Einfluß der generativen und vegetativen Verjüngung auf das Verbreitungsmuster der Bäume und die ökologische Dynamik im Waldgrenzbereich. Beobachtungen und Untersuchungen in den Hochgebirgen Nordamerikas und in den Alpen. *Geoökodynamik* **1993**, *14*, 153–182.
368. Marr, J.W. The development and movement of tree islands near the upper limit of tree growth in the southern Rocky Mountains. *Ecology* **1977**, *58*, 1159–1164. [CrossRef]
369. Kullman, L. Transplantation experiments with saplings of Betula pubescens ssp. tortuosa near the tree-limit in central Sweden. *Holarct. Ecol.* **1984**, *7*, 289–293. [CrossRef]
370. Kullman, L. Pine (Pinus sylvestris L.) tree limit surveillance during recent decades, Central Sweden. *Arct. Alp. Res.* **1993**, *25*, 24–31. [CrossRef]
371. Hobbie, S.E.; Chapin, F.S. An experimental test of limits to tree establishment in Arctic tundra. *J. Ecol.* **1998**, *86*, 449–461. [CrossRef]
372. Germino, M.J.; Smith, W.K.; Resor, A.C. Conifer seedling distribution and survival in an alpine-treeline ecotone. *Plant Ecol.* **2002**, *162*, 157–168. [CrossRef]
373. Dullinger, S.; Dirnböck, T.; Grabherr, G. Modelling climate change-driven treeline shifts. Relative effects of temperature increase, dispersal and invisibility. *J. Ecol.* **2004**, *92*, 241–252. [CrossRef]
374. Moen, J.; Cairns, D.M.; Lafon, C.W. Factors structuring the treeline ecotone in Fennoscandia. *Plant Ecol. Divers.* **2008**, *1*, 77–87. [CrossRef]
375. Batllori, E.; Camarero, J.J.; Ninot, J.M.; Gutiérrez, E. Seedling recruitment, survival and facilitation in Pinus uncinata tree line ecotones. Implications and potential response to climate warming. *Glob. Ecol. Biogeogr.* **2009**, *19*, 460–472. [CrossRef]
376. Holtmeier, F.-K.; Broll, G. Response of Scots pine (Pinus sylvestris) to warming climate at its altitudinal limit in northernmost Subarctic Finland. *Arctic* **2011**, *64*, 269–280. [CrossRef]
377. Treml, V.; Šenfeldr, M.; Chuman, T.; Ponocná, T.; Demková, K. Twentieth century treeline ecotone advance in the Sudetes Mountains (Central Europe) was induced by agricultural land abandonment rather than climate change. *J. Veg. Sci.* **2016**, *27*, 1207–1221. [CrossRef]
378. Bruening, J.M.; Bunn, A.G.; Salzer, M.W. A climate-driven tree line position model in the White Mountains of California over the past six millenia. *J. Biogeogr.* **2018**, *45*, 1067–1076. [CrossRef]
379. Harsh, M.A.; Buxton, R.; Duncan, R.P.; Hulme, P.E.; Wardle, P.; Wilmshurst, J. Causes of treeline stability: Stem growth, recruitment and mortality rates over 15 years at New Zealand Nothofagus tree lines. *J. Biogeogr.* **2012**, *29*, 2051–2071.
380. Vitali, A.; Camarero, J.J.; Garbarino, M.; Piermattei, A.; Urbinati, C. Deconstructing human-shaped treelines: Microsite topography and distance from the seed source control Pinus nigra colonization of treeless areas in the Italian Apennines. *For. Ecol. Manag.* **2017**, *406*, 37–45. [CrossRef]
381. Srur, A.M.; Villalba, R.; Rodriguez-Catón, M.; Amoroso, M.M.; Marcotti, E. Climate and Nothofagus pumilio establishment at upper treelines in the Patagonian Andes. *Front. Earth Sci.* **2018**. [CrossRef]
382. Moser, M. Die ektotrophe Ernährungsweise an der Waldgrenze. *Mitt. der forstl. Bundevers. Wien* **1967**, *75*, 357–380.
383. Schinner, F. Die subalpine Waldgrenze und Bedeutung der Mykotrophie im Gasteiner Tal. In *Ökologische Analysen von Almflächen im Gasteiner Tal*; Cernusca, A., Ed.; Veröffentlichungen des Österreichischen MaB-Hochgebirgsprogramms Hohe Tauern: Innsbruck, Austria, 1978; pp. 311–314.
384. Magnússon, S.H.; Magnússon, B. Effect of enhancement of willow (Salix spp.) on establishment of birch (Betula pubescens) on eroded soils in Iceland. Man and Biosphere Series. In *Nordic Birch Ecosystems*; Wielgolaski, F.E., Ed.; The Partenon Publishing Group: News York, NY, USA; Toronto, ON, Canada, 2001; pp. 317–329.

385. Haselwandter, K. Mycorrhiza in the alpine timberline ecotone: Nutritional implications. In *Trees at Their Upper Limit. Treelife Limitation at the Alpine Timberline. Plant Ecophysiology*; Wieser, G., Tausz, M., Eds.; Springer: Dordrecht, The Netherlands, 2007; pp. 57–66.

386. Reithmeier, L.; Kernaghan, G. Availability of ectomycorrhizal fungi to black spruce above the present treeline in eastern Labrador. *PLoS ONE* **2013**, *8*, e777527. [CrossRef]

387. Nara, K.; Hogetsu, T. Ecotomycorrhizal fungi on established shrubs facilitate subsequent seedling establishment of successional plant species. *Ecology* **2014**, *85*, 1700–1707. [CrossRef]

388. Nordfors, G.A. Något om den vegetative föryngringing i fjällskog. *Norrl. Skovårdsförbunds Tisk.* **1923**, 1–45.

389. Holtmeier, F.-K.; Broll, G. Layering in the Rocky Mountain treeline ecotonre. *Trees* **2017**, *31*, 953–965. [CrossRef]

390. Kuoch, R.; Amiet, R. Die Verjüngung im Bereich der oberen Waldgrenze in den Alpen. *Mitt. der Schweiz. Anst. für das Forstl. Vers.* **1970**, *46*, 159–328.

391. Stimm, B. Morphologisch-anatomische Untersuchungen zur Ablegerbildung und sproßbürtigen Bewurzelung der Fichte (Picea abies (L.) Karst. *Flora* **1987**, *179*, 421–443. [CrossRef]

392. Caccianiga, M.; Payette, S. Recent advance of white spruce (Picea glauca) in the coastal tundra of the eastern shore of Hudson Bay (Quebec, Canada). *J. Biogeogr.* **2006**, *33*, 2120–2135. [CrossRef]

393. Buckner, D.L. Ribbon Forest Development and Maintenance in the Central Rocky Mountains of Colorado. Ph.D. Thesis, University of Colorado, Boulder, CO, USA, 1977.

394. Daly, C. Snow distribution in the alpine krummholz zone. *Prog. Phys. Geogr.* **1984**, *8*, 157–175. [CrossRef]

395. Benedict, J.B. Rates of tree-island migration, Colorado Rocky Mountains. *Ecology* **1984**, *65*, 820–823. [CrossRef]

396. Kubisch, P.; Leuschner, C.; McNown, R.W.; Brownlee, A.H.; Sveinbjörnsson, B. Fine roots abundance and dynamics of stone pine (Pinus cembra) at the alpine treeline is not impaired by self-shading. *Front. Plant Sci.* **2017**, *9*, 602. [CrossRef]

397. Schröter, C. *Das Pflanzenleben der Alpen. Eine Schilderung der Hochgebirgsflora*; Raustein: Zürich, Switzerland, 1908.

398. Fries, T.E. Botanische Untersuchungen im nördlichen Schweden I. Ein Beitrag zur Kenntnis der alpinen und subalpinen Vegetation in Torne Lappmark. *Flora och Fauna* **1913**, *2*, 361.

399. Tengwall, T.Å. *Die Vegetation des Sarekgebirges. Naturwissenschaftliche Untersuchungen des Sarekgebirges in Schwedisch Lappland, geleitet von Dr. A. Hamberg 3:4*; Akademische Abhandlungen: Stockholm, Sweden, 1920.

400. Holtmeier, F.-K. Die klimatische Waldgrenze - Linie oder Übergangssaum (Ökoton)? Ein Diskussionsbeitrag unter besonderer Berücksichtigung der Waldgrenzen in den mittleren und hohen Breiten der Nordhalbkugel. *Erdkunde* **1985**, *39*, 271–285. [CrossRef]

401. Karlsson, P.S.; Weih, M. Soil temperatures near the distribution limit of the mountain birch (Betula pubescens ssp. czerepanovii): Implications for seedling nitrogen economy and survival. *Arct. Antarct. Alp. Res.* **2001**, *33*, 88–92. [CrossRef]

402. Kauhanen, H. On growth problems of mountain birch near its distributional limits. *Ungi Rep.* **1987**, *65*, 183–190.

403. Smith, W.K.; Germino, M.J.; Hancock, T.E.; Johnson, D.M. Another perspective on altitudinal limits of alpine timberlines. *Tree Physiol.* **2003**, *23*, 1101–1112. [CrossRef]

404. Slot, M.; Wirth, C.; Schumacher, J.; Mohren, G.M.J.; Shibistova, O.; Lloyd, J.; Ensminger, J. Regeneration pattern in boreal Scots pine glades linked to cold-inducted photoinhibition. *Tree Physiol.* **2005**, *25*, 1139–1150. [CrossRef]

405. Ball, M.C.; Hodges, V.S.; Loughlin, G.P. Cold-induced photoinhibition limits regeneration of snow gum at treeline. *Ecology* **1991**, *5*, 663–668.

406. Germino, M.J.; Smith, W.K. Sky exposue, crown architecture, and low-temperature photoinhibition in confier seedlings at alpine treeline. *Plant Cell Environ.* **1999**, *22*, 407–415. [CrossRef]

407. Johnson, D.M.; Germino, M.J.; Smith, W.K. Abiotic factors limiting photosynthesis in Abies lasiocarpa and Picea engelmannii seedlings below and above the alpine timberline. *Tree Physiol.* **2004**, *24*, 377–386. [CrossRef] [PubMed]

408. Tausz, M. Photo-oxidative stress at the timberline. Trees at their upper limit. Treelife limitation at the alpine timberline. *Plant Ecophysiol.* **2007**, *5*, 181–195.

409. Bader, M.Y.; van Geloof, I.; Rietkerk, M. High solar radiation hinders tree regeneration above the alpine treeline in Ecuador. *Plant Ecol.* **2008**, *191*, 33–45. [CrossRef]

410. Takahashi, K.; Hirosawam, T.; Morishima, R. How the timberline formed: Altitudinal changes in stand structure and dynamics around the timberline in central Japan. *Ann. Bot.* **2012**, *109*, 1165–1174. [CrossRef]

411. Sveinbjörnsson, B.; Hofgaard, A.; Lloyd, A. Natural causes of the tundra-taiga boundary. *Ambio Spec. Rep.* **2002**, *12*, 23–29.

412. Wang, Y.; Pederson, N.; Ellison, A.M.; Buckley, H.; Case, B.S.; Liang, E.; Camarero, J.J. Increased stem density and competition may diminish positive effects of warming at alpine treeline. *Ecology* **2016**, *97*, 1668–1679. [CrossRef]

413. Auer, C. Untersuchungen über die natürliche Verjüngung der Lärche im Arven-Lärchenwald des Oberengadins. *Mitt. der Schweiz. Anst. für das Forstl. Vers.* **1947**, *25*, 3–140.

414. Schönenberger, W. Standortseinflüsse auf Versuchsaufforstungen an der alpinen Waldgrenze. *Mitt. Eidgenössische Anst. für das Forstl. Vers.* **1975**, *51*, 359–428.

415. Weih, M.; Karlsson, P.S. The nitrogen economy of mountain birch seedlings: implications for winter survival. *J. Ecol.* **1999**, *87*, 211–219. [CrossRef]

416. Liang, E.; Wang, Y.; Piao, S.; Lu, X.; Camarero, J.J.; Zhu, H.; Zhu, L.; Ellison, A.M.; Ciais, P. Specis interactions slow warming-induced upward shifts of treelines on the Tibetan Plateau. *Proc. Natl. Acad. Sci. USA* **2016**, *113*, 4380–4385. [CrossRef] [PubMed]

417. Vuorinen, K.E.M.; Oksanen, L.; Oksanen, T.; Pyykönen, A.; Olofson, J.; Virtanen, R. Open tundra persist, but arctic features decline—Vegetation changes in the warming Fennoscandian tundra. *Glob. Chang. Biol.* **2017**, *23*, 3794–3807. [CrossRef] [PubMed]

418. Chhetri, P.K.; Cairns, D.M. Low recruitment above treeline indicates treeline stability under changing climate in Dhorpatan Hunting Reserve, Western Nepal. *Phys. Geogr.* **2018**, *39*, 324–342. [CrossRef]

419. Frei, E.R.; Bianchi, E.; Bernareggi, G.; Bebi, P.; Dawes, M.A.; Brown, C.D.; Trant, A.J.; Mamet, S.D.; Rixen, C. Biotic and abiotic drivers of tree seedling recruitment across an alpine treeline ecotone. *Sci. Rep.* **2018**, *8*, 10894. [CrossRef]

420. Cramer, W. Modeling the possible impact of climate change in broad-scale vegetation structure: Examples from northern Europe. In *Global Change and Arctic Terrestrial Ecological Studies*; Oechel, W.C., Callaghan, T., Gilmore, T., Holten, J., Maxwell, B., Molau, U., Sveinbjörnsson, B., Eds.; Springer: New York, NY, USA, 1997.

421. Bekker, M.F.; Malanson, G.P.; Alftine, K.J.; Cairns, D.M. Feedback and pattern in computer simulations of the alpine treeline ecotone. In *GIS and Remote Sensing Applications in Biogeography and Ecology*; Millington, A.C., Walsh, S.J., Osborne, P.E., Eds.; Springer International Series in Engineering and Computer Science Book Series: New York, NY, USA, 2001.

422. Rupp, T.S.; Chapin III, F.S.; Starfield, A.M. Modeling the influence of topographic barriers on treeline advance at the forest-tundra ecotone in northwestern Alaska. *Clim. Chang.* **2001**, *48*, 399–416. [CrossRef]

423. Rickebusch, S.; Lischke, H.; Guisan, A.; Zimmermann, N.E. Understanding the low-temperature limitations to forest growth through calibration of a forest dynamics model with tree-ring data. *For. Ecol. Manag.* **2007**, *246*, 251–263. [CrossRef]

424. Zeng, Y.; Malanson, G.P. Endogenous fractal dynamics at alpine treeline ecotones. *Geogr. Anal.* **2008**, *38*, 271–287. [CrossRef]

425. Diaz-Varela, R.A.; Colombo, R.; Meroni, M.; Calvo-Iglesias, M.S.; Buffoni, A.; Tagliaferri, A. Spatio-temporal analysis of alpine ecotones: A spatial explicit model targeting altitudinal vegetation shifts. *Ecol. Model.* **2010**, *221*, 621–633. [CrossRef]

426. Macias-Fauria, M.; Johnson, E.A. Warming-induced upslope advance of subalpine forests is severely limited by geomorphic processes. *Proc. Natl. Acad. Sci. USA* **2013**, *110*, 8117–8122. [CrossRef]

427. Case, B.S.; Hale, J. Using novel metrics to access biogeographic patterns of abrupt treelines in relation to biotic influences. *Prog. Phys. Geogr.* **2005**, *39*, 310–335. [CrossRef]

428. Weiss, D.J.; Malanson, G.P.; Walsh, S.J. Multiscale relationships between alpine treeline elevation and hypothesized environmental controls in the Western United States. *Ann. Assoc. Am. Geogr.* **2015**, *105*, 437–453. [CrossRef]

429. Alatalo, J.M.; Ferrarini, A. Braking effect of climate and topography on global change-induced upslope forest expansion. *Int. J. Biometeorol.* **2016**, *61*, 541–548. [CrossRef] [PubMed]

430. Bobrowski, M.; Gerlitz, L.; Schickhoff, U. Modelling the potential distribution of Betula utilis in the Himalaya. *Glob. Chang. Conserv.* **2017**, *11*, 69–83. [CrossRef]

431. Kruse, S.; Gerdes, A.; Kath, N.J.; Herzschuh, U. Implementring spatially explicit wind-driven seed and pollen dispersal in the individual-based larch simulation model: LAVESI-WIND 1.0. *Geosci. Model Dev.* **2018**, *11*, 4451–4456. [CrossRef]

432. Thuiller, W. Patterns and uncertainties of species's range shifts under climate change. *Glob. Chang. Biol.* **2004**, *10*, 2020–2027. [CrossRef]

433. Araújo, M.B.; Pearson, R.G.; Thuiller, W.; Erhard, M. Validation of species-climate impact models under climate change. *Glob. Chang. Biol.* **2005**, *11*, 1504–1513. [CrossRef]

434. Elith, J.; Graham, C.H.; Anderson, R.P.; Dudík, M.; Ferrier, S.; Guisan, A.; Hijmans, R.J.; Huetmann, F.; Leathwick, J.R.; Lehmann, A.; et al. Novel methods improve prediction of species' distribution from occurrence data. *Ecography* **2006**, *29*, 129–151. [CrossRef]

MDPI

St. Alban-Anlage 66

4052 Basel

Switzerland

Tel. +41 61 683 77 34

Fax +41 61 302 89 18

www.mdpi.com

Forests Editorial Office

E-mail: forests@mdpi.com

www.mdpi.com/journal/forests